Waste Plastics 2019

AA009305

Topical Conference at the 2019 AIChE Annual Meeting

Orlando, Florida, USA
10 - 15 November 2019

ISBN: 978-1-7138-0566-3

Printed from e-media with permission by:

Curran Associates, Inc.
57 Morehouse Lane
Red Hook, NY 12571

Some format issues inherent in the e-media version may also appear in this print version.

Printed with permission by Curran Associates, Inc. (2020)

For permission requests, please contact AIChE
at the address below.

AIChE
120 Wall Street, FL 23
New York, NY 10005-4020

Phone: (800) 242-4363
Fax: (203) 775-5177

www.aiche.org

Additional copies of this publication are available from:

Curran Associates, Inc.
57 Morehouse Lane
Red Hook, NY 12571 USA
Phone: 845-758-0400
Fax: 845-758-2633
Email: curran@proceedings.com
Web: www.proceedings.com

TABLE OF CONTENTS

582663 RAPID Manufacturing Institute and AIChE

Process intensification is of worldwide interest as a type of advanced manufacturing. It aims for new, more compact, and more efficient processes, especially by applying component integration, novel driving forces, and modular deployment of these units. Join the PAIC Town Hall to hear about the leadership role AIChE plays via RAPID: a US Department of Energy advanced-manufacturing institute for process intensification and the development of process and operation principles and their applications.

582664 The Importance of Credentialing and Licensure

How do additional credentials strengthen an engineer's competitive position? What are the credentials that a chemical engineer may find helpful? How does a credential differ from additional education, licensure and advanced degrees? Information will be shared that will hopefully open up additional avenues for engineers of all ages to consider in their career planning.

582665 Legal Developments on Plastics: Basel Convention and the Circular Plastics Economy

The PAIC Legal Developments Subcommittee follows legal issues arising in: energy, food, water and other resources; climate change; chemical regulation; anti-terrorism, physical and cybersecurity, and chemical process safety; ethics, manufacturing and business law. At the PAIC Town Hall, we will review the May 2019 addition of plastic waste to the Basel Convention and its provisions regarding the development of a worldwide circular plastics economy; and the progress of the Expert Working Group pursuant to the Convention.

582666 Responding to Paradigm Shifts

Paradigm shifts generally lead to winners and losers. Manufacturing, chemicals, and energy illustrate three important areas of change driven by technology innovation and a shift from product to integrated system thinking. At the PAIC Town Hall, we will discuss the need for new frameworks that mesh technology, societal, and environmental dimensions and for new ways to respond to on-going shifts.

565108 Aqueous Rheology of Hydrophobe-Substituted Cellulose Ethers

Hydrophobe-substituted cellulose ethers are well-known as bio-derived associative thickeners in a range of applications, including latex paints, oil-field fluids, and personal care products. Using dynamic oscillatory rheology, we examine the effect of molecular weight, associative substituent length and degree of substitution on the yield stress and critical associative concentration of their aqueous solutions.

568052 My Wonderful and Random Walk through Polymer Science in Industry

My Wonderful and Random Walk through Polymer Science in Industry

Ann Fornof

3M Company

Having spent nine years at 3M with experience with IBM prior, I have had the opportunity to work on projects large and small—successful, unsuccessful, and somewhere in between. In this talk, I will discuss how my tortuous path from undergraduate to graduate school followed by two postdocs provided the foundation upon which I have built my career in industry. The fundamental polymer science that was critical to my PhD work is the same fundamental polymer science that I now apply to industrial problems. In this talk, I will focus on those projects based on adhesion science—both at the low T_g (pressure sensitive) and high T_g (structural) ranges. These materials are used in applications ranging from window film to dental materials to on-demand adhesives.

558078 Design and Synthesis of Emulsion Polymer Dispersions and Resins for Industrial Products with Improved Eco-Footprint

Emulsion polymerization is a powerful technology for the commercial production of a wide range of polymer dispersions for an even wider range of industrial applications. This technique can produce polymers with diverse chemical functionality, molecular weight, morphology, and viscosity. The versatility of emulsion polymerization allows synthetic chemists and process engineers to design an almost infinite variety of polymers with a diverse balance of properties and applications. Consequently emulsion polymer dispersions and resins have found broad utility in markets such as coatings, plastics additives, adhesives, home and personal care, and construction chemicals among others.

One essential application for emulsion polymers is the waterborne architectural coatings market. In the United States over 800 million gallons of house paint are produced each year, very often containing significant amounts of TiO_2 to provide the opacity required by consumers. This raw material has a large impact on the environmental carbon footprint and is typically used inefficiently due to multi-particle interactions. To mitigate this, polymeric pre-composite emulsion particles have been designed to interact with coatings grade TiO_2, improving the eco-profile and performance while lowering the cost of paint by increasing the efficiency of the relatively expensive TiO_2. This novel technology utilizes a smarter material which reacts with TiO_2 to form a self-assemble pre-composite during the paint making process. The resultant pigment/polymer composites create a highly ordered distribution of TiO_2 in the paint film, improving not only its efficiency but also barrier and mechanical properties. This innovative technology provides cost-efficient improvements in key sustainability metrics as demonstrated by a third-party validated life cycle assessment. Production of this technology is measured by millions of pounds and was recently awarded R&D 100 award and EPA Presidential Green Chemistry Challenge Award.

Another challenge for this market is that environmental regulations continue to drive the architectural coatings market towards lower volatile organic compounds (VOC), it is vital for eco and environmentally friendly architectural coatings to possess properties such as low odor, zero emission, and formaldehyde abating while providing ever higher performance. To overcome these challenges, novel emulsion polymers were designed with post film formation crosslinking chemistries and optimized latex particle morphology. Through incorporation of new functionalities, this emulsion technology also provides formaldehyde abating abilities, greatly reducing air pollution due to impurities in coatings.

570390 Recent Advances in Low Temperature Cure Electrodeposition Coatings

PPG has been at the forefront of innovation in electrocoat, an automotive industry standard for improved corrosion performance, for over fifty years. Recently, low temperature cure electrodeposition coatings have emerged as a means to enable the introduction of lightweight substrates, such as aluminum or carbon fiber composites. Additional benefits of this technology include improved sustainability and a reduction in the process footprint for the application of automotive coating stacks. Presented herein are advances in structural attributes and characterization techniques to establish the structure–property relationships in these low temperature cure coatings. The effect of crosslinking chemistry on key properties such as cure response, appearance and mechanical properties will be discussed.

561668 Sustainable Plastics for Food Packaging

Plastics are critical for food packaging, storage, and distribution, but challenges exist around plastic waste and sustainability. This presentation will give an overview of several works aimed at improving the sustainability of polyethylene-based food packaging. First, compatibilization of blends improves plastic mechanical properties for lightweighting (ACS Macro Letters, 4, 30, 2015). Second, compatibilization methods are presented for mixed stream polyethylene and polypropylene recycling (Science, 355, 814, 2017). Lastly, high-throughput polymer processing research tools to enable next-generation food packaging materials will be discussed.

574929 Understanding Formation of Nitrogen Heterocycles during Catalytic Hydrothermal Liquefaction

Hydrothermal liquefaction (HTL) is a promising technology for conversion of wet wastes into fuel precursor oils. In this work, we studied catalytic-HTL of food waste using a series of acid and base catalysts. While base catalysts are generally more effective at increasing oil yield and decreasing its oxygen content, acidic catalysts are especially effective at removing nitrogen content. Removing nitrogen from the bio-oil is an important challenge for the industrial application of the process. Heterocyclic nitrogen compounds are especially problematic, as they are difficult to remove using hydrodenitrogenation. Moreover, in the context of fixing nitrogen, HTL has potential for re-utilization of fixed nitrogen, reducing the need for costly and energy intensive nitrogen reduction processes. Unfortunately, the complex reaction mixture resulting from catalytic HTL of food waste and other realistic feed streams is too complicated to understand mechanisms. Computational studies can overcome this challenge, providing a guide for targeted experimental work and forming the basis of science-based reactor models. To further simplify, we studied homogeneous formation of heterocyclic under HTL conditions. In this work, a reaction network that involves interactions between reducing sugars, amino acids, and dienes is chosen as a basis for the formation of undesirable nitrogen heterocycles. Specific attention is given to aza-Diels-Alder cycloadditions and Maillard reactions, as both of these are potential pathways to formation of organo-nitrogen compounds. Quantum chemical calculations are carried in Gaussian 16 on reactants, intermediates, and products using the CBS-QB3 method. Conformers of each structure are identified by scanning the potential energy surface. Transition states of each pathway are determined and validated using intrinsic reaction coordinate calculations. These results are used in CanTherm to calculate thermochemistry and rate coefficients while treating hindered rotors. These results suggest new mechanisms for formation of undesirable nitrogen heterocycles that can guide selection of HTL processes and conditions to minimize their formation.

574908 Biochar and Activated Carbons Derived from Food Waste: Properties and Applications

Biochar and activated carbons derived from food waste: properties and applications

Lei Yu [a], David P. Gamliel [b], Katherine Saltzgiver [a], Brianna Markunas [a] and Julia A. Valla [a]

[a] Department of Chemical and Biomolecular Engineering, University of Connecticut, Storrs 191 Auditorium Road, Unit 3222, Storrs, CT 06269-4602, United States

[b] Physical Sciences Incorporated, 20 New England Business Center Rd. Andover, MA, 01810, United States

Food waste can be considered as a useful, renewable resource for energy and/or materials production. In this study, biochars were prepared via pyrolysis of food waste and then used as precursors for production of activated carbons via physical and chemical activation. Pyrolysis and activation conditions were varied to optimize the properties of biochar and activated carbons, respectively. Activated carbons were characterized by N_2 sorption-desorption, scanning electron microscope (SEM), energy-dispersive X-ray spectroscopy (EDX), elemental analysis, inductively coupled plasma atomic emission spectroscopy (ICP-OES), X-ray powder diffraction (XRD), ^{13}C nuclear magnetic resonance (NMR), Fourier-transform infrared spectroscopy (FTIR) and Raman spectroscopy. Adsorption experiments of aromatic hydrocarbons were conducted to evaluate the feasibility of using the biochars and activated carbons derived from food waste for the adsorption of aromatic hydrocarbons (benzene and naphthalene) from water. The activated carbon with the highest adsorption capacity was prepared via steam activation at 950 °C for 1 h. The BET surface area, micropore volume and total pore volume of the activated carbon was 745 m^2/g, 0.185 cm^3/g and 0.594 cm^3/g, respectively. The benzene and naphthalene sorption capacity was 460 mg/g and 150 mg/g, respectively. The physicochemical properties and adsorption capacities of the activated carbons prepared from food waste were also compared to the corresponding properties and capacities of the activated carbons derived from miscanthus, as well as commercial activated carbons. Our study revealed that activated carbons prepared by physical activation of biochars derived from food waste resources are very promising candidates for water purification with capacities for aromatic hydrocarbons similar to the ones observed using commercial activated carbons. The results are presented in the Figure 1 below.

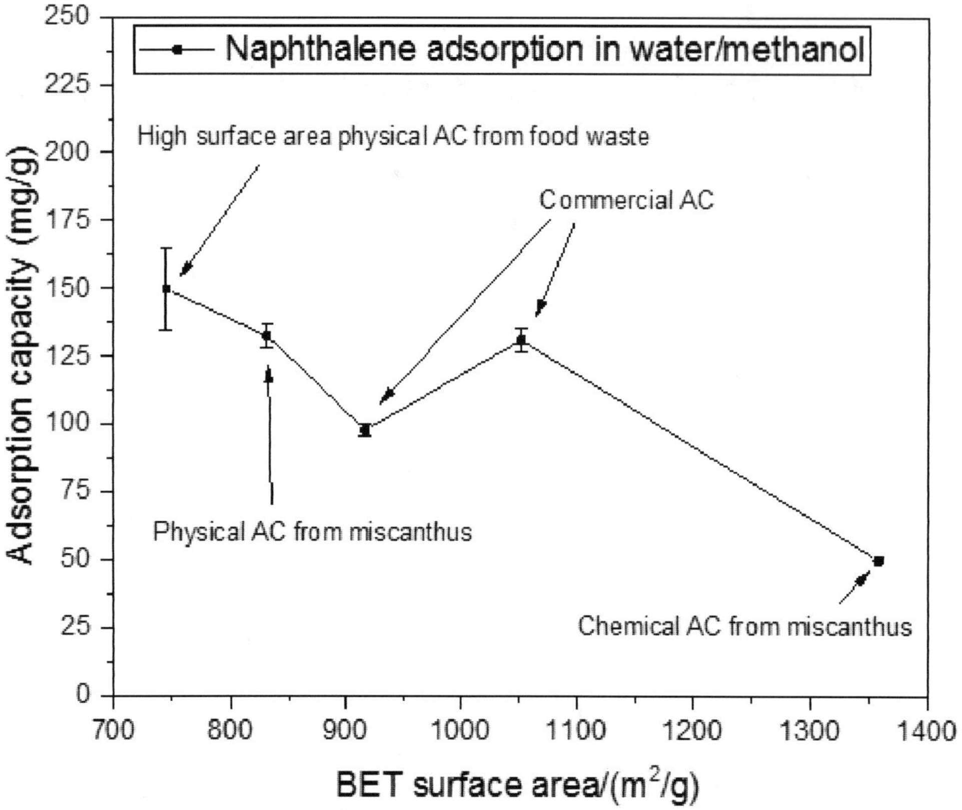

Figure 1: Adsorption capacity of naphthalene in activated carbons derived from various resources

566272 A Framework for Assessing Solid Waste Energy Recovery Systems for Small-Scale on-Site Applications

The generation of solid waste, which includes wood, paper, plastics, organic wastes (e.g., food wastes) and other debris can be problematic for isolated, small-scale communities. While several known technologies, such as pyrolysis or gasification, can be employed for on-site applications, there is no waste-to-energy technology that can accept all types of solid waste. At present, small-scale communities must decide which technology to use (if any) based on their unique site and situation. The purpose of this study is to develop a framework for assessing available technologies for the destruction of solid wastes with the simultaneous production of energy for small-scale operations (i.e., for populations of 100 to 1,000 people). To assess the feasibility of several existing technologies, we first surveyed several available technologies (by contacting company representatives and/or conducting on-line research) to characterize the technology and obtain the following performance metrics: rate of waste treated, type of waste treated, pre-processing requirements, transportability, number of operators, physical footprint requirements, support requirements (fuel, electricity), permitting and military testing & evaluation, environmental control limitations, safety, operator skills and training, specialty equipment, operational and down time, production of heat & electricity, production of residuals, noise & odor, and air emissions. Second, we developed a framework for assessing characteristics of each technology considering the typical solid waste generation rates (based on characteristics from U.S. military forward operating bases) for population sizes ranging from 100 to 1,000 people. Within the framework, individual metrics where binned into three overarching categories: "waste inputs", "support requirements", and "products / residuals". Each technology was assigned a baseline score considering each individual metric and the binned categories for comparison. Depending on the need, the framework further allowed for weighted analysis of each binned category. The developed framework was employed to assess several pyrolysis or gasification technologies for the U.S. military, but could be modified and employed for numerous small-scale on-site waste-to-energy applications.

576722 Microplastic Environmental Effect on the Aquatic Ecosystem of the Thermaic Gulf

The aim of this study is to quantify the abundance of floating microplastics in sediments and surface water of Thermaic Gulf, Greece and to identify the spatial distribution of the microplastic particles and their polymeric characterization. Moreover, the frequency of occurrence of ingested microplastics in marine biota also was examined. Microplastics are now found worldwide in all aquatic compartments such as surface water, water column, and sediments (Ivar do Sul and Costa, 2014). Recently, it has been estimated that 5 trillion microparticles are floating globally, while up to 890,000 plastic particles/ km^2 were predicted in the Mediterranean surface waters. (Eriksen et al., 2014).

Thermaic Gulf is a gulf constituting the northwest corner of the Aegean Sea. The city of Thessaloniki is at its northeastern tip, and it is bounded by Pieria and Imathia on the west and the Chalkidiki peninsula on the east, with Cape Kassandra at the southeasternmost corner. It is about 100 km long. Water surface samples were realized during the winter (February–March 2018). A total of 25 samples were collected at 500 m, 5 km and 10 km distance from the coast. Surface samples were collected using neuston net with cod end (300 μm mesh size) equipped with a flowmeter and towed on the water surface at 2–3 knots for 20 min. At the end of the samplings, the nets were rinsed thoroughly from the outside to ensure that both plankton and debris were washed into the end of the net and to prevent any contamination by rinsing water. Organic matter attached to the surface of microplastics was removed to allow clear identification of the type of plastics, using a solution of 30% H_2O_2. For sediment sampling, boxcorer grabs were used and to ensure representativeness of the samples, at least five sample replicates (5 m between replicates) were collected. Density separation of microplastics were carried out with a Munich Plastic Sediment Separator. After separation and purification, target microplastics were sorted from the remaining matrix. Large plastics were sorted out directly, while smaller-sized ones needed further observation under a dissection microscope. For plastic particle analysis, samples were observed under a stereo-microscope with a LED light and measured with micrometer ocular lens. Plastic particles were characterized and classified by colour, size and shape. The particles were dried at room temperature and weighted in analytical balance. Additionally, the polymer fingerprint was detected using Fourier Transform Infrared (FT-IR) spectroscopy.

A total of 15 sea bass was sampled from Thermaic Gulf. The fish were collected during the winter of 2018. The total length (TL) and total wet weight (TWW) of fish were recorded. The fish were stored at −20 °C prior microplastic detection. Also, the tissues from sea bass were analyzed.

A total of 1200 microplastics have been isolated from all 25 surface water samples, with a range of 13 to 520 total particles per sample. The minimum concentration was 13752 particles/km^2 and the maximum was 420513 particles/km^2. Regarding the sediments the minimum concentration was 0.02 MPs / g dry sediment and the maximum was 0.1 MPs/g dry sediment. Analysis of the size distribution of microplastics in surface samples showed that the most abundant size class is 1–2.5 mm both in the whole dataset. This finding suggests a potentially higher hazard or zooplankton organisms (Fossi et al., 2012). Plastic fragments were found to be primarily isolated plastic particles. Polyethylene (PE) and Polypropylene (PP) were the most abundant chemical species, in good agreement with the literature where these polymers account for the majority of the plastic particles floating in Mediterranean Sea. The results showed a higher abundance of microplastics at 5-10 km distance from the coast. Microplastics were also detected in in the guts, skin and muscles of sea bass collected at an amount that ranged from 0.08 g^{-1} to 1.1 g^{-1}.

This study reports the results of a monitoring campaign of microplastics in three different media. These data represent the first attempt to present the current state of microplastics in the Thermaic Gulf and to establish the baseline level for a future study. Being a Thermaic Gulf an important port within the European countries and because of its morphological features generate important quantities of waste originating from the South European counties so the evaluation of the presence of plastics is mandatory to apply strategies to reduce the quantity of microplastics in Thermaic Gulf.

REFERENCES

ERIKSEN, M., LEBRETON, L. C. M., CARSON, H. S., THIEL, M., MOORE, C. J., BORERRO, J. C., GALGANI, F., RYAN, P. G. & REISSER, J. 2014. Plastic Pollution in the World's Oceans: More than 5 Trillion Plastic Pieces Weighing over 250,000 Tons Afloat at Sea. *PLOS ONE,* 9, e111913.

FOSSI, M. C., PANTI, C., GUERRANTI, C., COPPOLA, D., GIANNETTI, M., MARSILI, L. & MINUTOLI, R. 2012. Are baleen whales exposed to the threat of microplastics? A case study of the Mediterranean fin whale (Balaenoptera physalus). *Marine Pollution Bulletin,* 64, 2374-2379.

IVAR DO SUL, J. A. & COSTA, M. F. 2014. The present and future of microplastic pollution in the marine environment. *Environmental Pollution,* 185, 352-364.

564081 Study and Optimization of the Supply Chain of the Recollection and Harnessing of Used Cooking Oil in the City of Bogotá

Currently, waste disposal is a major challenge for most mega-populated cities worldwide. Particularly in underdeveloped and developing countries, the lack of regulations and law enforcement, together with a generalized unconscious behavior regarding waste management had generated a large variety of impacts. Among the different urban residues, used cooking oil (UCO) is a highly problematic waste that is generally disposed through the sewage system. As a consequence of this improper disposal, there is blockage of sewage pipes, which in turn generates flooding during rainy seasons, proliferation of vectors (rats, cockroaches, bacteria, etc.), and bad odors while decomposing. In some cases, the waste oil reaches surface or underground waters generating a major ecological impact.

All these problems have been increasingly experienced in the capital city of Colombia, Bogota. The city and its surrounding metropolitan area concentrate nearly 20% of the country's population (~ 10 million), and it is expected to grow at 2.8% annually in the coming years. Then, the increasing population will boost impacts associated to UCOs mishandling. Nowadays the city expends nearly 6 to 8 million dollar per year on corrective sewage maintenance related to UCOs mismanagement.

In order to reduce UCO disposal problem, major cooking oil companies in Colombia have started responsible consumption programs to promote UCOs recollection and valorization. Currently, UCOs are collected to be used as feedstock for soaps and biodiesel production. However, as these are low-added value commodities, sustainability of the recollections and exploitation chain is challenging. Thus, all the constituent steps of the collection, treatment and transformation processes must be optimized to ensure sustainability. Particularly, supply chain optimization is a necessary task to accomplish during the sustainability assessment.

In this work, the analysis and optimization of a UCO recollection program, implemented in the city of Bogota by a major fat and oils, was accomplished. An analysis of available UCOs volumes, source locations, recollections frequencies and routes, and the corresponding impacts (e.g. fuel consumption) was carried out. Subsequently, the optimization of the supply chain was focused on maximizing recollection volumes while minimizing time, costs and the associated impacts (i.e. environmental). Thus, the supply chain was modelled using a Travelling Salesman Problem like formulation with combinatorial Genetic Algorithm as solution algorithm for the route optimization. The formulation is based on a graph where each collection location was represented by a node and the route is formed by changing the paths between them. Restrictions were incorporated to take into account shift times, truck capacities, and recollections times. The optimization problem was formulated and solved in Python.

572109 Economic Feasibility of Evaporation-Based Liquid Digestate Treatment for Biogas Plants

Biogas plants (BGP) are important renewable energy sources and they contribute to sustainable development. In BGP, organic waste is used to produce biogas which is either utilized for on-site electricity generation or is upgraded to biomethane and then distributed to a natural gas network or used as a fuel (compressed biomethane - CBG).

The core technology is the anaerobic digestion unit. Its development has currently achieved maturity. However, the current frontier of biogas technologies research is focused on liquid digestate processing which supports circular economy of biogas industry.

Evaporation-based systems are very suitable for further liquid digestate treatment especially when storage and transportation are associated with significant costs for operators. It reduces liquid digestate volume by thickening and at the same time preserves its fertilizing potential. Further, it can produce clean water suitable for discharge into the environment, i.e. is able to prevent ammonia from entering the outgoing freshwater stream.

On the other hand, such treatment unit has higher investment cost and reasonable payback period is not always achievable. It depends on many parameters such size of the plant, liquid digestate production, transportation and storage cost, waste heat availability, thickening-related parameters, etc.

This contribution provides detailed sensitivity analysis to various parameters in order to show what is important to consider when making decision about an investment into an evaporation-based liquid digestate treatment system. Since BGPs are costly, supporting tools are frequently applied by governments. These include investment support, higher feed-in tariffs for electricity, CHP incentives, etc. The conditions provided by such tools are also considered.

Tools supporting the biogas plant operation and investments are crucial. Reasonability of investment into an evaporation-based system for liquid digestate significantly decrease without them. However it is not the necessary condition. Beside the investment cost, the parameters that play key role are transportation cost, dry matter content in liquid digestate and cost of chemicals. Less important parameters, but still significant, are application cost of the liquid digestate, energy consumption of an evaporation unit, dry matter content in thickened concentrate and how much waste heat is available. For a specific combination of values of these parameters, the evaporation-based system for liquid digestate treatment is feasible.

Strategies for improvement of anaerobic digestion from agricultural waste: a critical review

Safenaz Shaaban, Doaa Bassuney, Mennat Allah Mohamed, and Mahmoud Nasr*

Sanitary Engineering Department, Faculty of Engineering, Alexandria University, 21544, Alexandria, Egypt

*Corresponding author: Dr. Mahmoud Nasr

Tele. (+20) 100 63 90 400

Email: mahmmoudsaid@gmail.com; mahmoud-nasr@alexu.edu.eg;

https://orcid.org/0000-0001-5115-135X

Abstract

Agricultural wastes, also known as agro-wastes, refer to the residual products discarded from various farming practices. Agro-wastes consist of multiple organic and non-biodegradable substances, and they include large quantities of crop residues such as rice straw, sugarcane bagasse, wheat shell, soybean hulls, and corn stalk. Agricultural wastes also comprise a wide range of animal residues such as livestock feces and urine, poultry litter and excreta, and carcasses (or parts) of animals, as well as, agro-industrial byproducts such as sugarcane bagasse ash (a byproduct of sugar factories), cassava bagasse, soy hulls, and sugar-beet pulp. Moreover, agricultural wastes contain nitrogen and phosphorus species released from crop fertilization, as well as, toxic elements due to the utilization of pesticides (herbicides and insecticides). Furthermore, agricultural wastes include lignocellulosic biomass such as wood and straw residues, which is considered a renewable resource for the 2nd generation of biofuels production. The fractions of cellulose, hemicellulose, and lignin vary among agricultural wastes (e.g., corn stover, rice husk, wheat straw, and sweet sorghum stalk). Based on the nature of farming systems, agro-wastes can be collected in the forms of solid, liquid, or slurry. The annual quantity of agricultural wastes has been estimated as 998 million tonnes, in which the organic fraction can reach up to 80% of the total solid wastes.

Anaerobic digestion process

Anaerobic digestion has found successful applications as an environmentally friendly approach for the conversion of agro-wastes into valuable products and renewable fuels (Shaaban and Nasr, 2020). The input agro-wastes (i.e., as a substrate) include wet organic substances, animal slurries, biosolids, and livestock manure (pig and cattle). The anaerobic bioconversion process reduces the volume and mass of agro-wastes, and the residual matter after digestion can be used as a soil amendment and as secondary fertilizer compounds (Mthethwa et al., 2019). In anaerobic digestion, biodegradable materials are stabilized and broken down by the action of microbial organisms in an oxygen-deprived environment. The microbial community includes fermentative bacteria, acetogens, and methanogens. The biological degradation reactions are complex, consecutive, and parallel, as they undergo multiple steps of hydrolysis, acidogenesis, acetogenesis, and methanogenesis. During these processes, the waste feedstock is decomposed into simple chemical components such as sugars, which are converted to organic acids. Finally, the organic acids are converted to biogas, *viz.*, about 55–75% CH_4, CO_2, and traces of other gases. The percentage of CH_4 varies according to the operating temperature and the amount of organic fraction in the feed substrate. These steps can be summarized as follows:

In the hydrolysis step, hydrolytic bacteria convert the complex organic compounds (e.g., proteins, lipids, and carbohydrates) to monomers (low-molecular-weight matters) including amino acids, long chain fatty acids, and simple sugars.

In the acidogenesis phase, acid bacteria stimulate the fermentation process, producing intermediary products such as volatile fatty acids (VFAs), alcohols, hydrogen, and carbon dioxide (Mthethwa et al., 2018).

In the acetogenesis process, acid-forming bacteria or acetogens utilize VFAs to obtain acetic acid, carbon dioxide, and hydrogen.

In the methanogenesis phase, methanogenic bacteria transform the intermediary products to CH_4 gas via two pathways:

Pathway-I: Methanogenic bacteria, converting CH_3COOH to CH_4 and CO_2.

Pathway-II: Methanogenic bacteria, converting CO_2 and H_2 to CH_4 and H_2O.

Anaerobic digestion is sustainable management of agricultural waste as it converts the organic matter into renewable energy that can replace conventional fossil fuels. Anaerobic digestion has the advantage of producing biogas and fertilizer, leading to minimizing the net energy input of the system. Moreover, the amount of waste sludge yielded from the anaerobic digestion system is lower than that of the aerobic treatment mechanisms. However, anaerobic digestion requires a high level of investment for a commercial or large scale application. Moreover, the environmental conditions should be strictly optimized to avoid the accumulation of volatile fatty acids and the inhibition of bacterial growth. Additionally, the fluctuation of input variables (e.g., flowrate, organic load, and chemical composition) provides a negative impact on the efficiency, control, and stability of the process. Furthermore, a lack of knowledge about the activities of living organisms and the phenomena occurring in the bioprocess offers serious challenges for the wide implementation of anaerobic digestion (Kumari et al., 2017).

Anaerobic digestion of agricultural wastes is influenced by various factors, which can be summarized as follows:

Carbon to nitrogen (C/N) ratio

A C/N ratio of 20 – 30 has been recognized to provide the optimum condition for most anaerobic digestion applications. The unbalanced C/N ratio results in the accumulation of ammonia, leading to the inhibition of microbial activities. Accordingly, different types of chemical components are mixed with the feedstock wastes to provide the favorable C/N ratio that can improve gas productivity.

Organic loading rate

Organic loading rate (OLR) refers to the mass of volatile solids (VS) or chemical oxygen demand (COD) per cubic meter of reactor's volume per day. OLR is used to estimate the ability of a digester to convert the feedstock waste into biogas and other end products. The high OLR requires large digester's size, leading to an increase in the initial cost of investment.

Medium pH

pH is an important environmental factor in the anaerobic digestion process, as it affects the growth rate of microorganisms and the biological conversion pathways (Nasr et al., 2015). The optimum pH range for the biomethanation processes has been reported as 6.6 – 7.0.

Alkalinity

The accumulation of volatile fatty acids in the digester during the acidogenesis phase can lead to the deactivation of methanogenic bacteria. Hence, alkalinity (e.g., ability to resist acidification via the addition of calcium carbonate) is an essential factor in the anaerobic digestion process as it safeguards the buffer capacity and stability of the digester.

Inoculum (Seeding)

A stable methanogenic community is required for the efficient degradation of organic compounds to produce biogas. Hence, the anaerobic digestion system should be enriched with sufficient quantities of micro-organisms. The acclimatization stage allows the bacterial cells to adapt to the new environmental conditions (Nasr et al., 2015).

Substrate pretreatment

Agricultural wastes may include a heterogeneous complex of biodegradables and non-biodegradables. Accordingly, the input substrate for anaerobic digestion should be subjected to a pretreatment process (e.g., physicochemical, and biological treatments) (Reddy et al., 2017). This stage is used to (a) form high amounts of readily biodegradable substrates, (b) reduce the formation of inhibitory products, (c) minimize operational costs (e.g., energy demands), and (d) avoid the loss of sugars.

Other factors

Anaerobic digestion of agricultural wastes and metabolic conditions for microorganisms' growth are also influenced by several factors including activators (e.g., enzymatic additives), retention time, and temperature regime (mesophilic and thermophilic) (Nasr et al., 2013).

Because agricultural wastes are abundant (available), low cost, and renewable, they can be used as a valuable input for bioenergy generation. Hence, this study presents a comprehensive review on the application of anaerobic digestion for biogas production from agricultural wastes. The review illustrates the optimum operating conditions (e.g., culture temperature, medium pH, nutrients availability, organic load, and digestion time) that maximize biogas productivity and improve organic biodegradability. The reactor's configurations, viz., suspended bed reactor, fixed bed reactor, fluidized bed reactor, two-stage processes, and hybrid reactors employed for the enhancement of digester stability are discussed. The types of agricultural wastes such as lignocellulose biomass, livestock manure, crop residues, poultry litter, and cattle wastes used for biomethane production are illustrated. These objectives are represented regarding recent literature studies. Recommendations for future works are also suggested.

Acknowledgements

The authors would like to acknowledge Nasr Academy for Environment Protection

References

Kumari, S., Nasr, M. & Kumar, S., 2017. Technological advances in biohydrogen production from microalgae. *Algal Biofuels: Recent Advances and Future Prospects,* pp. 347-360.

Mthethwa, N., Nasr, M., Bux, F. & Kumari, S., 2018. Utilization of *Pistia stratiotes* (aquatic weed) for fermentative biohydrogen: Electron-equivalent balance, stoichiometry, and cost estimation. *International Journal of Hydrogen Energy,* 43(17), pp. 8243-8255.

Mthethwa, N.P., Nasr, M., Kiambi, S.L., Bux, F., Kumari, S. 2019. Biohydrogen fermentation from *Pistia stratiotes* (aquatic weed) using mixed and pure bacterial cultures. *International Journal of Hydrogen Energy,* 44(33), pp. 17720-17731.

Nasr, M., Tawfik, A., Ookawara, S. & Suzuki, M., 2013. Biological hydrogen production from starch wastewater using a novel up-flow anaerobic staged reactor. *BioResources,* 8(4), pp. 4951-4968.

Nasr, M., Tawfik, A., Ookawara, S., Suzuki, M., Kumari, S., Bux, F., 2015. Continuous biohydrogen production from starch wastewater via sequential dark-photo fermentation with emphasize on maghemite nanoparticles. *Journal of Industrial and Engineering Chemistry,* Volume 21, pp. 500-506.

Nasr, M., Tawfik, A., Suzuki, M. & Ookawara, S., 2015. Mathematical modeling of bio-hydrogen production from starch wastewater via up-flow anaerobic staged reactor. *Desalination and Water Treatment,* 54(1), pp. 50-58.

Reddy, K., Nasr, M., Kumari, S., Kumar, S., Gupta, S.K., Enitan, A.M., Bux, F. 2017. Biohydrogen production from sugarcane bagasse hydrolysate: effects of pH, S/X, Fe^{2+}, and magnetite nanoparticles. *Environmental Science and Pollution Research,* 24(9), pp. 8790-8804.

Shaaban, S. & Nasr, M., 2020. Toward Three R's Agricultural Waste in MENA: Reduce, Reuse, and Recycle. *In: Negm A., Shareef N. (eds) Waste Management in MENA Regions. Springer Water. Springer, Cham,* pp. 337-353.

581709 Filament Fabrication and 3D Printability of Poly(3-hydroxybutyrate-co-3-hydroxyvalerate) (PHBV)/Poly(lactic acid) (PLA) blends using a Commercial Chain Extender

A series of poly(3-hydroxybutyrate-*co*-3-hydroxyvalerate) (PHBV) and poly(lactic acid) (PLA) (60:40), (50:50), and (40:60) %wt. based blends were successfully 3D printed using a commercial chain extender (CE). Breakage and buckling behaviour of the filaments during melt extrusion and 3D printing were addressed accordingly. The incorporation of the CE allowed high printing temperatures (220°C) of PHBV, resulting in good bed adhesion and warpage prevention. When investigating the 3D printing parameters of the blends, a bed temperature of 60°C and a layer height of 0.25mm were found to exhibit the highest mechanical properties. Scanning electron microscopy of tensile fractured specimens revealed that good weld formation was present for all samples that were printed at the optimal printing parameter. Poor adhesion between layers was observed when printing the blends with a bed temperature of 90°C. This study highlights the effect of the CE on the printability of PHBV/PLA based blends as well as the importance of appropriate printing parameters in the performance of 3D printed parts.

Acknowledgements: This research is financially supported by the Ontario Ministry of Agriculture, Food and Rural Affairs (OMAFRA), University of Guelph, Bioeconomy Industrial Uses Research Program Theme Project #030252; the Natural Sciences and Engineering Research Council (NSERC), Canada Discovery Grants Project #400320; and the Ontario Ministry of Economic Development, Job Creation and Trade ORF-RE09-078 (Project #053970);.

Keywords: 3D printing, poly(3-hydroxybutyrate-3-*co*-3-hydroxyvalerate), poly(lactic acid), fused deposition modeling, biopolymers.

557689 Evaluation of Binderless Board Made from Composted Rice Straw As Substrate for Rice Seedling Production

China is the largest rice producer and consumer in the world; approximately 20% of arable China's lands is annually devoted to planting this valuable crop. With the increase of China's population, the rice yield will increase to 7850kg/ha by 2030, resulting in a total of 200 million kg of rice per year to feed the Chinese people. Therefore, China has the potential to produce a large amount of rice straw. Today, compost plays the major role as a peat substitute. Composts straw have beneficial attributes, such as providing plant nutrients, suppressing disease, promoting the growth of beneficial microbes. In this paper, the straw was composted under controlled condition to form the rice seedling substrate. After the composted straw was beated for an hour, the substrate block with dimension of 27 cm ×55 cm × 1 cm were formed by vacuum adsorption from water.

The nutrient in the substrate block, including organic matter, nitrogen, phosphorus and potassium, was improved significantly in comparation with the traditional substrate. The results of rice seedling grown in straw-based substrate block showed that the seedling height, stem diameter, root length, leaf area, SPAD of rice produced from the substrate block are nearly the same as the normal substrate. And after transplanted, the effective tillering number improved 7.4%, the yield of the rice improved 5.7%, the ratio of straw to grain decreased 11.5%. That is to say, the straw based substrate can reduce ineffective tillering, and decrease the amount of produced straw.

The substrate block can substitute the seedling tray and reduce the agricultural white pollution. Agricultural wastes in rice production can be utilized and recycled. In addition, the labor intensity and cost of rice seedling production can be reduced, and the rice production efficiency was improved.

572555 A Closed-Loop Biorefinery for Woody Biomass Conversion Using Lignin-Derived Deep Eutectic Solvents

Lignocellulosic biomass is an attractive energy source; however, conversion of the biomass is still technically and economically challenging due to its rigid and compact structure and recalcitrant nature toward conversion caused by chemical and physical features of the biomass. Therefore, effective and innovative conversion technologies are eagerly needed for converting the biomass to fuels and industrial chemicals.

Deep eutectic solvents (DES) pretreatment is an effective method for biomass conversion and/or fractionation. DESs have been proposed as alternatives to ionic liquids (ILs). These solvents do not only share the typical properties of ILs such as low volatility, wide liquid range, non-toxicity, and biocompatibility, but they also have some advantages over the traditional ILs like easy synthesis and relatively inexpensive components, thus large scale application is feasible. Tuning properties of DESs by replacing the hydrogen bond donor (HBD) with phenolics sharing structural similarities with lignin can facilitate better solubility of lignin during the fractionation. Relatively mild pretreatment conditions minimize the unwanted lignin modification and keep relatively intact structures during the pretreatment; therefore, it promotes subsequent lignin valorization.

Herein, a biomass conversion and fractionation process employing novel deep eutectic solvents (DES) prepared from lignin-derived phenolic compounds is studied. With choline chloride as HBA, the proposed process utilized the fractionated lignin-derived compounds as an HBD to enhance the total biomass utilization. Different phenolic acids such as 4-hydroxybenzoic acid and caffeic acid were investigated for effective lignin fractionation and sugar release from woody biomass (poplar). To evaluate the impact of DES pretreatment on biomass conversion, enzymatic digestibility was conducted. Diverse analytical techniques including 2D HSQC NMR, ^{31}P NMR, GPC, and wet chemistry using HPLC were performed to elucidate the structural changes during the reactions. Recyclability of DES was also studied to develop an economically feasible closed-loop system.

572703 Reuse of Sustainable Wastes for Fibreboard Production: The Case of Waste Paper and Water Hyacinth

In this early study, new cement bonded bi-composite fibreboard (FB) was made from water hyacinth (WaHy) and waste paper (WP). Ordinary Portland cement (PdCe) used as a binder was mixed with other two additives; gypsum plasters (GyPl) and wood ash (WdAh) in defined proportions to form bonding matrices. The WP and WaHy were pre-treated and a linked process developed for the mixing and consolidation steps. The FBs produced were based on different proportions of composites, binder and additives mixed. Equal amount of composites mixed with binder and additive in ratio 2:1 by weight resulted in FB having acceptable densities properties. Improved tensile strength was observed for bi-composite FB from WaHy mixed with WP. In general, FB having densities in the range of 0.50 – 0.57 gcm^{-3} compared favourably with the ASTM and ANSI standards (95%), thus making the FBs a potential alternative for building and construction purposes. More elaborate research with advanced analytical techniques are hereby suggested.

582073 Polyhydroxyalkanoate and Cellulose Ester Based Biodegradable Plastic Blends for Sustainable Packaging

Poly(3-hydroxybutyrate-co-3-hydroxyvalerate) (PHBV) has positive biodegradable, non-toxic, sustainable, and compostable properties, making it attractive for use in packaging. However, the applications in flexible packaging are limited due to its brittleness and narrow processing temperature range [1]. To mitigate the brittleness of PHBV, a compostable polymer such as cellulose acetate (CA) can be used with a 2.45 degree of substitution to positively influence the toughness of PHBV [2]. Furthermore, based on the solubility parameters, PHBV and cellulose acetate are potentially miscible [3], [4].

In this research, PHBV was blended with CA to investigate the overall mechanical properties of the blend; however, CA requires plasticization in order to be melt compounded. Prior to processing, CA powder was plasticized with triethyl citrate (TEC) and rested at room temperature for 12 hrs. Subsequently, PHBV pellets were melt compounded with the plasticized CA (pCA) powder in a twin-screw extruder followed by injection moulding. Our study found 70/30 wt.% PHBV/pCA blend ratios had improved impact strength, exceeding virgin PHBV and pCA which is attributed to PHBV degradation from the high processing temperature. TEC plasticizer migrated during processing into the PHBV, partially plasticizing it, thus reducing its melt temperature and forming secondary crystallite structures in the form of a double melt peak from differential scanning calorimetry (DSC) study. Using CA as a filler in PHBV without TEC reduced the tensile and flexural strength, and had poor interfacial adhesion from scanning electron microscopy (SEM) study. The impact strength however, was improved at a 70/30 wt.% PHBV/CA blend ratio with 50% higher tensile modulus and strength, and flexural strength compared to PHBV/pCA 70/30 wt.% blends.

Acknowledgment

This research is financially supported by the Natural Sciences and Engineering Research Council (NSERC), Canada Discovery Grants (Project # 400320); the Ontario Ministry of Agriculture, Food and Rural Affairs (OMAFRA) – University of Guelph, the Bioeconomy Industerial Uses Research Program Theme (Project # 030255); and the Ontario Ministry of Economic Development, Job Creation and Trade ORF-RE09-078 (Project #053970)

References

[1] K. Khosravi-Darani and D. Z. Bucci, "Application of Poly(hydroxyalkanoate) In Food Packaging: Improvements by Nanotechnology," *Chem. Biochem. Eng. Q.*, vol. 29, no. 2, pp. 275–285, 2015.

[2] R. M. Gardner, C. M. Buchanan, R. Komarek, D. Dorschel, C. Boggs, and A. W. White, "Compostability of cellulose acetate films," *J. Appl. Polym. Sci.*, vol. 52, no. 10, pp. 1477–1488, 1994.

[3] B. Wang, J. Chen, H. Peng, J. Gai, J. Kang, and Y. Cao, "Investigation on Changes in the Miscibility, Morphology, Rheology and Mechanical Behavior of Melt Processed Cellulose Acetate through Adding Polyethylene Glycol as a Plasticizer," *J. Macromol. Sci. Part B Phys.*, vol. 55, no. 9, pp. 894–907, 2016.

[4] J. S. Choi and W. H. Park, "Effect of biodegradable plasticizers on thermal and mechanical properties of poly(3-hydroxybutyrate)," *Polym. Test.*, vol. 23, no. 4, pp. 455–460, 2004.

582775 Extraction of Eucalyptus Globulus Leaves with Distinct Methods and Solvents. Comparison and Analysis of the Extracts

The increased awareness and research for improvements and valorization of by-products from established processes is marking the recent years as an important moment for biorefinery processes. Under the biorefinery scope, several biomasses are regaining interest for valorization, as the knowledge of sources and product certification rises. Essential oils are a classical example of a different approach for forestry residues valorization, coexisting with a different class of extracts obtained from biomass extraction, plant extracts [1].

In the present work, distinct methods were employed for *E. globulus* leaves extraction. Conventional techniques, namely Soxhlet, and solid-liquid extraction in batch mode (SLE-batch) at four different temperatures (20, 35, 55 and 75 °C), were performed with solvents of distinct polarity: dichloromethane, ethanol, methanol, *n*-hexane, petroleum ether, ethyl acetate and isopropanol. To obtain the essential oil of the *E. globulus* leaf, hydrodistillations were performed. Supercritical CO_2 extractions were also performed at 300 bar and 40 °C, with the addition of modifiers to CO_2, namely ethanol and ethyl acetate, in contents of 5-10 wt.%.

All the extracts produced, along with the ground *E. globulus* leaf, commercial samples of *E. globulus* leaf oil obtained by steam-distillation from Portugal and India, pure eucalyptol and betulinic, oleanolic and ursolic acids, were analyzed with Fourier transform infrared – attenuated total reflectance (FTIR-ATR) technique.

To avail from all the information of the FTIR-ATR spectra, a statistical tool, multidimensional scaling (MDS), was employed. MDS enables the representation of similarities/dissimilarities between complex data as distances in a low-dimensional space, allowing its visual interpretation and analysis [2].

The extractions performance varied significantly with the method and solvent employed, along with the temperature in the SLE-batch runs. The yields of extraction obtained ranged from 0.14 wt.% to 37.80 wt.% for hydrodistillation and Soxhlet extraction with methanol, respectively. All SLE-batch experiments scored lower than the Soxhlet ones. Furthermore, within the SLE-batch experiments, temperature had a clear effect, as lower temperatures yielded less and *vice-versa*. The supercritical extractions yields were similar to the less polar solvents, but higher than those of hydrodistillation.

MDS provided a scale reduction from 1000 D to 2 D with a stress factor of 9.9 %, considered good [2]. From its visualization, a clear distinction of the extraction methods employed can be observed, along with solvent polarity and temperature effect.

In this work, an experimental and modeling study on *E. globulus* leaf was performed, aiming at mapping the similarities/dissimilarities between natural products that can be produced by extraction from this biomass.On the whole, this study is expected to shed light on the biorefinery opportunities for this biomass part, including a valid enrichment of product portfolio options or market segments.

References

1. 1. G. Vânia G. Zuin, L.Z. Ramin, Green and Sustainable Separation of Natural Products from Agro-Industrial Waste: Challenges, Potentialities, and Perspectives on Emerging Approaches (Springer, 2018)
2. 2. Borg, P.J.F. Groenen, P. Mair, Applied Multidimensional Scaling (Springer, Berlin, 2013)

570267 Towards a Comprehensive Decision-Support Framework for Sustainable Manufacturing

From the sustainability science point of view, industrial sustainability is a triple-bottom-line based, discretized system state transition problem, which requires continuous sustainability assessment and decision analysis in order to derive best possible decisions for implementation However, put sustainability in practice is not a trivial task and it needs a systems approach with a suitable framework that enables decision makers to integrate sustainability thinking into their business strategies successfully.

To provide a fundamental definition for sustainability, we defined the sustainability as vector function [1]. Using this definition, the sustainability status of a system and its transition toward sustainability can be measured, monitored, and interpreted. As sustainability is a continuous process, we formulated the problem as a multi stage decision problem in [2]. To achieve the sustainability goals, we formulated the sustainability problem as a multi-objective control problem and proposed a decision support system in [3] with focusing more on tactical layer.

In this paper, we propose a new comprehensive decision support framework for manufacturing sustainability enhancement that is an extension to our recent work in [3]. This two-layer (strategic and tactical) framework is designed based on the sustainability fundamentals and control theory. The two layer that are executed at different time scales are integrated to implement sustainability development strategies successfully. A model predictive control (MPC) strategy is used at the strategic control layer to find and update optimal sustainability strategic plans continuously (e.g. at each year) for archiving short-to-long term sustainability goals with considering external information changes (e.g. new regulations, supplies and demands) provided by external monitoring component and internal information provided by sustainability assessor at the tactical layer. The tactical layer, on the other hand, is responsible for implementing sustainability strategic plans successfully by minimizing the projects' implementation costs and time. The components of each layer, implementation mechanism, and methods are presented. The proposed framework provides a structured step-by-step guide to industry for achieving economic, environmental, and social sustainability goals with considering uncertainties related to external and internal parameters that affect the business. A case study on biodiesel manufacturing is illustrated to demonstrate methodological efficacy.

References:

[1]Moradi-Aliabadi, M. and Y. Huang, "Vector-Based Sustainability Analytics: A Methodological Study on System Transition toward Sustainability," Industrial and Engineering Chemistry Research, 55(12), 3239-3252, 2016.

[2] Moradi-Aliabadi, M. and Y. Huang, "Multistage Optimization for Chemical Process Sustainability Enhancement under Uncertainty," ACS Sustainable Chemistry and Engineering, 4(11), 6133-6143, 2016.

[3] Moradi-Aliabadi, M. and Y. Huang, "Decision Support for Enhancement of Manufacturing Sustainability: A Hierarchical Control Approach," ACS Sustainable Chem. Eng., 6(4), 4809–4820, 2018.

---

* All correspondence should be addressed to Prof. Yinlun Huang (Phone: 313-577-3771; E-mail: yhuang@wayne.edu).

577723 Evaluating Dynamic Resilience of Intermittent and Uncontrollable Techno-Ecological Systems

Engineering sciences have made a constant effort towards the goal of sustainable development. In particular, chemical engineering has developed multiple techniques like heat and mass integration, waste reduction and eco-industrial parks aimed at reducing the ecological impact. Using the framework of Techno-Ecological Synergy (TES), Bakshi and co-workers [1, 2, 3] have incorporated ecological systems in process and supply chain design space to obtain novel ecologically and economically win-win solution. These novel systems aim to improve sustainability by matching the ecological goods or service utilization rate with its regeneration rate.

Unlike technological systems, the dynamics of ecosystems are not "controllable". In order to ensure sustainability of TES system, the design and operation should be "viable" over long periods of time i.e. the design should be resilient. The term "resilience" has been defined across multiple disciplines with varying definition. Resilience can be used to describe a system’s resistance capacity, coping capacity, adaptive preference or its transformative capacity to avoid irreversible changes while sustaining large shocks [4].

In this work, we employ the mathematical framework of viability theory [5] to analyze the viable operational domain of TES systems. We employ various dynamic resilience metrics to analyze and design control strategy for ensuring a sustainable operation of the novel TES systems. We compare the resilience of TES systems with that of conventional technocentric systems. This study helps us integrate the intermittent and uncontrollable dynamics of ecosystem with the control of technological system’s paradigm and design "resilient" sustainable systems.

References

[1] Bhavik R. Bakshi. Toward Sustainable Chemical Engineering: The Role of Process Systems Engineering. *Annual Reviews in Chemical and Biomolecular Engineering*, Accepted, 2019.

[2] Utkarsh Shah and Bhavik R. Bakshi. Accounting for nature’s intermittency and growth while mitigating no 2 emissions by technoecological synergistic design-application to a chloralkali process. *Journal of Advanced Manufacturing and Processing*, 0(0):e10013, Mar 28, 2019.

[3] Tapajyoti Ghosh, Xinyu Liu, and Bhavik R Bakshi. Including ecosystem services in sustainable process design across multiple spatial scales. In *Computer Aided Chemical Engineering*, volume 44, pages 1837–1842. Elsevier, 2018.

[4] Christophe Béné and Luc Doyen. From resistance to transformation: a generic metric of resilience through viability. *Earth’s Future*, 6(7):979–996, 2018.

[5] Jean-Pierre Aubin. *Viability theory*. Springer Science & Business Media, 2009.

574791 A Multi-Layer Process Control Framework for Sustainability: Application to Biomass/Coal Co-Gasification System

Industry, government, and society have begun to shift from economic stand-alone focus to the inclusion of sustainability in the decision-making process. This shift is due to the adverse environmental impact and unsustainable development caused by human activities, including chemical industry releases. As a result, process systems engineering (PSE) approaches have been developed, predominantly for incorporating sustainability into chemical process design and optimization. However, the development of sustainability-oriented control schemes is still scarce, especially when compared to the number of available efforts on steady-state sustainability assessment and optimization. Previous work has shown that process control plays a critical role in the transient sustainability and emission performance, especially for systems with different operating conditions (e.g., steady-state multiplicity [1]). The main challenges that have been mentioned in previous literature are: 1) existence of multiple objectives for optimization involving social, economic and environmental issues (e.g., pollution prevention and abatement); and 2) difficulty of integrating the sustainability indicators into a process control framework [2].

To address these challenges, a novel multi-layer process control framework is proposed to drive the system to a sustainable operating point that is defined using a multi-objective optimization algorithm. In particular, a sustainable Model Predictive Control (MPC) scheme is formulated based on dimensionless sustainability performance indicators from the U.S. EPA's GREENSCOPE (Gauging Reaction Effectiveness for the ENvironmental Sustainability of Chemistries with a Multi-Objective Process Evaluator) tool [3]. Such indicators are associated with the process state variables and can capture the sustainability information of the current process condition, including economic, environmental and social aspects. The selected sustainability indicators are employed as constraints in the designed controller to maintain the process operation within a pre-defined sustainable zone, where the performance indicator values are higher than the desired thresholds. To characterize this sustainable control zone, dynamic radar plots are developed to visualize the multidimensional sustainability indicators during a transient period.

The developed method is illustrated via a biomass/coal co-gasification process for syngas production with the end goal of methanol manufacturing. For this application, the whole process model is developed in Aspen Hysys based on existing literature information [4-5]. With the established models in Hysys and a link for data communication between Hysys and MATLAB, a multi-objective optimization problem is solved to maximize profit and optimize the process sustainability performance (e.g., environmental release and resource use minimization), by employing a genetic algorithm-based approach developed in MATLAB. A reduced-order model for the process is also derived in MATLAB for control implementation purposes. In this presentation, the details on the application results of this multi-layer control framework for improving sustainability are discussed, focusing on the process dynamic sustainability performance. These results demonstrate the capability of the proposed sustainable control scheme for maintaining the process within sustainable zones during transient periods.

References:

1. Li S, Mirlekar G, Ruiz-Mercado GJ, Lima FV. Development of chemical process design and control for sustainability. Processes. 2016;4(3):23.
2. Daoutidis P, Zachar M, Jogwar SS. Sustainability and process control: A survey and perspective. J. Process Control. 2016;44:184–206.
3. Ruiz-Mercado GJ, Gonzalez MA, Smith RL. Sustainability indicators for chemical processes: III. biodiesel case study. Ind. Eng. Chem. Res. 2013;52(20):6747–60.
4. Li S, Feliachi Y, Agbleze S, Ruiz-Mercado GJ, Smith RL, Meyer DE, Gonzalez MA, Lima FV. A process systems framework for rapid generation of life cycle inventories for pollution control and sustainability evaluation. Clean Technol. Environ. Policy. 2018;20: 1543-1561.
5. Robinson PJ, Luyben WL. Simple dynamic gasifier model that runs in Aspen Dynamics. Ind. Eng. Chem. Res. 2008;7784–92.

567732 Process Engineering Inspired Comparison of Sustainability and Resilience Specific Features of Natural and Human-Built Process Systems

From chemical and biological engineering points of views, sustainability and resilience means long term planning, design and control of robust large scale processes. In the past decades mankind built "artificial" processes against the principles of the natural (e.g. biological) processes, to achieve more profit in a short time horizon. In order to keep balance in the space of finite resources and reservoirs, we have to pay more attention to the computer assisted engineering of large water-food-energy- ecosystem related process networks.

The structure and functionalities of the complex natural and human-built processes are more sophisticated, than the networks in sense of network science, on the one hand. However, the case-specific local functionalities of the individual building blocks in these structures are less complex, than the various specific mathematical methods, on the other. As an intermediate solution, we have developed Programmable Structures that can be generated from two functional meta-prototypes and from the declaration of the process network. The meta-prototypes and the generated process structures are prepared for the semantically distinguished, but syntactically uniform representation of the "*model specific conservation law based, additive*" measures, and of the "*over-writable*" signals.

Applying this methodology, we studied many different process systems from cellular signaling, through technological systems to the ecosystem involved agricultural process networks. The results from the analysis of these quite different processes, studied by a unified approach, oriented our theoretical interest toward model based comparison of sustainability- and resilience-specific features of natural and human-built processes.

According to the preliminary investigations on this comparison, it became obvious that we can learn useful (chemical and biological engineering based) sustainability related design and control principles. The most important lessons can be summarized as follows:

1. Both natural and human-built processes are determined by "model specific conservation law based" stoichiometric balance processes, but the natural processes are inherently self-determined, while human-built ones are determined by short term decisions. As a consequence of the self-determinedness, natural processes prefer local solutions, in contrary to the more and more global character of human-built processes.
2. Another common feature is that the primary (controlling) information is carried by conservation law based vehicle processes in both natural and artificial processes. However, in natural processes a direct feedback exists between the signal and the vehicle process. In the lack of this feedback, human generated information processes are often characterized by the (sometimes useful, sometimes harmful) far reaching effects. The basic architecture of natural processes is characterized by the mutual cooperative control feedback between the functionally connected neighbors. However, the artificial processes are controlled by the hierarchically organized human decisions.
3. The secondary (evaluating) information appears only in the human-built processes, in form of objectives (i.e. interests). Accordingly, the evaluation feedback in the artificial processes are manifested in the various forms of usually hierarchical optimization. Nevertheless, consciously designed cooperative evaluation feedback amongst the functionally connected neighbors could also be applied in the human-built process architectures. Engineering designed and controlled cooperation means that the functionally connected neighbors must tend to evolve mutually suboptimal, compromise functioning.

Programmable Structure of process models helped to learn the clear separation of the chemistry invented "model specific conservation laws based" additive measures from the over-writable signs. This, combined with biology invented cooperative feedback, suggested design and operation principles for sustainable complex process systems, as follows:

- to consider long term balances of the model specific conservation laws based measures;

- to understand "information process" as a sub-process of "conservation process" that consumes less measures, but effects more on the functioning of the complementing sub-processes and the environment, than *'vice versa'*; and

- to understand cooperative process as a mutual (control or evaluation) feedback between the functionally connected neighbors.

The existence and utilization of these principles will be illustrated by some recently studied examples for biosystem, ecosystem, (bio)technological and agro-environmental process networks.

578082 An Input-Output Framework to Assess the Importance of Insect Pollinators to US Industry Sectors: A Case Study of Apple Production

Ecosystem services remain highly critical and largely undervalued in industrial communities. Although the supply of these services is often considered boundless, many ecosystem services, including insect-mediated pollination, have suffered severe degradation and face overconsumption beyond rate of replenishment. While insects provide other valuable ecosystem services beyond pollination, both wild and managed species of bees and other pollinating insects are instrumental in yield and/or quality of approximately 75 pollination-dependent crops in the United States of America. In forty-five years, production of pollination-dependent crops has increased more rapidly (50-62%) than the global population of managed bees (45%) while simultaneously, wild flying insects in all regions (including protected natural lands) have declined up to and exceeding 75 percent over only the last twenty-seven years, both in biomass and with decreasing biodiversity. We determine that insect-mediated pollination of crops accounts for over 30 billion US dollars of pollination-dependent crop production alone making the ecosystem service a highly valuable asset to the economic welfare of the agricultural sector. In addition, there are related agricultural sectors that rely upon pollination-dependent crop production (fertilizers, pesticides, equipment) as well as non-agricultural industry sectors (pharmaceutical, fuel, real estate) that share complex linkages with crop production and agricultural sectors, leading to intricate indirect dependence upon pollination service mediated by both commercially-managed and wild species of bees and other insects throughout the economy.

We present an input-output (I-O) modeling based framework for quantifying the economic value of industry sectors attributable to insect-mediated pollination. Of pollination-dependent crops, each has a specific dependence upon insect-mediated pollination service. Using available field data, we first reconcile and update estimates of pollinator dependence coefficients for individual crops while explicitly quantifying uncertainty. Using publicly available data from the USDA (United States Department of Agriculture), NASS (National Agricultural Statistics Service), and reconciled pollinator dependence coefficients, we establish bounds on economic value of agricultural crops attributable to insect pollinators. This information is then used in conjunction with the 2012 Benchmark I-O account data to quantify the economic dependence of non-agricultural sectors on insect-mediated pollination. We focus specifically on the apple farming industry and recreate the apple farming supply chain. We disaggregate the fruit and nut farming sector to represent the apple farming sector with greater detail.

While the direct economic value of apples attributable to insect pollinators is 3 billion dollars, indirect economic dependence of non- agricultural sectors on insect pollinators is equally as significant, though lacking previous quantification. In addition, we further identify the industry sectors that are most vulnerable to loss of insect pollinators. This model allows for specificity in analyses of indirect economic impacts of pollination service mediated by insects by providing a framework for the incorporation of crop-specific dependence coefficients into sector expenditures. The results are compelling and highlight the need for better understanding the role of insect pollinators in life cycles of agricultural and non-agricultural products and processes.

567111 Rapid Estimation of Release Inventories and Evaluation: A U.S. Metal Coating Industry Case Study

Industrial facilities might release many chemical substances during the production of value-added products. These chemical releases may cause damage to a receptor, either occupational and non-occupational population, and the environment. Therefore, to determine whether a chemical poses a considerable risk to human health and the environment, several tasks must be completed. These include determining the releases to the environment from conditions of use. Therefore, this work aims to develop a process-oriented, decision-making, and learning-from-data framework which integrates: (a) estimation of chemical releases from an industrial facility taking into consideration the type of processes, the condition of use, and the physical-chemical properties of a substance; (b) environmental sustainability indicators based on U.S. Environmental Protection Agency (USEPA)'s GREENSCOPE; (c) a decision-making procedure based on fuzzy techniques to include stakeholders' opinions; and (d) an aggregation method to incorporate the existing trade-off between the USEPA's GREENSCOPE environmental indicators. A case of study for the U.S. metal coating industry is employed for applying the proposed framework as a tool for supporting environmental decision-making efforts and chemical release minimization. In addition, this framework may be suitable for chemical risk evaluation and environmental sustainability decision-making for industrial systems.

The views expressed in this abstract are those of the authors and do not necessarily represent the views or policies of the U.S. Environmental Protection Agency.

580488 Mechanistic Modeling of the (Bio)Conversion of (Bio)Macromolecules

Complex reacting systems are ubiquitous, yet it is still challenging to develop quantitative models to describe them. Two main challenges are to assemble the reaction mechanism or network and to specify the quantitative parameters governing the species and reactions that comprise it. To address the first challenge, we have developed methods for automated generation of reaction mechanisms of complex systems that allow kinetic models of substantive detail to be built. Molecules are represented as graphs and matrices, and operations on these representations allow reaction to be carried out, molecule uniqueness to be determined, and properties to be calculated. We have applied our methodology to a wide range of different problems, including production of silicon nanoparticles, biochemical transformations, polymerization and depolymerization, acid-catalyzed olefin oligomerization, and tropospheric ozone formation. While the chemistries we have studied are seemingly very disparate, applying a consistent methodology to study them reveals that there are many common elements of complex reaction networks. To conquer the second issue, we have developed a hierarchical approach for specifying rate coefficients and thermodynamic properties that relies on experimental data, structure/reactivity relationships, group additivity approaches, and quantum chemical calculations. This talk will focus on the application of these approaches to a variety of different chemistries

578934 Cyclooctene-Facilitated Low Temperature Aerobic Oxidation of Ethylbenzene with Solubilized Au Clusters Catalysts

Traditionally, a catalyst functions by direct interaction with reactants. In a new Extended Catalyst System (ECS), an intermediate produced by one catalytic reaction enables an independent reaction to proceed. An example is the coupled Au-initiated selective epoxidation of cyclooctene with the selective oxidation of ethylbenzene. Although oxidation of ethylbenzene could not proceed in the presence of either solubilized Au nanoclusters or cyclooctene, it occurred readily when both were present simultaneously. This combined system effectively extended the catalytic effect of Au. Reaction kinetics and spin trap experiments suggested that the cyclooctenyl peroxy and oxy radicals were responsible for initiating the ethylbenzene oxidation. ECS enables parallel reactions to proceed without constraints by stoichiometric relationships, offering new degrees of freedom in industrial hydrocarbon co-oxidation processes.

570382 Stability of Supported Metal Catalyst in Aqueous Phase Reactions

A significant challenge associated with the conversion of biomass-derived molecules is the need to perform the reactions in the condensed phase. This challenge is magnified when the solvent is water where metal oxide supports are typically not stable. Carbon supports are effective alternatives, but the chemistry of the carbon surfaces can be difficult to characterize. Additionally, control of the pore structures of carbon supports is less well developed than with metal oxides. To overcome these issues, metal catalysts on carbon over-coated silica were synthesized in which the chemistry of the carbon layer was controlled by the calcination conditions. Advanced solid state NMR techniques were utilized to extensively characterize the carbon over-layers allowing for systematic assessment of the impact of either oxygen or nitrogen heteroatoms on the hydrothermal stability of supported Pd nanoparticles. Utilizing the insights from those studies, extended aqueous-phase reaction testing of supported Pt and Ru catalysts was performed to assess the potential for their stable operation.

580267 Chemical Recycling - the Answer to Plastic Waste?

The complexities of recycling of plastic waste are many. First of all the complexity of the feedstock and the fact that many reactions are occurring simultaneously makes it very challenging for chemical engineers. Also on the reactor level no consensus exist on what is the optimal design for each feed type, not even on what is the best "chemical recycling" option because of the varying economics and different government incentives Worldwide. If we could predict how the product distribution and conversion depends on the feedstock composition, reaction conditions and assess process operation issues, we could rationally design and optimize the system on the computer. The first step is to identify all the important species and reactions; this can be done using automatically or by hand with different levels of complexity. Unfortunately, we seldom have enough experimental data to determine the rate parameters for all the important reactions. Moreover the data that is typically available does not allow to determine intrinsic kinetics, not even to close to mass balances. With novel experimental and computational techniques we currently try to improve our understanding of chemical recycling, but still a huge step for mankind is needed to come to truly fundamental understanding of chemical recycling.

578653 A View of Cancer Immunology through a Reaction Engineering Lens

The reactor plays a central role in chemical processes, where state variables are controlled to create a favorable landscape that encourages a desirable reactive outcome. Yet, despite our best intentions, the realities of reacting systems - like inferential uncertainty, heterogeneity, and non-ideal conditions - can limit our ability to achieve our goals. To combat these complexities, the reaction engineering community have improved analytical and analysis tools and redesigned reactors to mitigate their impact. In some regard, using the immune system to treat cancer has some of the same aims: creating favorable conditions that encourage a patient's immune cells to attack and kill malignant cells. In this talk, I will discuss some techniques developed to overcome challenges in designing and operating chemical reactors and their analogies in engaging host immunity to control tumor growth, specifically related to identifying reaction networks.

574095 Abiotic Degradation with Reactive Iron Mineral Coatings in Anaerobic Sediment

Abiotic dehalogenation with reactive iron minerals plays an important role in the natural attenuation of chlorinated organic solvents. These reactive iron minerals are most significant in redox transition zones. However, there are a number of challenges in collecting anaerobic cores representative of the subsurface environment. In this study, an anaerobic sediment core was collected from an industrial site with 1,4-dichlorobenzene and 4-chloroaniline plumes. With the redox condition preserved, a number of approaches including sequential extraction were applied to quantify reactive iron mineral coatings. Results revealed an abundance of siderite, pyrite, mackinawite, magnetite, and iron oxyhydroxides in three redox transition zones. Experiments were carried out to evaluate abiotic dehalogenation of 1,4-dichlorobenzene. The reaction rate followed a second order rate expression that was a function of the 1,4-dichlorobenzene and the reactive mineral coatings present in the sediment; for the three transition zones rate, constants for the dominant reactive mineral coatings were obtained. The time to achieve 90% removal ranged from 50 to 200 days for the zones studied. To degrade 4-chloroanaline, persulfate was applied with the reduced iron mineral coatings in the sediments. Persulfate forms $SO_4^{-\cdot}$, a radical, and a strong and effective oxidizing agent. Degradation rates revealed first-order rate constants with 90% removals requiring 9 and 22 hours. From these studies, the degradation pathways and models for target constituents of concern can be used to enhance these processes in contaminated subsurface systems.

579690 Nanoparticles for Energy and Environment Applications – Wastewater Treatment

Oil extraction involves thermal processes that use large amounts of water which also results in large amounts of oil sands process-affected water (OSPW). Because of the zero-discharge policy, this water cannot be discharged back in rivers or lakes so it is stored in tailing ponds. One of the challenges faced with produced water and interfered with their effective water management is particle settling. Without the introduction of a process or material, it would take these particles thousands of years to settle. Despite the fact that many existing processes are promising for fine particle settling, they lack several properties that can be improved by nanoparticle technology.

In this study, nanoparticles are prepared and modified to promote interaction with suspended particles in the wastewater. The prepared material was fully characterized using FTIR, TGA, BET, XRD and SEM. Nanoparticles provide promising results for the enhancement of fine particle settling making it an environmentally friendly and cost-effective process.

Nanoparticles is a new innovative technology with many applications for energy and environment sustainability. They are effective in enhancing fine particles settling in tailing ponds that would otherwise remain suspended for decades. They also promote dewatering in alignment to the Canadian government emphasis on active management of oil sands tailings and land reclamation.

573928 Laboratory Investigation of Combined Chemical/Biological Treatment of Hexavalent Chromium, Carbon Tetrachloride and Trichloroethylene in Aqueous System

Among the hundreds of chromium contaminated sites in the United States are Hanford and Savannah River Sites, which are under the purview of the U.S. Department of Energy (DOE). At the Hanford Site, which is the most contaminated nuclear site in the United States, aquifer and groundwater contamination represents a complex mixture of inorganics including hexavalent chromium , Cr(VI), radionuclides and nitrates, and organic contaminants, e.g., trichloroethylene (TCE) and carbon tetrachloride (CCl_4).

Conventional treatment of chromium contamination focuses on immobilization of the highly toxic and water-soluble form Cr(VI), by reducing it to the relatively less toxic and less mobile trivalent form, Cr(III) while organic chlorinated compounds and nitrates can also be reduced using ferrous sulfate. Both types of contaminants (inorganic and organic) also respond to biodegradation. In this paper, results will be presented from a two-step comprehensive approach: (i) to chemically reduce hexavalent chromium, carbon tetrachloride and trichloroethylene using ferrous sulfate and zero valent iron (ZVI) amendments, which leads to nontoxic compounds, but also results in oxidation of ferrous to ferric iron. (ii) In the second step, ferric iron is regenerated to ferrous through biotic transformation using metal-reducing bacteria, *Shewanella oneidensis* MR-1. The microorganisms support biodegradation of contaminants, and convert ferric to ferrous iron in the process, leading to a self-containing sustainable process. The results can be used in designing an at a hazardous waste site.

571321 Rapid Dechlorination of Polychlorinated Biphenyls (PCBs) in an Engineered System

Polychlorinated biphenyls (PCBs) are posing great environmental problems due to their toxicity. Removal of PCB in polluted sites are challenged by the high physical/chemical treatment costs, while microbial dichlorination of PCBs are largely impeded by difficulties in growing PCB dechlorinators to a high cell abundance. In this study, we accelerated the onset of PCB dechlorination by pre-cultivating functional microbes on alternative electron acceptors in an engineered system. Extensive dechlorination of Aroclor 1260 was achieved within 1 week by culture CG4 pre-grown tetrachloroethene (PCE). Compared to control cultures fed solely with Aroclor 1260, the PCB dechlorination rate was up to 30 times greater in cultures pre-cultivated with alternative electron acceptors. However, when culture CG4 was simultaneously exposed to PCE, and PCBs, PCB dechlorination was not observed until complete depletion of PCE. Expression of the bifunctional reductive dehalogenase (RDase) gene *pcbA4* was inhibited by the presence of both substrates and continued with only one substrate in the medium. Utilization of chloroethenes and PCBs as substrates by culture CG4 was found to be largely sequential rather than concurrent, suggesting that PCE is a preferred substrate for the RDase responsible for PCE and PCB dechlorination.

570464 Adsorption on NDMA on Biomass-Derived Activated Carbons in Lab-Scale Columns

The purpose of this research is to test the ability of biochars made from local agroforestry and forestry wastes, to adsorb N-nitrosodimethylamine (NDMA) at very low concentrations out of water. According to the EPA, NDMA is a probable human carcinogen found in ground and drinking water, especially in ground water near rocket testing sites that used unsymmetrical dimethylhydrazine fuel. The treatment standards for NDMA typically range from 1 to a few ng/L (parts per trillion), which creates analytical challenges for monitoring and development of remediation methods.

In this study, biochars produced from pecan shells and local forestry waste were tested in a lab-scale column for adsorption of NDMA. Concentrations of carbon-14-labeled NDMA in well water obtained near a NASA rocket-test site were measured at regular time intervals over the course of the flow through the biochar-filled column for several adsorbents. Liquid scintillation counting was used for quantification to enable faster and most consistent results with smaller sample sizes. Data from the breakthrough curves was used to calculate column design parameters to estimate the cost of adsorption treatment.

562651 Transforming Plastic Waste into Microporous Adsorbents for Environmental Remediation

By 2050 plastics will outweigh fish if 8 million tonnes of plastics continue to flow into oceans each year and damage marine ecosystems at an annual cost of ~ £9.9 bn. This can be overcome with better plastic recycling technologies. Existing plastic recycling methods are focused on energy-intensive mechanical and chemical methods. In this work, we employed the facile method of solvent-recycling to transform plastic waste into microporous materials with high specific surface areas of up to 1000 m^2/g. The specific surface areas and morphologies of these microporous recyclates could be tailored to suit various environmental remediation applications using different combinations of reaction conditions such as reagent type, ratios and temperature, benefitting carbon capture, dye removal and inorganic toxicants sequestration from water. The adsorption capacities of these recyclate materials were comparable to state-of-the-art nanomaterials such as metal organic frameworks (MOFs), polymers synthesized from expensive fine substrates and commercial adsorbents. Our approach simultaneously benefits the living environment by tackling plastic pollution via the transformation of plastic waste into adsorbents that can potentially be employed to sequester a variety of environmental pollutants – greenhouse gases, organic and inorganic molecules.

569836 Study of Foam Injection Technology for Soil Remediation Contaminated with Hydrocarbons

Soil flushing and soil washing are two remediation techniques preferred for cleaning soil contaminated with hydrocarbons due to their technical simplicity and cost. The efficiency of these techniques when injecting only water may be moderate due to the interaction of the contaminants with the soil components (organic, inorganic). The efficiency of soil flushing techniques may be increased with the use of surfactants. Surfactants decrease the interfacial tension between oil and water leading to the drag of contaminants in the flushing solution; also, they may interact with the soil surface, detaching the adsorbed hydrocarbon components by wettability exchange mechanisms.

The use of surfactants seem a good alternative to increase the efficiency of soil remediation processes, however, the injection of surfactant solutions may lead to different problems such as dragging of contaminant to clean sites even to water reservoirs causing secondary pollution. In the same line, surfactants adsorb to the minerals in the soil causing modification of the natural chemistry of the soil avoiding the natural recomposition of the system. Different strategies have been proposed: the implementation of surfactant solutions at low concentration, the use of bio-surfactants, and recently the injection of foams. Foams are colloid comprising on air trapped in a water matrix, this colloidal system can carry very low amount of surfactant in a very huge volume and possess a high viscosity, which can give control to the flooding process. In this work, we present a review on the latest technologies of foam injection for soil remediation, emphasizing the key variables to consider during soil flooding techniques. In the same line, we present the results we have obtained with a formulation comprising a mixture of synthetic surfactant (SDS) and a bio-surfactant in a test for removing hydrocarbons from soil, foam stability and its relation to hydrocarbon removal.

556406 Facile Ball Milling Method for Developing Surface Functional Groups on Carbon-Based Adsorbents and Their Roles in the Removal of Heavy Metals in Aqueous Solution

Water pollution by heavy metals poses a serious threat to both human health and the environment due to their toxic properties and tend to accumulate in living organisms. Carbon-based adsorbents such as graphene and its derivatives, carbon nanotubes, activated carbon, and biochar are often used to remove heavy metals from aqueous solutions. This work provides a comprehensive review of recent research on various carbon adsorbents in terms of the governing removal mechanisms of their surface functional groups and the associated removal behaviors and performances to heavy metals in aqueous solutions. It then categorizes various synthesis methods that are commonly used to functionalized carbon adsorbent for enhanced heavy metals removal. Facile ball milling method is successfully developed in this study to introduce additional heteroatoms, primarily oxygen and nitrogen onto biochar surfaces for enhanced material and sorptive properties to heavy metals in aqueous solutions. Ball milled biochar had greater internal and external specific surface area, larger pore volume, nano-size properties, stronger negative zeta potential for both O and N functionalizing methods. Faster adsorption kinetics and 6 -10 times enhanced Ni (II) adsorption were displayed. Ball milling thus has great potential to increase the efficiency of environmentally friendly biochar for various environmental applications.

571889 Environmentally-Responsible Conversions from Gasification Waste to Energy Storage Devices: Synthesis of Waste-Derived Nanomaterials and Environmental Evaluation

Waste gasification technology is considered as a sustainable solution to reduce waste volumes and recover energy by producing syngas. However, the leftover of gasification become another waste which needs combustion and landfill. Herein, with the concept of waste-to-resource, an environmentally-responsible pathway has been developed to extract vanadium from a gasification waste, carbon soot, and further converted into vanadium pentoxide. In this study, V_2O_5 nanosheets were synthesized from the leachate of gasification waste, which show comparable specific capacitance as the one from commercial V_2O_5. A flexible supercapacitor was designed by using waste-derived nanosheet and carbon nanoparticle as active materials. The whole process successfully synthesizes V_2O_5 from carbon soot and explores the potential application of carbon waste. Compared to the traditional process of V_2O_5 production, this process demonstrated significantly lower negative environmental impact in terms of human health, ecosystem diversity and resource availability, in a ReCiPe 2016 environmental impact assessment. Moreover, the interesting physical and chemical properties of carbon soot raise the possibility of utilizing carbon soot as a readily-available resource in a large scope of application of carbon materials.

AMELIORATED PHENOL UPTAKE FROM PAPER MILL EFFLUENT ONTO SPENT BREWER'S *SACCHAROMYCES CEREVISIAE* IN THE PRESENCE OF CETYLTRIMETHYL AMMONIUM BROMIDE

Natarajan Saravani, V.S.B. Engineering College, Karur, Tamilnadu,
M.Manimozhi, School of Electrical Engineering, Vellore Institute of Technology, Vellore,
**M. Arulmozhi, Bharathidasan Institute of technology (BIT) Campus, Anna University, Tiruchirappalli, Tamilnadu,*
Subburayalu Ramesh, Annamalai University, Chidambaram, Tamilnadu.

Abstract

The prime purpose of this research is to evaluate the influence of foam separation, an adsorptive bubble separation technique, employed as a finer treatment technique after biosorption. Cetyl Trimethyl Ammonium Bromide (CTAB) played the multiple roles of foam production, phenol elimination, and microbial collection and hence acts as a percentage promoter. The process has its own merits of lower capital investment, less operating and maintenance cost, less power consumption, relatively faster ad eco friendly technique. The effect of several working parameters, such as initial feed concentration, equilibrium time, biosorbent dose, pH, liquid pool height, surfactant concentration, and air flow rate, were experimentally explored. The higher floatability of *Saccharomyces cerevisiae* was obtained in a maximum time of 10 min. Among the models tested, Fowler- Guggenheim model (R^2=0.9703) was the most adequate. The experimental data fitted pseudo second order kinetics with a kinetic constant of $5.5425X10^{-5}hr^{-1}$. Fourier transform infrared spectroscopy (FTIR) spectroscopy revealed the active participation of imidazole in the biosorptive foam separation process. Electro kinetic measurements were carried out to determine the iso electric point (IEP) of the *Saccharomyces cerevisiae* cells. The zeta potential profile of the baker's yeast cell was affected by the presence of phenol at different pH values. The recovery of phenol loaded biomass and final traces of phenol by flotation were found to be 99.75%.

Corresponding author
*Dr. M. Arulmozhi, Professor and Head, Department of Petrochemical Technology, Bharathidasan Institute of technology (BIT) Campus, Anna University, Tiruchirapalli-620 024, Tamilnadu, India.
E-mail id: arulmozhisenthilkumar@gmail.com Mobile: +919364467846
Phone: +91 431 2407946 Fax: +91 431-2407333

Keywords: Cetyl Trimethyl Ammonium Bromide; Fowler- Guggenheim isotherm; Isoelectric point (IEP);*Saccharomyces cerevisiae*; Scanning Electron Microscopy; Zeta potential.

Abbreviations: CTAB, cetyl trimethyl ammonium bromide; 4-AAP, 4-amino anti pyrine; IEP, Isoelectric point; FTIR, Fourier transform infrared spectroscopy; SEM, Scanning Electron Micro

1. Introduction

A massive industrial growth has occurred wide over the globe in the past few decades. The growth is so immense that the environment has entirely changed from what it had been before to what it is today!!!(Virendra Kumar et al, 2014, Abdelwahab et al 2013). Pollution being the major rationale for this drastic change has its consequences in air, water, soil etc. Water, being a chief natural resource, is used in nearly all applications. Paper making, a hefty industry, uses a large extent of water, and consequently the discharge of wastewater is also massive (Jacob Varughese Vadakenat et al, 2014). Pollution due to paper and pulp industry has been rated among the twelve top pollutants. Therefore the proper treatment before the release to the environment is essential. One chief drawbacks of the direct discharge is the incidence of persistent phenol. Higher concentration of phenol lead to multiple disadvantages in both aquatic and human respiratory systems, and may even cause skin disorders eventually leading to cancer. These compounds once discharged have the ability to cause genetic changes to the exposed organisms (Kaizar Hossaina et al, 2015, Bandyopadhyay et al 1998, Swamy AVVS & Syamala Devi K, 2012).

Industries encloses very high phenol concentration; e.g., plastic manufacturing (600-2000ppm), petrochemical manufacturing (50-600ppm) and stocking production (6000ppm). Phenol an aromatic pollutant is present in the wastewater of numerous industries. Effluent standard of phenolic wastewaters is less than 1 ppm for discharging to surface waters. Phenolic compounds are toxic even at low levels and pose a threat to the biosphere because of their recalcitrant nature. Phenol vapors acts as air pollutant and precipitate from the air into the soil. The need to remove these contaminations has led to the development of new technologies that emphasize the detoxification and destruction of the pollutants rather than the conventional approach of disposal (Girish CR &Ramachandra Murty V 2012).

Conventional methods of phenol wastewater treatment include both physical and chemical. These methods include chlorination, ozonation, solvent extraction, incineration, chemical oxidation, membrane process, coagulation, flocculation, adsorption, ion exchange, reverse osmosis, electrolysis, etc (Wei et al 2007, Zijun Xiao et al 2012, Prpich*et al.,* 2005). But these methods led to secondary problems due to formation of toxic materials such as cyanates, chlorinated phenols, hydrocarbons, etc. Solvent extraction poses a risk of contamination of treated water by the solvent & high cost of solvent is yet another disadvantage. The solvents used for phenol recovery are benzene, isopropyl ethyl and butyl acetate. Activated carbon (frequently used adsorbent) is disposed by incineration generating many new compounds such as dioxins and furans. Chemical oxidation operates at high temperature and high pressure. The realization of the high cost of phenol waste treatment by physico-chemical methods has opened the door for the investigation of new methods. The requirement to remove these contaminants has led to the development of new technologies for the fine treatment of the phenolic compounds. Biological treatment, an alternative method is versatile, inexpensive, receiving the most attention due to its environmentally friendly approach & high biodegradation abilities.Thus treatment of phenolic wastewaters is necessary and biological methods are the most appropriate techniques due to mineralization of toxic organic compounds and inexpensiveness (Prpich and Daugulis *et al.,* 2005).

Biosorption is a surface phenomenon, where physical or chemical interactions between phenol molecules and the functional group present in the microbial cell wall plays an important role. The structure of a microbial cell wall is much intricate, thus allowing utmost options for biosorption to occur. Many sorbents are being in use including a range of bacteria, algae fungi etc. (Aditi Banerjee et al, 2010, Yang CF & Lee CM 2007, Gurusamy Annadurai et al 2002, Kumar et al. 2005, Juan & Han-Qing 2006, Jiang Yan et al 2006, Juan Wu & Han-Qing Yu 2008, Jegan et al. 2010, Aditi Banerjee et al 2010, Yu Zhang et al 2011, Aditi Banerjee et al. 2011 Riham Surkatti & Muftah H El-Naas 2014, Neetu Singh &Chandrajit Balomajumder 2016) Natarajan Saravani (2016) recognized that *Micrococcus lylae,* a Gram positive bacteria, is an efficient biosorbent for phenol removal. My Linh Nguyen & Ruey-Shin Juang (2015) investigated the ability of chitosan beads modified with histidine and *Saccharomyces cerevisiae* using glutaraldehyde as a cross linking agent (SC-HIS-CCBs) for the biosorption of phenol. They performed batch biosorption experiments using the SC-HIS-CCBs at a pH (2.0 - 7.0), and initial phenol concentration (25 -250 mg/L). The biosorption capacity of phenol was 28.4 mg/g at a pH 2.0 and 313 K.

The purpose of this study is to investigate the biosorptive foam separation capabilities of *Saccharomyces cerevisiae* to treat paper and pulp mill industry effluent. The uptake and accumulation of hazardous organics by microorganism offers an economical, practical and efficient alternative to existing methods. CTAB a cationic surfactant efficiently fascinates anionic phenolate. Isotherm and Kinetic studies was conceded out by scheming which curve fits the most. About ten models for isotherm and five models for kinetic fit were analysed.The qualitative aspects of infrared spectroscopy is one of the most prevailing attributes of this diverse and multipurpose analytical technique. Over the years, much has been published in conditions of the fundamental absorption frequencies which are the key to unlocking the structure–spectral relationships of the associated molecular vibrations. FTIR is widely used to study about the functional groups present in the outer surface of the cell. In flotation as well as in solid/liquid separation the surface properties of the minerals, i.e. the specific surface charge is of essential importance. In the phenol-water interface electrical forces are also acting. Charged solid particle is surrounded by liquid ions of opposite charge to that of particle surface. As the solid particle moves through the liquid, a liquid layer surrounding the particle will move together with it. Consequently the ionized particle will carry a thin liquid shell making with it an integral whole. The potential difference between the external surface of the shell and the liquid as a whole is called electro kinetic potential or zeta potential. A zeta-potential study is used in colloid chemistry for observing the behavior of dispersive systems in liquids & characterizes the electrical double layer on the solid liquid interface, a very important fact in flotation and flocculation processes. Surface morphology was characterized using Scanning Electron Microscope

2. Materials and methods

2.1. Materials and instruments

Phenolic effluent was collected from nearby paper mill industry and its physicochemical parameters were analyzed. Sodium hydroxide flakes (98%), Hydrochloric acid (HCl), Cetyl Trimethyl Ammonium Bromide (CTAB) was bought from SDFCL (India). Reagents used in 4-amino anti pyrine (4-AAP) procedure for spectrophotometric determination (4-AAP, K_3Fe (CN)

6, K_2HPO_4, KH_2PO_4 and ammonium hydroxide) were of analytical grade and supplied by Merck (Germany).

Dried yeast cells *Saccharomyces cerevisiae,* procured from a local brewery were ground to fine powder using an agate pestle and mortar in order to increase the surface area of the bio mass. It is then stored in a desiccator for future use. Dried biomass was used as an absorbent.

Sartorius basic meter PB-11 (Germany) was used for measuring pH of feed solution and UV-Vis spectrophotometer Shimadzu-UV-2450 (Japan) used for measuring absorbance of the phenolic solution.

2.2 Biosorption experiments

The biosorption experiments were carried out in a series of 250 mL Erlenmeyer flasks containing 50 mL phenol. The phenol biosorption performance at different concentration varying from 100-250ppm was investigated. 0.10g biosorbent was transferred into flasks with an initial phenol concentration ranging from 100-250ppm. Additional experiments were carried out with various biosorbent dosages ranging from 0.05-0.2 g. The pH studies were carried out for different acidic and basic conditions (6, 7, 8, and 9). All the flasks were sealed with cotton stoppers and shaken in a rotary shaker with a speed of 120 rpm. Samples of the flasks were separately taken at appropriate times to observe residual phenol using 4-amino anti pyrine method.

2.3 Biosorptive foam separation experiments

The biosorptive foam separation process scheme is shown in (Fig.1 & Fig.2).The biosorptive foam separation column is a cylindrical glass tube (1000mm in height and 160mm diameter) with a curved opening & a sparger of pore size 40 microns, fixed at the bottom, functioning as a gas distributor. Airflow is metered using a rotameter. Foamate was collected continuously at the top of the column. The liquid obtained by breaking the foam and the residual solution were analyzed using a UV spectrophotometer at a wavelength of 510 nm for phenol concentration. Feed solutions with different surfactant concentrations were used.

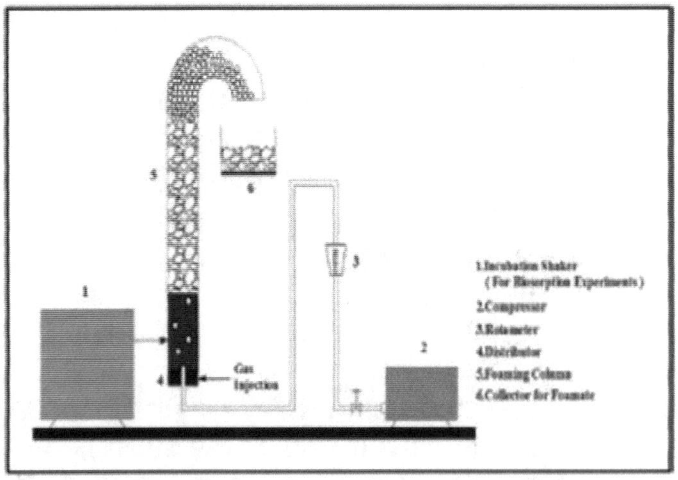

Fig.1 Biosorptive Foam Separation Column

The surfactant to be used in the process must be capable of providing electrostatic interaction with the sample to be analyzed. Hence, the surfactant used here is of cationic Cetyl trimethyl ammonium bromide (CTAB) to attract anionic phenolate.

A known volume of the phenol biosorbed solution, having predetermined concentration of phenol was taken into the foam column. 0.1 % (w/v) Surfactant was added. Air was then passed through the phenol biosorbed solution at a controlled flow rate. After allowing drainage for some specific time, the foam and the residual liquid were separated. The liquid obtained by breaking the foam and the residual pool liquid was then analysed to determine the phenol concentrations using UV Spectrophotometer. The above procedure was repeated for various surfactant concentration, flow rates and height of liquid pool height (Natarajan Saravani et al, 2016).

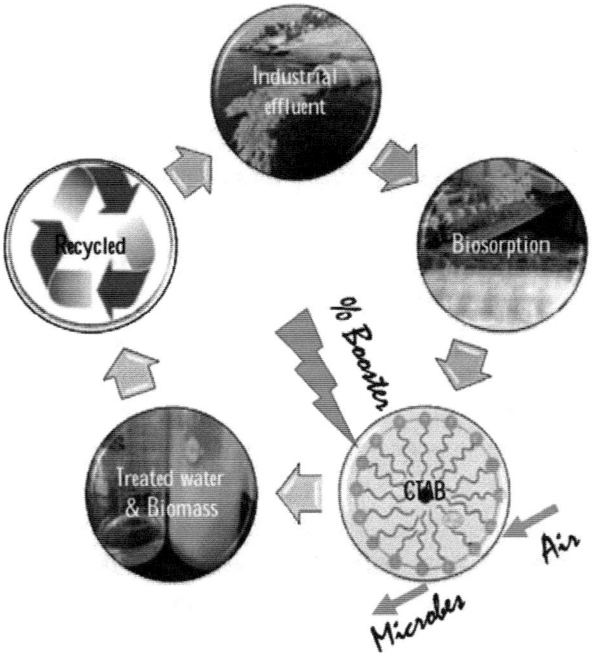

Fig.2 Schematic representation of biosorptive foam separation process

2.4 Phenol estimation

The residual phenol concentration in the culture medium was estimated using 4-aminoantipyrene method based on rapid condensation with 4-aminoantipyrine, followed by oxidation with potassium ferricyanide under alkaline conditions to give a red colour product. The concentration of phenol was calculated from the absorbance values at 510 nm using standards. Percentage removal and enrichment ratio were determined by Eq. (1), and (2):

$$\text{Removal of phenol (\%)} = \frac{C_o - C_f}{C_o} \times 100 \qquad (1)$$

$$\text{Enrichment ratio} = \frac{C_{cf}}{C_o} \qquad (2)$$

Where C_o and C_f are the phenol concentration (mg/L) in the influent and effluent streams respectively, and C_{cf} is phenol concentration in the collapsed foam solution (mg/L), (Natarajan Saravani et al, 2016).

2.5 Characterization experiments

The FTIR spectra of the *Saccharomyces cerevisiae* were recorded on a Perkin Elmer spectrum version 10.03.09, instrument model spectrum two and instrument serial number 92626.

The zeta (ζ) potential of the *Saccharomyces cerevisiae* cells were performed as a function of pH, for both fresh and after biosorptive foam separation using the micro electrophoretic apparatus Zeta Plus (Brookhaven Instruments Corporation, USA).

Scanning electron microscopic images were caught and hence the morphology was characterized using VEGA3 SBH version 4.1.20.3 model, and the instrument serial number VG98912771N.

3. Results & Discussion

3.1 Effects of feed concentration, equilibrium time, biosorbent dose, and pH on biosorption

Batch Experiments was ensued at room temperature, using various concentrations ranging from 100 to 250 mg/l to find the equilibrium time and optimum concentration of phenolic content in pulp and paper mill effluent. The conditions smeared to precede with the batch experiments were; speed of agitation 120 rpm, neutral pH and 0.1g dosage in 50 ml. With an increase in agitation time from 0 to 60 hr, the percentage removal reached up to 99.0%. With the increase in phenol concentration ranging from 100-250 mg/l, the percentage removal increased till 99.0% for 100 mg/l in 60 hrs. There was a rapid adsorption comprehended in the start, and becomes bit slower while the equilibrium state was achieved. Equilibrium was attained before the expected time and it was also agreed by other authors. Initial concentration difference provides the driving force. Higher the difference between the aqueous and solid phase, there is an increase in the rate of adsorbate molecule in bulk solution that passes through to reach the adsorbent surface (Juan et al, 2006).

The biosorbent dosage effects were tested for a range of 0.05g-0.20g. The results for the entire range of dose 0.05g-0.20g examined showed a very similar trend of very rapid removal at the initial, followed by a final stability and hence equilibrium. An increase of biosorbent dosage from 0.05g-0.20g leads an increase from 33.1 % to 99.0 % of removal of phenol. This increase in phenol loading capacity for a corresponding increase in dosage may be due to the accessibility of greater amounts of active sites of the sorbent remaining unsaturated. Utmost removal in the pulp and paper effluent was accomplished at a reduced dose of 0.1 g. A further increase shows decreased response due to the binding of almost all the phenol to the sorbent and establishment of equilibrium. That is, the available phenol molecules are inadequate to cover all the exchangeable sites on the sorbent, resulting in a lower phenol uptake (N. Saravani and M. Arulmozhi, 2016).

With varying pH ranging from acidic to base (6-9), pH optimization experiments were carried out using the optimum concentration (100 mg/l), dose of 0.1 g to find out the optimum. The pH is an important parameter that affects phenol solution chemistry and the surface characteristics of *Saccharomyces cerevisiae*. Response remains almost constant throughout the range of pH examined. According to the results obtained it was seen that both acidic and basic pH attain constant equilibrium and errand biosorption, yet biosorption of phenolic compound from pulp and paper was most favored at neutral pH 7. At acidic pH, the cell wall of *Saccharomyces cerevisiae* (biosorbent) was closely related with Hydronium ions competing for negative active sites and, it was caused due to the repulsive force that restricts the approach of phenolate ions. Amide groups and imidazole are positively charged, when protonated and play a vital role in phenolics adsorption. Moreover the amino and carboxyl groups, nitrogen and oxygen of the peptide bonds may also participate in the scene, which may mainly depends on the degree of displacement of protons, determined by the pH (Saifuddin, N. and Raziah, 2007). The maximum equilibrium biosorptive capacity of phenol compounds in pulp and paper effluent was achieved at pH 7. When the pH was increased further the degree of protonation decreases and hence the percentage removal decreases.

3.2 Effects of liquid pool height, surfactant concentration, and air flow rate on biosorptive foam separation performance

Biosorptive foam separation is a hybrid procedure, which in this context was used to enhance phenolics removal in the treatment of the paper and pulp industry effluent. Percentage removal of phenol in the effluent was noted to be 99.0% at the end of biosorption. Considering the aspect that even minute traces of phenol could be harmful, the next stage involves the removal of phenol loaded biomass and the minimal traces of phenol left behind, by the addition of a cationic surfactant. The exhausted cells and the final traces of phenol were eradicated along with the foam produced during the process. The influence of flotation parameters on percentage removal and enrichment ratio were studied and optimized during the process.

The effect of liquid pool height on the percentage removal of phenol and enrichment ratio using Cetyl Trimethyl Ammonium Bromide was studied at a surfactant concentration of 0.2 (w/v %) and an airflow rate of 4.5 lpm. As the liquid pool height increases from 10 to 30 cm the percentage removal of phenol also increases from 37.5 to 50.4%. Similarly the enrichment ratio also increased from 9.0 to 13.5. This consequence is mainly due to the concept that as the residence time of the bubbles is high, when the liquid pool height is high. The enrichment of the phenol on the surface of the bubble is also high until it reaches a stage of equilibrium. When the height is increased beyond 30 cm the percentage and the enrichment ratio decreases to 48.35% and 5.61 respectively. This is because the equilibrium has been already reached and further addition had no effect (Arulmozhi et al 2013).

The effect of Cetyl Trimethyl Ammonium Bromide on spent aqueous phenol under constant conditions of liquid pool height of 30 cm and air flow rate of 4.5 lpm & a variety of surfactant concentrations ranging from 0.1 to 0.4 (w/v %) of water were studied. The ability of surfactants in the treatment of effluents has been reported by Afzal Shah et al (Afzal Shah et al, 2016). The recovery of phenol using CTAB as the surfactant was found high due to its cationic nature as phenol is slightly acidic and anionic. As the concentration of the surfactant increased the percentage removal of phenol was also found to increase from 9.8 to 50.92%

and the enrichment ratio decreased from 10.78 to 9.73 respectively. On further increase of the surfactant concentration from 0.3 to 0.4% the percentage removal of phenol was found to decrease from 23.2 to 22.85% with an increase in the enrichment ratio from 11.96 to 12.27, in agreement to the report by Arulmozhi et al (Arulmozhi et al, 2013). The increase in the percentage removal was due to the fact that the effluent containing phenol binds to the cationic surfactant and removed to the foam. Once the level of saturation is attained the percentage of phenol decreased as surplus micelles were formed with high surfactant concentration (on increasing beyond 0.2 (w/v %).

The effect of air flow rate on enrichment ratio and percentage removal of the phenol containing effluent is at an optimum conditions liquid pool height of 30 cm and surfactant concentration of 0.2 (w/v %). As the air flow rate increases from 4.5 to 5 lpm, the percentage removal increased from 50.0% to 60.64% and the enrichment ratio increased from 1.73 to 4.85. This occurred due to the basis that initially at low flow rates, the quantity of phenol to be transported to the foam bubble surfaces are more. But the residence time of the bubble in the column is low which in turn results in more water content on the foam thereby enhancing the removal of phenol in the suspension. On further increase of air flow rate to 5.5 and 6, it was found that the percentage removal of phenol decreased to 27.8 and 25.0% and the enrichment ratio reduced to 3.6 to 1.5 due to higher air flow the bubbles reached the top of the column causing removal of surfactant rather than the trace level phenol. These results were in accordance with the results by Arulmozhi et al (Arulmozhi et al 2010).

3.3. Isotherm studies, Fowler Guggenheim isotherm model

In order to provide quantitative information on the adsorption capacity of the sorbent, experiments were carried out under the experimental conditions of 30^{0}C, 2 g/L of the sorbent and phenolic effluent of different initial concentration (100-250 mg/L). These flasks were placed on a rotating incubation shaker with a speed of 120 rpm till equilibrium. The amount adsorbed at equilibrium w_e is calculated by the Eq. (3)

$$q_e = \frac{(Co - Ce)V}{m} \tag{3}$$

Where C_o is the initial phenol concentration, V is the volume of solution (L) and m is the dry weight of the biosorbent (g) Equilibrium data were fitted using ten different isotherm models, namely Langmuir, Freundlich, Tempkin, Redlich Peterson, BET, Dubinin Radushkevich, Elovich, Halsey, Harkins Jura, and Fowler Guggenheim isotherm models. Results showed that Fowler Guggenheim isotherm fitted the data well with an R^2 value of 0.9703 (Fig. 2).

Fowler- Guggenheim isotherm model equation explains the lateral interactions of the adsorbed molecules. The equation is expressed as Eq. (4);

$$\frac{\theta}{C_o} = K_{FG}\,(1-\theta)^{n_{FG}} \tag{4}$$

And the linearized form is given by Eq. (5)

$$ln\ \frac{C_e(1-\theta)}{\theta} = -lnK_{FG} + \frac{2W\theta}{RT} \tag{5}$$

Where K_{FG} is Fowler- Guggenheim equilibrium constant (L mg^{-1})

n_{FG} is model exponent

$\theta = 1 - \frac{C_e}{C_o}$ is the degree of surface coverage.

W is the interaction between the adsorbed molecules (kJ mol^{-1})

R is the gas constant (8.314 J $mol^{-1} K^{-1}$)
T is the absolute temperature (K)

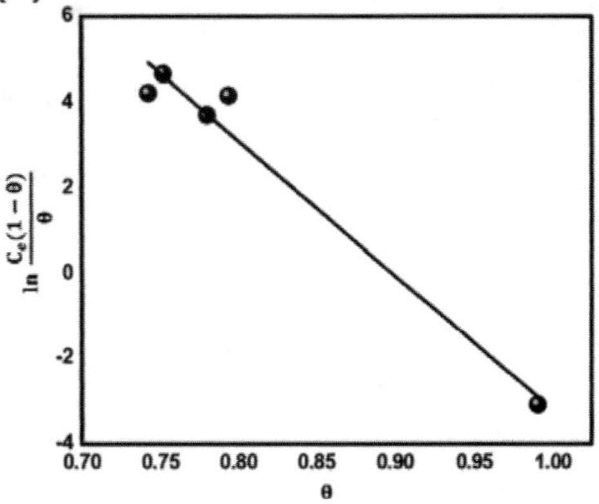

Fig. 2 Linearized Fowler Guggenheim isotherm model for phenol removal (temperature of 30°C, dose of 0.1g/50 mlbiosorbent, agitation speed of 200 rpm and pH 7).

3.4. Biosorption kinetics & Pseudo second order model

Biosorption kinetics was studied at 30^0 C and the rate was monitored till a contact time of sixty hours and the results were compared to theoretical models. Adsorbed amount of phenol at any time q_t is calculated by Eq. (6)

$$q_t = \frac{(Co-Ct)V}{m} \qquad (6)$$

Where q_t is the amount of biosorption at time t (mg/g), C_o is the initial concentration of the phenolic compound in the solution (mg/L), C_t is the concentration of the phenolic compound in the solution at time t (mg/L), V is the volume of solution (L) and m is the mass of the biosorbent (g).

Pseudo first order, Pseudo second order, Elovich, Intraparticle diffusion and Mass transfer diffusion models (Natarajan Saravani et al, 2016) were tested in this study and best fit shown in Fig. (3). Results showed that the Pseudo second order model fitted the data well. The pseudo-second-order based on adsorption equilibrium capacity is expressed in Eq. (7) and linear form of pseudo second order is given in Eq. (8)

$$\frac{dq_t}{dt} = K_{2ad}(q_e - q_t)^2 \qquad (7)$$

$$\frac{t}{q_t} = \frac{1}{K_{2ad}q_e^2} + \frac{t}{q_e} \qquad (8)$$

Where K_{2ad} Pseudo-second-order rate constant (1/mol.hr).
A linear plot of t/q_t vs. t gives a slope of the equilibrium adsorption capacity q_e

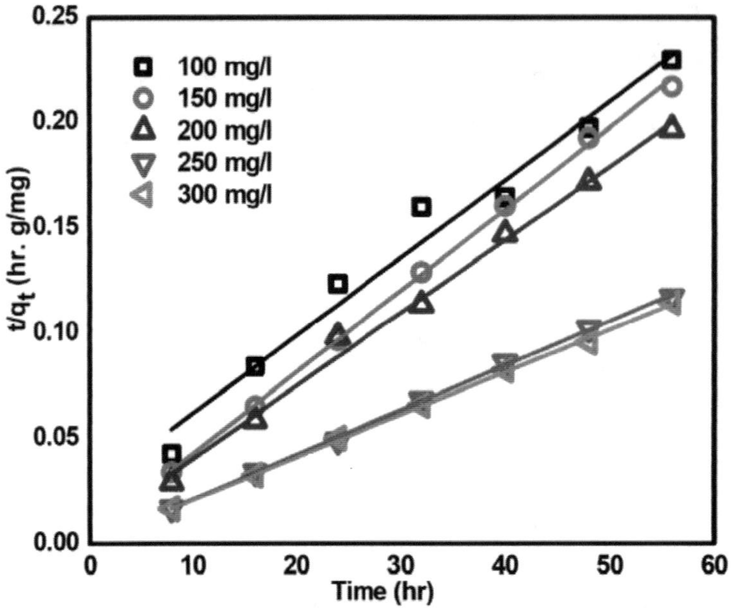

Fig. 3 Linearized Pseudo second-order model of phenol removal (temperature of 30°C, dose of 0.1g/50 mlbiosorbent, agitation speed of 200 rpm and pH 7).

3.5. Fourier transform infrared (FTIR) analysis

The FTIR spectra of raw, phenol and, phenol CTAB treated yeast in the range of 4000–400 cm^{-1} were taken to acquire information on the nature of the cell wall and phenol interaction, were shown in Fig.4. Chemical characteristics of the samples were analyzed by FTIR spectrometry after the samples were prepared as KBr discs. FTIR spectrum of the fresh biosorbent exhibited broad and strong absorption bands around 3200-3500 cm^{-1} indicating the presence of –OH stretching vibrations. Other peaks around 1691.27cm^{-1}(amide band), 1335.46 cm^{-1} (aromatic nitro compound NO_2 symmetric stretching; imidazole) and 1213.97 cm^{-1} (C–H stretching) have also been observed. Peaks around 724.14 cm^{-1} (NH bending), 523.58 cm^{-1} (-CH stretching of aromatic ring) and 438.73cm^{-1} (C–X stretching) have also been observed. The phenol treated yeast showed a broad peak at around 3600–3000 cm^{-1}, which could be attributed to sorption of the phenol onto the yeast. A shift in peak from 1691.27cm^{-1} to 1690.30 cm^{-1} and 1335.46 cm^{-1} to 1334.50 cm^{-1} is an indicative of active participation of amide band and the aromatic nitro compound NO_2 symmetric stretching (imidazole) in phenol adsorption. The peaks at 1213.97 cm^{-1} can be designated as the C-H stretching in carboxyl or amine groups. Other peaks around 724.14 cm^{-1}(NH bending), 473.44 cm^{-1} (S-S stretching) and 432.94 cm^{-1} (C–X stretching) have also been observed. The phenol CTAB treated yeast showed a new peak at 3743.15 cm^{-1} confirming the presence –OH stretching and bending and also showed a broad peak at around 3600–3000 cm^{-1}, which could be attributed to sorption of the phenol onto the yeast. A peak at 1688.37 cm^{-1} confirms the presence of amide band. A peak at 1621.84 cm^{-1} confirms the presence of intermolecular β structure of amide band. A peak at 1334.5 cm^{-1}(aromatic nitro compound NO_2 symmetric stretching). The peak at 1219.97 cm^{-1} is indicative of aliphatic C-N stretching. Other peaks around 720.28 cm^{-1}(NH bending), 472.47 cm^{-1} (S-S stretching) and 435.83 cm^{-1} (C–X stretching) have also been observed (Stuart).

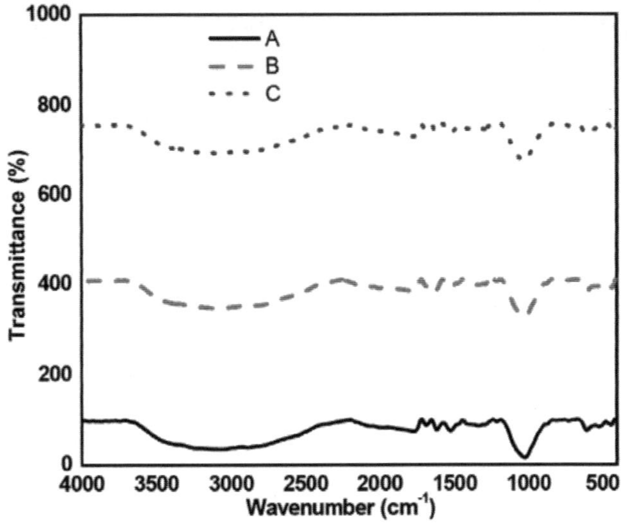

Fig. 4 FTIR spectra of *Saccharomyces cerevisiae*(A) fresh biosorbent (B) after biosorption and (C) after biosorptive foam separation,

3.6. Electro-kinetic studies

This instrument determines the electrophoretic mobility and converts to ζ potential using the Smoluchowski approximation (Natarajan Saravani et al, 2016). The reported values are an average of 10 measurements. The ζ potential measurements provide information regarding the surface charge of fresh and phenol CTAB loaded cells (Fig. 5).The pH values were adjusted with the addition of NaOH and HCl. Electrophoretic mobility measurements were also performed and the respective ζ potential values were calculated, in order to examine the possible biosorption mechanism. The electro kinetic study provides preliminary information regarding the type of surfactant to be added for the separation of fine particles by flotation. The experimental negative ζ values, right through the pH values examined, shows an isoelectric point (IEP) at a pH of around 2, proved the application of the positive surfactant. This renders hydrophobicity and hence, floatability. The hydrophobicity of yeast mainly depends on the presence of various polysaccharides, proteins, lipids, and carboxylate, amines, and imidazole, phosphate, sulfhydryl, sulfate and hydroxyl group moieties. When examined alone, the zeta potential values going on decreases till a pH of 7 and the charge reverses after that, which corresponds to maximum sorption capacity. The surface charge of the non-living biomass becomes less negative till the neutral pH and becomes more negative towards basic region in the presence of cationic surfactant. At its extreme end, the suspension is quite unstable. Microbial walls are anionic due to the presence of carboxyl, hydroxyl, phosphoryl and other negatively charged sites. The phenol binding properties of yeast are largely due to positively charged amino and imidazole groups, which are part of the cell wall structure. Due to this fixed cation content, they may exhibit high sorption capacities (Saifuddin, N. and Raziah, 2007). From the above discussions it is evident that the zeta-potential changes according to changes in the solution pH, and zeta-potential measurements makes it possible to follow the changes. After biosorptive foam separation the zeta potential of biomass becomes less negative. This could be attributed to the electrostatic interaction between the biomass, CTAB and phenolate ions. This effect is caused by the dissociation of the functional groups at high pH values, which will promote an electrostatic attraction toward negatively charged phenolate ions in the solution.

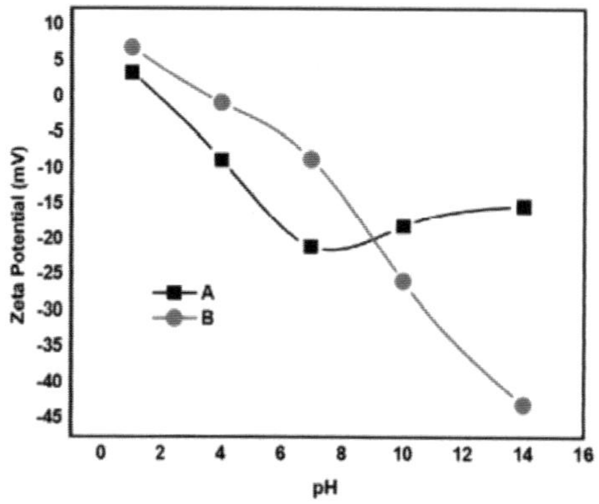

Fig. 5 ζ-Potential-pH curves of (A) fresh and (B) phenol CTAB loaded *Saccharomyces cerevisiae.*

3.7 SEM Analysis

It is evident from the Fig.6 that the *Saccharomyces cerevisiae* dead cells were of oval shaped aggregates with a smooth and glossy finish. They stay apart each other with almost no interaction. The change in surface morphology of the *Saccharomyces cerevisiae,* before and after biosorptive foam separation was observed at a magnification of 5 µm, using an accelerating voltage of 10 kV. All the samples were sputter coated with a gold ultrathin film prior to its mounting on a sample stub for SEM imaging. All the *Saccharomyces cerevisiae* dead cells were perfectly oval both before and after biosorptive foam separation. *Saccharomyces cerevisiae* cells were bit heftier indicating the presence of phenolic compounds after biosorptive foam separation.

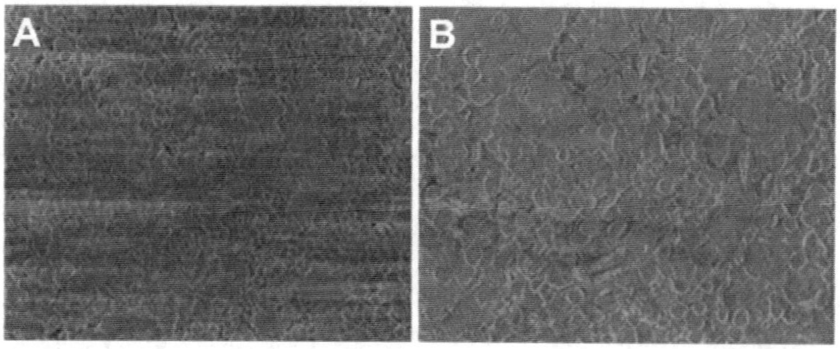

Fig.6 SEM images of *Saccharomyces cerevisiae***(A)** fresh biosorbent**(B)** after biosorptive foam separation.

4. CONCLUSION

The results of biosorptive foam separation of phenolic compounds by *Saccharomyces cerevisiae*, has been substantiated to be a most favorable technique. It is a two stage process where sorption onto dead cells of *Saccharomyces cerevisiae* constitutes the first stage and separation of phenol loaded biomass, and phenol lean solution without *Saccharomyces cerevisiae* cells constitutes the next. The conditions for maximum phenol removal are 100 mg/L of initial phenol concentration, sixty hours of contact time period, 0.1 g of cells, and a pH of 7, and the conditions for the second stage of biosorptive foam separation are a liquid pool height of 30 cm, cationic surfactant concentration of 0.2% (w/v), and an air flow rate of 5 L/m. Isotherm and kinetic studies revealed that the biosorption process followed the Fowler-Guggenheim isotherm model (R^2 = 0. 9703) and pseudo-second order kinetics with a kinetic constant of $5.5425X10^{-5}$ hr^{-1}. FTIR spectra for the fresh biosorbent, after biosorption, and after biosorptive foam separation were generated to indicate the interactions and bonding. The ζ-potential of the *Saccharomyces cerevisiae* cells was affected by the presence of phenol at different pH values showing isoelectric point at a pH of around 2. The results obtained from this study showed that *Saccharomyces cerevisiae* can be a good sorbent for the removal of phenol from an aqueous solution and CTAB can be a most appropriate surfactant. The recovery of phenol loaded biomass and final traces of phenol by floatation were found to be of the order of 99.75%.

Acknowledgements

The authors express their gratitude for the financial support extended by the Department of Science and Technology (DST-Government of India).Grant. No: SR/FT/ETA-0017/2012 and Technical Education Quality Improvement program (TEQIP II).

References

1. Abdelwahab O, Amin NK (2013), 'Biosorption of phenol from aqueous solutions by Luffacylindrica fibers: Kinetics, isotherm, and thermodynamic studies'. Egyptian Journal of Aquatic Research vol. 39, pp.215–223.
2. Aditi Banerjee, AlokeGhoshal K (2010); 'Phenol degradation by Bacillus cereus: Pathway and kinetic modeling', *Bioresource Technology*, 101, 5501-5507.
3. Aditi Banerjee, Aloke K. Ghoshal (2010), 'Isolation and characterization of hyper phenol tolerant Bacillus sp. from oil refinery and exploration sites', Journal of Hazardous Materials, vol.176, pp. 85–91
4. Aditi Banerjee, Aloke K. Ghoshal (2011), 'Phenol degradation performance by isolated *Bacillus cereus* immobilized in alginate', International Bio deterioration & Biodegradation, vol. 65, pp.1052-1060.
5. Afzal Shah,SuniyaShahzad,AzeemaMunir,Mallikarjuna N. Nadagouda,GulShahzada Khan,DilawarFarhan Shams,Dionysios D. Dionysiou, Usman Ali Rana(2016); 'Micelles as Soil and Water Decontamination Agents', *Journal of Chemical Reviews,* DOI: 10.1021/acs.chemrev.6b00132.

570638 Biosurfactant Fermented from Waste Cooking Oil and Its Application on Oil Sludge Celaning

The maximum biosurfactant concentration of 5.98 g L^{-1} was achieved from kitchen waste oil by *Pseudomonas aeruginosa*. Then biosurfactant was extracted from the broth and applied to wash oil-based drilling cuttings. Effect rules of six parameters on oil recovery were investigated separately, and the main influencing factors of concentration, pH, and temperature were screened out by Plackett-Burman design. And then the oil washing conditions were optimized by response surface method. The maximum recovery rate was predicted to be 70.48%, when rhamnolipid concentration was 325 mg L^{-1}, pH was 10.70, temperature was 63 °C, washing time was 35 min, stirring speed was 450 r min^{-1}, and liquid/solid ratio was 5:1. Finally, the optimal washing parameters and corresponding maximum recovery rate were verified through experiment, with the relative error less than 0.006%. The results indicating promising prospect of biosurfactant application in industrial oil recovery.

571053 Liquid-Liquid Extraction of Chlorophenols from Wastewater Using Hydrophobic Ionic Liquids

In this work, liquid-liquid extraction of 3-chlorophenol, 2, 5-dichlorophenol, 2, 4, 6-trichlorophenol and pentachlorophenol from wastewater was conducted using five hydrophobic ionic liquids (ILs). Moreover, the impact of key parameters like molecular structure, pH, temperature, phase ratio and initial concentration of feed on the removal of the chlorophenols (CPs) from their aqueous solution were studied. The results showed that the molecular structure of the IL has an important impact on the removal of CPs. The evaluation of different ILs with the same CPs revealed that piperidinium and pyridinium based ILs have greater extraction efficiency than sulfonium and imidazolium based ILs. It was found that the extraction efficiency decreased with the decrease of IL: wastewater mass ratio, with a maximum extraction at a ratio of 1:1. The results showed that temperature and initial concentrations in the range of 25-45°C and 4-1000 mg/L, respectively had no significant impact on the extraction for all studied systems.

576142 Overview of Waste Valorization Concepts and Principles

AIChE defines waste valorization as "the process of reusing, recycling or composting waste materials and converting them into more useful products including materials, chemicals, fuels or other sources of energy". "Waste-to-energy" aspects are becoming more prominent due to the rapid depletion of natural resources and increase in waste generation.

Waste valorization is the process of taking waste and changing it into constituent parts that can be utilized, and have value beyond the cost of the energy needed to process the transformation. The concept of waste valorization ties in with recycling and reuse technologies. Namely "enhancing the value", valorization is a process of changing residues into products at a much greater value. The results can include quality chemicals, materials, fuels, and energy, as well as many other products beneficial for a local economy.

It requires not only technical knowledge but also a global approach to accomplish this goal. Residues and products often need to be seen from a wider perspective that just one industry. Creating a closed system of valorization, industries may need to communicate with each other in the global economy pattern.

This presentation presents a technical overview on waste valorization addressing the principles and concepts involved.

571173 Conversion of Polyvinylchloride Containing Solid Wastes into Value-Added Products

Conversion of polyvinylchloride containing solid wastes into value-added products

Xianglan Bai, Joel Braden

Department of Mechanical Engineering, Iowa State University, Ames, IA, USA, 50011

Municipal Solid Waste (MSW) is a mixture of organic and synthetic materials. The organic material has the potential to be converted into valuable fuels through thermochemical conversion. On the other hand, main plastic wastes, such as PE and PP, can be an excellent source of chemicals or high-density liquid fuels. Previous studies by our group and others have shown that co-conversion of biomass and hydrocarbon-based plastics through catalytic or non-catalytic pyrolysis can be benefited from synergistic effects among the different materials. Nevertheless, MSW also contains significant quantities of halogenated plastics like polyvinyl chloride (PVC) that are difficult to recycle using traditional methods. When MSW is thermally converted (i.e., pyrolysis, gasification and incineration), PVC can not only produce corrosive hydrochloride gas, but also react with organic waste-derived products to form organohalogens and dioxins that create various environmental and health hazards. Our group has investigated a new approach to utilize PVC as an asset to promote MSW conversion to value-added products. We first co-converted cellulose and PVC in heated tetrahydrofuran (THF) solution. It was found that levoglucosan was the major monomer in liquid product. The highest monomeric sugar yield was 42% at 335 °C and 8 minutes with 2.5% cellulose weight equivalent PVC. The liquid products also contained up to 46% of soluble oligomeric sugars and small amounts of 5-HMF and furfural. Chlorinated organics were nearly absent in either liquid product or the pyrolyzates of the solid residue. Therefore, the results confirmed our hypothesis that hydrochloric acid released during thermal decomposition of PVC acts as an effective catalyst in THF to promote cellulose decomposition to soluble sugars. Encouraged by the result, we also converted a mixture of cardboard, PVC and PE in THF solution. As a result, not only PVC greatly facilitated rapid decomposition of cardboard to liquid product, but the presence of PE also helped to increase the conversion. Up to 68% of cardboard was converted into liquid product containing various organic compounds, such as furans, sugars and phenolics. The remaining solid residues could be easily separated into two fractions. The first fraction was fine char derived from cardboard and it had a HHV of 20.8 MJ/kg, compared to 11.7 MJ/kg with the cardboard. The second fraction of the residue was a waxy material. Upon pyrolyzing, the waxy residue produced a serious of aliphatic hydrocarbons identical to that are produced from pyrolysis of pure PE. More significantly, Cl atom was nearly theoretically converted to HCl and thus can be recovered from the solvent. Overall, we explored an environmentally friendly approach to convert mixed solid wastes into value added products.

574480 Development of High Performance Composites Using Recycled Materials

The purpose of the work is to develop and characterize high performance polymer matrix composites using recycled materials that otherwise are sent to landfills. Recycled polymer and recycled fibers are blended and molded to form a composite that possesses both high specific modulus and specific strength. Preliminary result of mechanical tests and thermomechanical analysis on the high performance composite material are presented.

576718 Sustainable Mass-Burn Refuse-Derived-Fuel-to-Energy System: A Municipal Solid Waste Management Case Study

Waste-to-energy (WtE) or energy-from-waste (EfW) is the process of generating energy in the form of electricity and/or heat from the primary treatment of waste. WtE is a form of energy recovery. Most WtE processes produce electricity and/or heat directly through combustion, or produce a combustible fuel commodity, such as methane, methanol, ethanol or synthetic fuels.

Gasification and pyrolysis processes have been known and used for centuries and for coal as early as the 18th century.... Development technologies for processing [residual solid mixed waste] has only become a focus of attention in recent years stimulated by the search for more efficient energy recovery.

Waste-to-energy can significantly reduce the amount of waste needing disposal, generate energy for a community, and also reduce municipal solid waste (MSW) transportation costs. It becomes more favorable for wastes that have high-energy content, low-moisture content, and low-ash content. These wastes include paper, plastics, textiles, rubber, leather, and wood. Therefore, MSW to a large extent is of biological origin (biogenic). Typically, more than half of the energy content in MSW is from biogenic material. Consequently, this energy is often recognized as renewable energy according to the waste input.

A case study of a WtE facility, operating at 200,000 tons of refuse annually, will be presented with pertinent information including but not limited to design and operation, annual electricity generation via turbine, life-cycle costs and overall profitability, total ash handling, emission control of PM, NO_x, SO_2, mercury, dioxins, and protection of health and environmental.

569146 Investigation of Reactions at the Gas/Solid Interface of Municipal Solid Waste Residues

Reactions at the gas-solid interfaces are present during the treatment/combustion of municipal solid waste (MSW) in waste-to-energy facilities, leaving a solid residue, ash, that has significant potential to valorize. Storage and transport of MSW ash after combustion in waste-to-energy facilities allows for reactions at the gas-solid interface such as corrosion and hydration. These reactions can potentially lead to issues such as production of undesirable gases, rise in temperature above regulation limits, and the consumption of recoverable metals and minerals from the ash. The issues of crucial concern in the waste management industry are elevated temperature in landfills, and the production of hydrogen from ash.

In order to minimize these issues in MSW management, it is necessary to understand the effect temperature, pressure, moisture content, and ash content on the rates of reaction and the products of the reactions. It is also important to study the characteristics of MSW ash and the its contribution in the reactions occurring in a landfill.

The first part of the talk focuses on understanding the impact of pressure, moisture, and temperature on MSW pyrolysis as a possible cause for elevated temperature in landfills (ETLFs). MSW provided from landfills is tested under simulated landfill conditions in controlled temperature and pressure environment. The effect on landfill gas concentrations (CH_4 and CO_2) in the reactors over a range of temperatures (49 to 177 °C) and moisture content (40 to 60 wt.%) in the presence of biological inhibitors are studied. Results from these tests identify a range of moisture content that impact the gas concentration during MSW decomposition. Gas analysis using microGC (gas chromatography) from these experiments identify the presence of hydrogen during MSW decomposition, which is also observed in elevated temperature landfills.

The second part of the talk focuses on characterization of MSW ash, and the reactions at the gas-solid interface of waste-to-energy ash after metal recovery. The ferrous and non-ferrous metals recovered from waste-to-energy ash are characterized using scanning electron microscopy (SEM), energy-dispersive X-ray spectroscopy (EDS), and X-ray diffraction (XRD). Hydrogen evolution resulting from the gas-solid interface reactions in MSW ash are recorded over time in controlled environment using gas analysis. The hydrogen production rate shows correlation with the loss of aluminum from MSW ash. Results also show higher hydrogen concentration in ash samples containing higher metal fractions, which points to ash hydration and carbonation as plausible reactions causing the elevated temperature behavior in landfills with co-disposal of ash.

577816 Microwave Enhanced Advanced Oxidation Treatment of Anaerobic Digested Effluent

Microwave enhanced advanced oxidation treatment (MWAOP) is a thermo-chemical treatment process which has been used in several studies for effective treatment of organic wastes (Saha, M. 2018). In MWAOP, hydrogen peroxide oxidant is converted to hydroxyl radicals in presence of microwave heating. Hydroxyl radical is a much stronger oxidant than hydrogen peroxide. MWAOP oxidizes organic waste to final products such as volatile fatty acids (VFA) and carbon dioxide (CO_2) (Liao et al. 2007). This process disintegrates solids and helps to release nutrients from organic waste (Saha, M. 2018). In this study, MWAOP has been applied to treat anaerobic digested effluent. Anaerobic digestion is extensively used for renewable energy production from organic wastes. The anaerobic digested effluent is rich in nutrients and can be used as a potential source for nutrient recovery. This study was focused on investigating the effectiveness of MWAOP to treat anaerobic digested effluent to produce more suitable substrate for the subsequent nutrient recovery. The digested effluent was collected from a dairy farm. All experiments were performed in a batch scale microwave system of 2450 MHz. The digested effluent was treated at two different temperatures (90°C and 110°C) using MWAOP. The results showed that settleability of MWAOP treated digested effluent was improved and soluble COD was increased after the treatment. Hence, MWAOP treated digested effluent might improve the subsequent nutrient recovery process.

References

[1] Liao, P. H., Lo, K. V., Chan, W. I., & Wong, W. T. (2007). Sludge reduction and volatile fatty acid recovery using microwave advanced oxidation process. Journal of Environmental Science and Health Part A, 42(5), 633-639.

[2] Saha, M. (2018). Microwave enhanced advanced oxidation treatment of organic slurries (Master's dissertation, University of British Columbia).

564782 CO_2 Conversion to 3D Graphene for Clean-Energy Application

Abstract: In the recent year, we discovered several reactions for the synthesis of 3D graphene [1-5]. In the presentation, I will highlight recent findings for CO_2 conversion to 3D graphene in my group: (1) CO_2 can react with alkaline metals (Li, Na, or K) to form various types of 3D graphene; (2) the synthesized 3D graphene materials possesses unique properties (such as high electrical conductivity and large accessible surface area); and (3) the 3D graphene materials exhibited excellent performance as electrodes for solar cells, supercapacitors, and batteries.

References:

1. H. Wang, K. Sun, F. Tao, D. J. Stacchiola, Y. H. Hu, "3D Honeycomb-Like Structured Graphene and Its High Efficiency as Counter Electrode Catalysts for Dye-sensitized Solar Cell", *Angew. Chem. Int. Ed.* 52, 9210(2013).
2. W. Wei, K. Sun, L. Cheng, A. J. Pak, E. Paek, G. Hwang, Y. H. Hu, "The Bright Future for Electrode Materials of Energy Devices: Highly Conductive Porous Na-Embedded Carbon", *Nano Lett.* 16, 8029(2016).
3. L. Chang, D. J. Stacchiola, Y. H. Hu, "An Ideal Electrode Material, 3D Surface-Microporous Graphene for Supercapacitors with Ultrahigh Areal Capacitance", *ACS Appl. Mater. Interface* 9, 24655(2017).
4. W. Wei, B. Hu, F. Jin, Z. Jing, Y. Li, A. A. G. Blanco, D. J. Stacchiola, Y. H. Hu, "Potassium-Chemical Synthesis of 3D Graphene from CO_2 and Its Excellent Performance for HTM-free Perovskite Solar Cells", *J. Mater. Chem. A*, 5, 7749 (2017).
5. L. Chang, K. Sun, Y. H. Hu, "New Chemistry for New Material: Highly Dense Meso-Porous Carbon Electrode for Electric Double-layer Capacitors with Ultrahigh Areal, Gravimetric, and Volumetric Capacitances", *ACS Appl. Mater. Interface*, 10, 33162(2018).

570669 Waste Plastics to Energy Storing Carbon Anodes through Green Processing

For the methodical recycling of abiquitous plastic waste, greener manufacturing or chemical processes are required for the environmental and societal benefits[1]. Sustainable engineering is the process of designing a system(s) in which they use energy and resources sustainably; ultimately generating an ability of future generations to meet their own needs. Environmentally ***harmful plastic waste*** is a major concern currently throughout the world.[2] Though plastics improve our lives in countless ways, they also pose a **serious** threat to our environment as plastic waste. However, the major challenges faced in recycling are (i) plastic waste collection, segregation, cleaning with associated cost, (ii) chemically different polymers extrusion yields low quality product and (iii) restrictions and imparted duties by other countries on import of recyclable plastic waste.

During last decade, Pol's research team innovated and researched on an environmentally beneficial thermochemical upcycling[3] processes with possible important applications for the obtained products contributing to the sustainable engineering field. Developed one-step, low-energy, solvent-less process is called **"upcycling"** because it produces products[4] having *greater value* than the original plastics. Developed transformative processing method could **permanently** reduce pollution at its source, providing **greener reaction conditions.** Established **greener** processing system for the conversion of unsorted plastic waste into structurally tunable carbon microspheres[5] and carbon nanotubes[6] that have important tribological[7] and battery applications[5,8] will be discussed. Moreover, recent **invention** lead by PI transforms polymer and starch waste, including used packing peanuts, into a multi-functional carbon sheets powder via an **environmentally friendly**, inexpensive, and scalable production[9] route. The manufactured carbon sheet particles demonstrate a wide-range of potential applications including: **electrodes** for energy storage, toners for printers, activated carbons for **storing CO_2**, conductive additives for tires, lubrication additives, and many more. This industrially viable, facile **green process** utilizes trashed packing peanut feedstock that is heat treated to 600 °C under an inert atmosphere in a single-step attaining reproducible product as promising lithium[9] and sodium[10] ion battery anode. In summary, this talk is centered on the originated **green processing,** which applies economic and technologically feasible parameters to engineer a product decreasing pollution at its source, minimize hazardous exposures, and reduce energy consumption protecting human health.

References

1. United Nations Environment Assembly. *Towards a pollution-free planet.* Report No. UNEP/EA.3/L.19 (United Nations, 2017).
2. United Nations. *Transforming our World: The 2030 Agenda for Sustainable Development.* Report No. A/RES/70/1 (United Nations, 2015).
3. V. G. Pol, Upcycling: Converting Waste Plastics into Paramagnetic, Conducting, Solid, Pure Carbon Microspheres. *Environ. Sci. Technol.*, **2010**, *44*, 4753–4759.
4. V. G. Pol, J. Wen, K–C. Lau, S. Callear, D.T. Bowron, C–K. Lin, S. A. Deshmukh, S. Sankaranarayanan, L. A. Curtiss, W. I. F. David, D. J. Miller, M. M. Thackeray, Probing the Evolution and Morphology of Hard Carbon Spheres. *Carbon,* **2014**, *68,* 104-111
5. V. G. Pol, M. M. Thackeray, Spherical Carbon Particles and Carbon Nanotubes Prepared by Autogenic Reactions: Evaluation as Anodes in Lithium Electrochemical Cells, *Energy & Environ. Sci.*, **2011**, 4, 1904–1912
6. V. G. Pol, P. Thiyagarajan, "Remediating Plastic Waste into Carbon Nanotubes," *J. Environ. Monitoring*, 2010, 12, 455–459.
7. K. K. Mistry, V. G. Pol, M. M. Thackeray, J. Wen, D. J. Miller, A. Erdemir, Synthesis and Tribology of Micro-Carbon Sphere Additives for Enhanced Lubrication, *Tribology Transactions*, **2015**, 58, 3.
8. Pol, V. G.; Lee, E.; Zhou, D.; Dogan, F.; Calderon-Moreno, J. M.; Jonhson C. S. Spherical Carbon

as a New High-Rate Anode for Sodium-ion Batteries. *Electrochim. Acta*, **2014**, *127*, 61-67.

9. Etacheri, C. N. Hong, V. G. Pol, Upcycling of Packing-Peanuts into Carbon Microsheet Anodes for Lithium-Ion Batteries. *Environ. Sci. Technol.* **2015**, *49*, 11191.
10. J. Tang, J. Barker, V. G. Pol, "Sodium-ion Battery Anode Comprising Carbon Sheets: Stable Cycling in Half and Full-Pouch Cell Configuration", *Energy Technology*, 2018, 6, 213-220.

560332 Molecular Simulation and Experimental Study of CH4 Adsorption in Surfactant Solution

Molecular Simulation and Experimental Study of CH4 Adsorption in Surfactant solution
Yihan Xue
Beijing Key Laboratory of Urban Oil and Gas Distribution Technology, China University of Petroleum (Beijing), Beijing, China
Email address:
xyh3610@126.com(Yihan Xue)
Abstract: The research on the storage and transportation technology of green methane is a hot spot nowadays. In order to overcome the disadvantages of LNG and CNG, the effect and method of methane adsorption in surfactant aqueous solutions are studied. In order to compare the effect of the methane adsorption in surfactant aqueous solutions, five surfactant aqueous solutions of sodium dodecyl sulfate (SDS), polyoxyethylene sorbitan monooleate (Tween 80), octylphenol ethoxylate (TX-100), dodecyl dimethyl betaine (BS-12) and castor oil polyoxyethylene ether (EL-20) at the concentration range of 25 g/L-200 g/L are selected for the study. Molecular simulation and experimental research are adoped in the research process. The adsorption energy and adsorption amount in five surfactant systems with different concentrations is numerically simulated at normal pressure or low pressure with normal temperature, at the same time, experimental study is performed. The numerical simulation results agree well with the experimental results. The results show that SDS has the best adsorption effect on methane at the same concentration. According to the research results, suggestions are made for the industrial application of methane adsorption in surfactant aqueous solutions.
Keywords: Methane storage and transportation, Surfactant aqueous solution, Adsorption energy, Adsorption amount

564914 Reactivity of Ni-Co/γ-Al_2O_3 Catalysts for Hydrodeoxygenation of Guaiacol

Reactivity of Ni-Co/ γ-Al_2O_3 catalysts for hydrodeoxygenation of guaiacol

Deepak Raikwar [1], Saptarshi Majumdar [1], and Debaprasad Shee [1*]

[1]*Department of Chemical Engineering, Indian Institute of Technology Hyderabad, Kandi, Sangareddy, 502 285,*

Telangana, India

Email: ch13m15p000001@iith.ac.in, saptarshi@iith.ac.in, dshee@iith.ac.in

Abstract:

Keywords: **Guaiacol, Hydrodeoxygenation, Nickel-Cobalt, Cyclohexane, Benzene**

1. Introduction:

Lignin is one of the primary component of lignocellulosic biomass and the most abundant aromatic polymer available in nature. It has been considered as an alternative and renewable feedstock to petroleum-based fossil fuel for the production of platform chemicals such as benzene, xylene, toluene (BTX). However, the lignin depolymerized bio-oil primarily contains array of oxygenated aromatic monomers which makes bio-oil unsuitable for direct utilization. Therefore, further upgradation of lignin-derived bio-oil is essential to eliminate such oxygen functionalities and to refine their properties. Catalytic hydrodeoxygenation (HDO) is considered as the most effective route for upgrading lignin-derived bio-oil during which platform chemicals (aromatics) and blending agents (cycloalkanes) can be produced. Guaiacol is often selected as a representative model compound for upgradation studies mainly due to its presence in bio-oil in high fraction and existence of three types carbon-oxygen linkages: C(sp^3)-OAr, C(sp^2)-OCH_3 (methoxy), and C(sp^2)-OH (hydroxyl). The presence of three different linkages in molecular structure with different bond dissociation energies poses a specific challenge for development of active and selective catalyst to produce specific product from guaiacol.

In recent years, several catalysts have been utilized in the HDO of guaiacol which completely or partially removes oxygen functionality from the compound[1]. Supported monometallic/bimetallic noble metal catalysts (Rh, Ru, Pt, Pd) has been extensively studied for HDO of guaiacol. However, the application of these catalysts is limited due to their high cost and scarcity. Recently, Ni-based catalysts have been applied in HDO reactions mainly because of high hydrogenation activity and low cost[1]. The high hydrogenation activity of Ni can be tailored with the addition of a second metal. Therefore, a combination of Ni and various other transition metals such as Fe, Mo, and Cu have been evaluated for guaiacol HDO . The major drawbacks associated with these reports are either incapability of catalysts to break $C(sp^2)$–OH group giving partially deoxygenated products or excessive hydrogenation capability and coke formation. In present study, HDO of guaiacol over less expensive bimetallic nickel-cobalt supported catalyst was examined for the production of valuable chemicals. The main objective of present study was to: (i) Optimization of Ni and Co mole ratio for complete removal of oxygen functionality from guaiacol (ii) Understand the coordinative effect Ni and Co species on physiochemical characteristics and structural properties (iii) Elucidate the structure-reactivity relationship of Ni-Co bimetallic catalysts.

2. Experimental:

Promoted and non-promoted Ni, Co, and Ni-Co catalysts supported over pretreated γ-Al_2O_3 were synthesized using co-impregnation method. All prepared catalysts were characterized by BET, XRD, TPR, and FT-IR[3]. HDO of guaiacol was carried out in a 300 mL stainless steel make high-pressure and temperature batch reactor (Parr Instruments, USA, Model no:4566HT).

3. Result and Discussion:

BET surface area of calcined and reduced Ni, Co and Ni-Co catalysts were obtained in the range of 90-220 m^2g^{-1}. The powder XRD patterns of calcined $_xNi_yCoAl$ bimetallic catalyst with different Ni(x)/Co(y) mole ratio at a constant total metal loading of 9.2 $mmolg^{-1}{}_{cat}$ are displayed in Figure 1. The XRD patterns of oxide catalysts gave a clear indication for the formation of nickel-cobalt mixed oxide ($NiCo_2O_4$) due to shifting and weakening of diffraction peaks corresponding to Co_3O_4 and NiO species[5]. Therefore, subsequent formation of an oxidized bimetallic phase could provide an easier way for the formation of Ni-Co alloy during H_2 reduction. The H_2-TPR profile of alumina supported $_xNi_yCo$ bimetallic catalyst with various Ni/Co mole ratio calcined at 723 K for 6h is shown in Figure 2. The presence of a sharp reduction peak with high H_2 consumption at 573 K in H_2-TPR profile of bimetallic catalysts with Ni/Co mole ratio of 1:2 ($_{3.05}Ni_{6.1}CoAl$) further confirms the presence of pure $NiCo_2O_4$ spinel. Therefore, the possibility of formation of Ni-Co alloyed species during reduction is higher in $_{3.05}Ni_{6.1}CoAl$ as compared to other catalysts in the series at an optimal reduction temperature of 723 K. Metal alloy formation may be the reason for enhanced catalytic activity because of the change in electron density as a result of formation of heteronuclear metal-metal bond[6].

The effect of variation in Ni/Co mole ratio for $_xNi_yCoAl$ during guaiacol conversion and product distribution is demonstrated in Figure 3. A guaiacol conversion of 95% with cyclohexane (21%), cyclohexene (33%), phenol (20%) and benzene (10%) were obtained over monometallic $_{9.2}CoAl$. Whereas, monometallic $_{9.2}NiAl$ resulted in a guaiacol conversion of 79% with cyclohexane

(44%), phenol (17%) and benzene (13%) as major product. The higher selectivity of aromatic ring saturated product over $_{9.2}$NiAl clearly demonstrate high hydrogenation capability of metallic Ni than Co. The bimetallic catalyst had a better performance during guaiacol HDO for the formation of cyclohexane and benzene as compared to monometallic ones. The conversion of guaiacol was observed to increase with the increase in Ni content up to Ni/Co mole ratio of 1:2, but decreased with further replacement by Ni. The bimetallic catalyst $_{3.05}Ni_{6.1}$CoAl (Ni/Co mole ratio=1:2) showed substantially higher guaiacol conversion (99%) with cyclohexane (56%) and benzene (29%) as the major product. A closer analysis of variation in selectivity of aromatic product (benzene) represents that it passed through a maximum at the Ni/Co mole ratio of 1:2. From these results, it can be inferred that Ni/Co mole ratio influences molecular and electronic structure and catalytic activity of $_{x}Ni_{y}$CoAl. The higher activity of $_{3.05}Ni_{6.1}$CoAl suggests that there exists an optimal Ni/Co ratio close to the value required for the formation of $NiCo_2O_4$ spinel. Upon reduction in H_2, $NiCo_2O_4$ spinel transformed to Ni-Co alloy of different electronic properties. A further increase or decrease in mole ratio of Ni/Co breaks the stoichiometry required for the formation of $NiCo_2O_4$ spinel and therefore Ni-Co alloyed domains, as a result, activity of catalyst decreases. It is also consistent with XRD and H_2-TPR data. H_2-TPR profile of $_{3.05}Ni_{6.1}$CoAl suggests the presence of a higher proportion of $NiCo_2O_4$ spinel than NiO and Co_3O_4. Ni-Co alloyed domains provides a different adsorption site than Ni and Co with new electron density which in turn effects the chemisorption capacity of guaiacol and other intermediates during guaiacol HDO[7]. The metallic Ni activates H_2 molecule whereas metallic Co and Ni-Co alloyed domains provides oxophilic sites for the adsorption of guaiacol coordinating with an oxygen atom. The activated H_2 molecule then interacts with the oxygen atom and facilitates cleavage of C_{Ar}-O linkage. The Ni-Co alloyed species promote the activation of C_{Ar}-OCH_3 and C_{Ar}-OH linkage thereby enhances the selectivity towards aromatic products. The presence of higher fraction Ni-Co alloyed species on $_{3.05}Ni_{6.1}$CoAl is attributed to its superior activity during guaiacol HDO. The further optimization of reaction parameters such as reaction temperature, initial H_2 pressure, catalyst loading leads to a maximum selectivity of 57% and 32% for cyclohexane and benzene with complete conversion of guaiacol.

Therefore, in terms of selectivity of deoxygenated products (cyclohexane and benzene) and guaiacol conversion, Ni-Co bimetallic catalyst can be a suitable catalyst for HDO of compounds present in lignin-derived bio-oil.

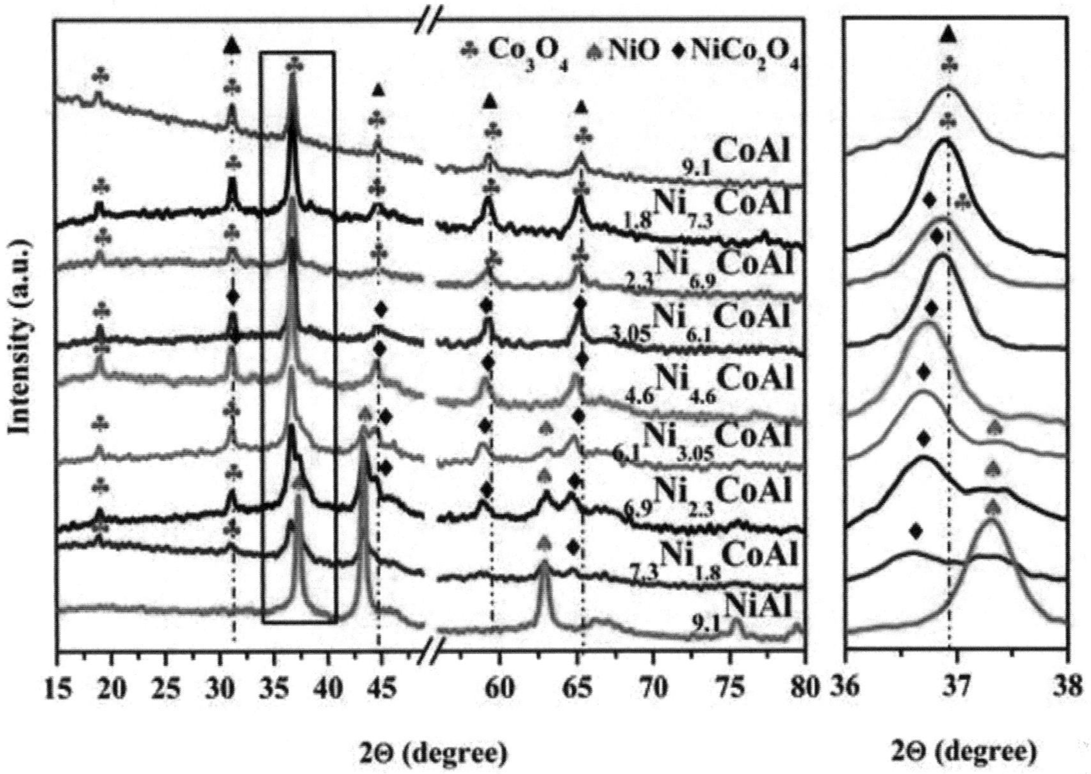

Figure 1 XRD patterns of oxidized ${}_{x}Ni_{y}CoAl$ catalysts of varying Ni/Co mole ratio.

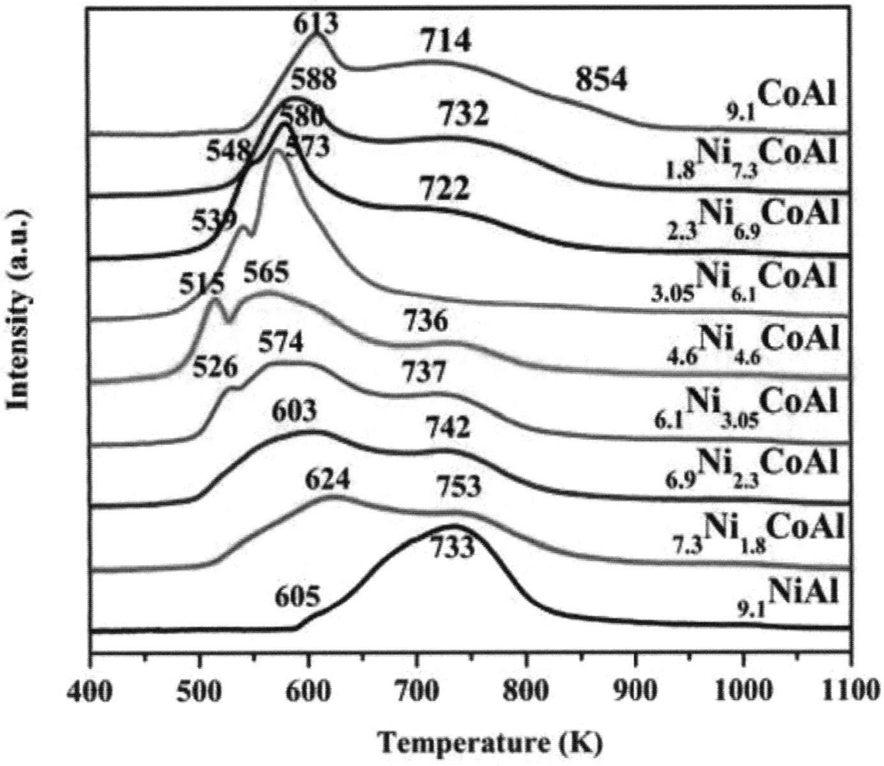

Figure 2 H_2-TPR profile of ${}_{x}Ni_{y}CoAl$ catalysts of varying Ni/Co mole ratio.

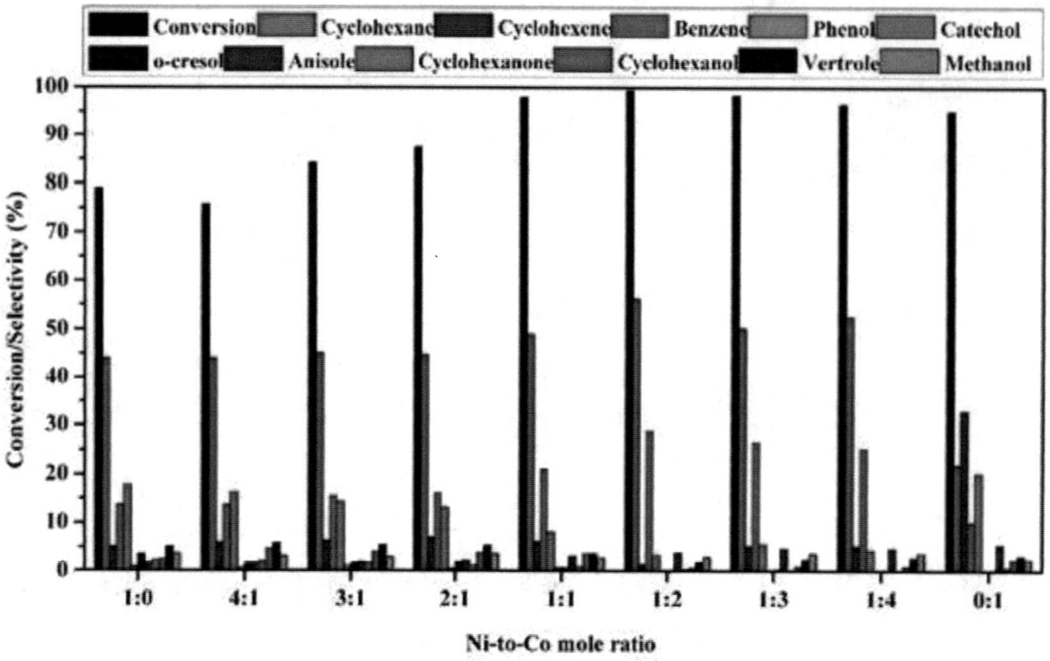

Figure 3 Effect of nickel to cobalt mole ratio on guaiacol conversion and product distribution.

Reaction conditions: Guaiacol= 5 ml, Tetraline= 95 ml, Catalyst loading= 0.25% (w/v), Temperature=

573 K, Initial hydrogen pressure= 10 bars, Reaction time= 4h.

4. Conclusion:

The following inference can be drawn from present study:

- Co has been used as a modifier with Ni to develop alumina supported non-noble bimetallic catalyst for controlled HDO of guaiacol in a high temperature and pressure batch reactor under a wide range of process parameter.

The molar ratio of Ni and Co species in supported$_x$Ni$_y$CoAl catalyst is governing factor for generation of catalytically active $NiCo_2O_4$ and Ni-Co alloyed species.

The characterization results suggest the presence of these species significantly in catalyst with a Ni/Co mole ratio of 1:2 ($_{3.05}Ni_{6.1}CoAl$).

A superior guaiacol conversion (99%) and significant selectivity of cyclohexane (57%) and benzene (32%) was obtained under optimal reaction conditions over $_{3.05}Ni_{6.1}CoAl$.

5. Acknowledgments

The authors are grateful for financial support from Department of Science and Technology (DST), Govt. Of India.

6. References

1 A. Ng, K. Lup, F. Abnisa, W. Mohd and A. Wan, *J. Ind. Eng. Chem.*, 2017, **56**, 1–34.

2 H. Fang, J. Zheng, X. Luo, J. Du, A. Roldan, S. Leoni and Y. Yuan, *Appl. Catal. A Gen.*, 2017, **529**, 20–31.

3 D. Raikwar, M. Munagala, S. Majumdar and D. Shee, *Catal. Today*, 2019, **325**, 117–130.

4 X. Zhang, T. Wang, L. Ma, Q. Zhang, Y. Yu and Q. Liu, *Catal. Commun.*, 2013, **33**, 15–19.

5 S. Yao, X. Wang, Y. Jiang, F. Wu, X. Chen and X. Mu, *ACS Sustain. Chem. Eng.*, 2014, **2**, 173–180.

6 D. W. Goodman and J. A. Rodriguez, *Science (80).*, 1992, **257**, 897–903.

7 T. Ishihara and H. Arai, *J. Catal.*, 1992, **241**, 232–241.

557505 "H_2 -Free" Hydrodeoxygenation of Guaiacol over Ni-Mo/CeO_2-C Nanocatalysts

H_2-free hydrodeoxygenation of guaiacol over Ni-Mo/CeO_2-C nanocatalysts

W. Jin [1], L. Pastor-Pérez [1,2]*, J.J. Villora-Picó[2], S.Gu, A. Sepúlveda-Escribano[2], T. R. Reina [1]*

[1] Chemical and Process Engineering Department, University of Surrey, Guildford, UK

[2] Departamento de Química Inorgánica, Instituto Universitario de Materiales de Alicante, Alicante, Spain

**Autor principal: l.pastorperez@surrey.ac.uk*

1. Introduction

Catalytic hydrodeoxygenation (HDO) is a fundamental technology to achieve the effective upgrading of bio-oil with low emission. Currently, the widespread implementation of HDO is limited by the supply of high pressure H_2 (4-20 MPa). To date, several methods (*e.g.* catalytic transfer hydrogenation (CTH), reforming followed by HDO, the combination of metal oxidation with water and HDO and non-thermal plasma (NTP) technology [1]) have been proposed which could achieve the deoxygenation with *in-situ* hydrogen generation. It is noteworthy that metal oxidation with H_2O (*i.e.* Zn) followed by subsequent HDO is an attractive approach due to cheap water is used as the hydrogen source. Nevertheless, the regeneration of metal requires high energy input [2-3] limiting its industrial application. Undoubtedly, it is ideal to use water as the hydrogen source during HDO process considering readily available and cheap advantages of water over other candidates. Accordingly, we propose a novel HDO method to realize the *in-situ* HDO suppressing the supply of external H_2. It is envisioned that water would undergo splitting on the catalytic surface to produce hydrogen. Hydrogen can further participate in the HDO of bio-oil to faciliate the oxygen removal purpose. Ni-based catalysts are promising candidates for *in-situ* HDO process by tuning the support composition and/or including promoters to adjust their catalytic functions. In this work, Ni-based CeO_2 and/or activated carbon supported catalysts were synthesized, characterized and tested in guaiacol HDO water-only reaction system. This novel method brings a new perspetive in the development of econimically viable bio-oil upgrading technologies.

2. Experimental

CeO_2-C (10wt.% CeO_2) was prepared in advance before the synthesis of Ni-based catalysts. Five catalysts namely, Ni/CeO_2, NiMo/CeO_2, Ni/C, Ni/CeO_2-C and NiMo/CeO_2-C (with 15 wt.% Ni and 2 wt.% Mo) were synthesized by wet impregnation method and characterized by H_2-TPR, XRD, N_2-adsorption, TEM, XPS and Raman analysis. Prior to the activity tests, catalysts were activated at 400 °C for 1h under H_2 atmosphere. The catalytic performances of catalysts for guaiacol HDO process were tested in a high pressure batch reactor (Parr Series 5500 HPCL Reactor) by feeding 50 mL of a 0.01 mg/mL guaiacol solution and 200 mg of catalyst at 250 °C for 4h. After the reaction, the catalyst was collected by filtration.

Products were separated from water by using ethyl acetate and then analyzed by GC/FID and GC/MS.

3. Results and discussion

The XRD patterns of Ni-based catalysts after activation were shown in Figure 1.It is clear that ceria and Ni dispersion are much better when activated carbon are used as primary support. The Ni and ceria particle sizes are smaller in the carbon supported materials. The textural properties summarised in Table 1 also indicate the suitability of activated carbon to achieve large surface areas. The catalytic performance of all studied catalysts is presented in Figure 2. C-containing catalysts exhibited higher activity compared to CeO_2 supported catalysts. For instance, 22 % conversion of guaiacol is obtained in HDO over Ni/CeO_2-C catalyst, which is a significant result in the case of water-only HDO reaction system. Regarding the characterization of spent samples, no metal sintering and minor concentration of carbon deposition were observed in the TEM images (Figure 3) and also by XRD and Raman.

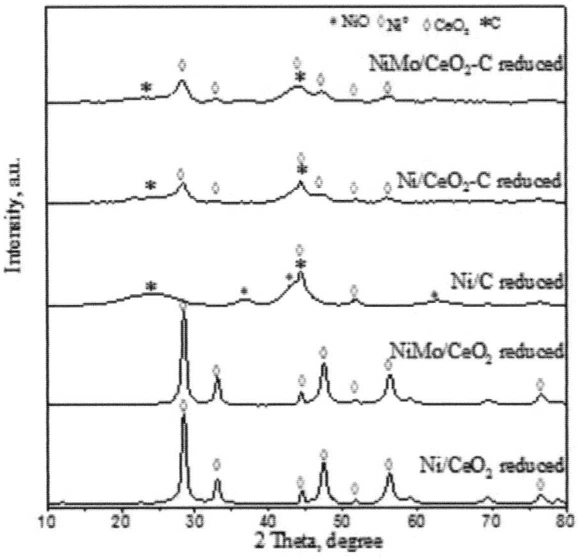

Figure 1 XRD patterns of reduced catalysts

Table 1 Textural properties of catalysts

	S_{BET} (m^2/g)	V_{micro} (cc/g)^a	V_{total} (cc/g)
Ni/CeO_2	78	0.029	0.064
NiMo/CeO_2	46	0.017	0.069
Ni/C	1062	0.378	0.85
Ni/CeO_2-C	1027	0.368	0.79
NiMo/CeO_2-C	1017	0.364	0.79

4. Conclusions

A novel approach consisting on the combination of water splitting and subsequent hydrodeoxygenation (HDO) is proposed in this work with the aim to suppress the supply of H_2 to the HDO reactor. CeO_2-C supported catalysts performed high HDO ability compared to CeO_2-supported catalysts in guaiacol *in-situ* HDO process. The high HDO activity of CeO_2-C supported catalysts is likely attributed to the high dispersion and small particle size of active phase along with the high surface area of catalysts. The proposed novel method by using available and cheap water as hydrogen source in HDO reaction is an interesting alternative which opens new research possibilities to achieve low-cost bio-oil upgrading process.

References

1. Jin, W.; Pastor-Perez, L.; Shen, D.; Sepúlveda-Escribano, A.; Gu, S.; Reina, T. R., *ChemCatChem* **2019**, 11, 924-960.

2. Cheng, S.; Wei, L.; Alsowij, M. R.; Corbin, F.; Julson, J.; Boakye, E.; Raynie, D., *Journal of the Energy Institute* **2018**, 91, 163-171.

3. Cheng, S.; Wei, L.; Julson, J.; Muthukumarappan, K.; Kharel, P. R.; Cao, Y.; Boakye, E.; Raynie, D.; Gu, Z., *Journal of the Taiwan Institute of Chemical Engineers* **2017**, 74,

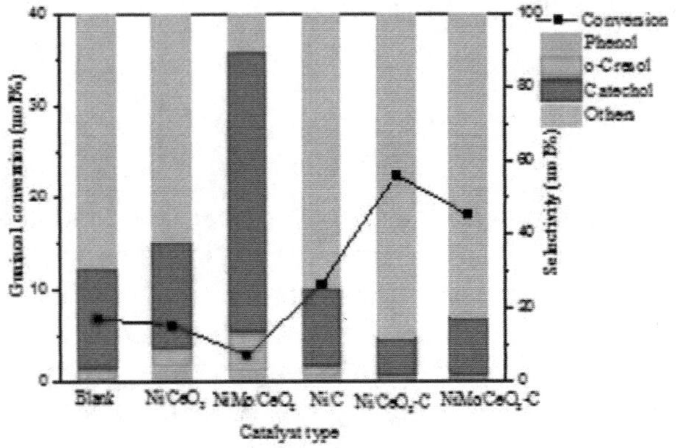

Figure 2. Activity of catalysts during *in-situ* HDO process

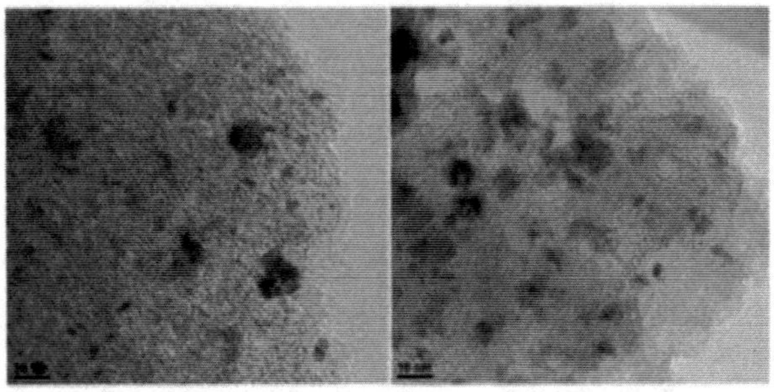

Figure 3. Selected TEM images of Ni/CeO_2-C

146-153.

568113 Microwave Carbocatalysis for Bioenergy and Biochemical Conversion

This work focuses on the utilization of the synergy of microwave irradiation and graphene-based catalysis as applied to various reactions related to biomass conversion processes including cellulose depolymerization, glucose conversion into platform chemicals, and transesterification/esterification/etherification of oil fractions and products present in biomass.

556465 The Green Production Process of e-Caprolactam with Vapor-Phase Beckmann Rearrangement

The Green Production Process of ¦Å-caprolactam with Vapor-Phase Beckmann Rearrangement

Hao Wang, Shibiao Cheng, Yingqi Fan, Li Xie, Yusheng Jiang, Keyong Yang, Baoning Zong

Process Develop Department, Research Institute of Petroleum Processing (RIPP), SINOPEC, Beijing 100083, P. R. China ¨C wanghao.ripp@sinopec.com;

Keywords: ¦Å-caprolactam, vapor-phase, Beckmann rearrangement

1 Introduction

Vapor-phase Beckmann rearrangement is a green, free-ammonium sulfate ¦Å-caprolactam production process. With the increasing demand for ¦Å-caprolactam, this process has been widely concerned. The existing liquid-phase method, as shown in Eq.(1), uses sulfuric acid as a catalyst, and produces about 1.5 tons of ammonium sulfate per ton of ¦Å-caprolactam, which will be a great burden of ¦Å-caprolactam enterprises in the future. Compared with the existing processes, the vapor-phase rearrangement process, as shown in Eq.(2), has the advantages of low equipment corrosion, low waste water discharge, high product quality, safe and stable operation, and thus has a broad market development prospect as a new generation of ¦Å-caprolactam production process.

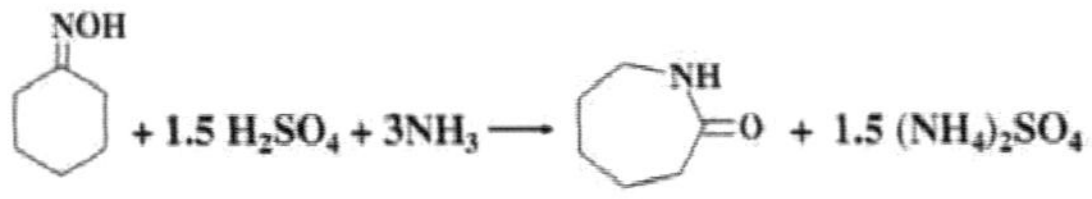

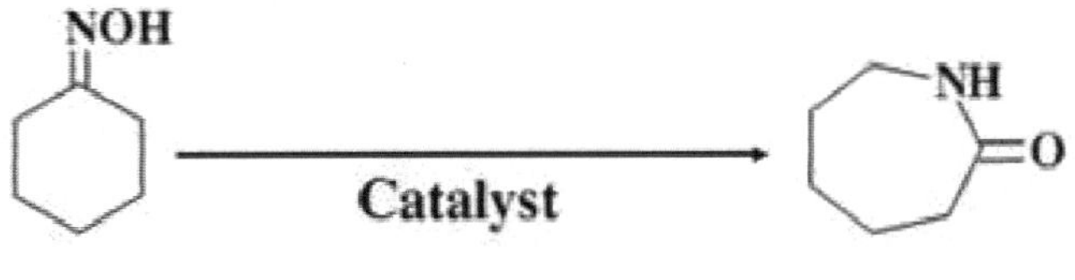

2 The production process of ¦Å-caprolactam with vapor-phase Beckmann rearrangement

The process consists of vapor-phase Beckmann rearrangement reaction unit and crystallization refining unit. The determined process conditions and process flow are quite different from those of the existing HPO process.

The vapor-phase Beckmann rearrangement process is a gas phase reaction process, which requires cyclohexanone oxime to participate in the reaction with gas phase. However, due to its heat-sensitive characteristics, its vaporization problem is the primary problem in the development of vapor-phase Beckmann rearrangement process. Secondly, since the reaction proceeds at a high temperature, the product distribution characteristics determine that the refining process requires a crystallization method with high-efficiency separation method. In the process of crystallization process development, the high quality product was been obtained. RIPP-SINOPEC relies on 800 tons/year cyclohexanone oxime vapor-phase Beckmann rearrangement pilot equipment device to solve lot problems after years of process development. A total of 5000 hours of experimental investigation was completed in this device. Time-on-stream of the catalyst was greater than 700 hours, the conversion of cyclohexanone oxime was greater than 99.9%, and the selectivity of caprolactam was greater than 96.5%. Meanwhile, the high quality ¦Å-caprolactam was produced with crystallization refining in this pilot plant. At present, the process has already been industrialized and promoted.

3 Conclusion

Vapor-phase Beckmann rearrangement process of produced ¦Å-caprolactam was developed by RIPP-SINOPEC, which is a green process and high atom economic. The technology has completed pilot plant verification and will realize industrial production in recent years.

571592 Vapor-Phase Hydrotreating of Furfural over Ni-M/SiO_2 Catalysts (M= Fe, Cu, Pt, Pd) for Propylene Production

Furfural can be mass-produced from lignocellulose biomass and can be a platform chemical for producing valuable chemicals. The propylene is one of the valuable chemicals, and it has a prospect from the furfural hydrotreating. In previous work, we examined 5wt% Ni/SiO_2 catalysts for the conversion of furfural under a hydrogen atmosphere. The propylene selectivity was governed by the higher temperatures (>225 ^oC) and edge sites of Ni. In this study, the Fe, Cu, Pt, and Pd (Ni: M= 5:1 in molar ratio) dope in the Ni/SiO_2 catalyst to change the Ni structure. The evaluation of catalytic performance is not only by reaction activity but also by coking analysis over these Ni-base catalysts. Finally, it concludes how the second metal changes the Ni structure and the optimized Ni-Pd/SiO_2 improves the reaction temperature to 175 ^oC for the best performance of propylene production in these catalysts, including Ni/SiO_2 and Pd/SiO_2 catalysts, too.

574432 Hydrogenation of 2-Methylnaphthalene over Bi-Functional Supported Nickel Catalysts

The products of thermal biomass conversion processes, such as pyrolysis, tend to be mixtures of aromatic and poly-aromatic compounds. While aromatics are desirable for gasoline (boiling point cuts below 200°C), they are not desirable in middle distillates. Instead, middle distillate fractions need to be hydrogenated and have naphthenic rings opened to improve the combustion characteristics for kerosene and diesel applications. Thermal DeOxygenation (TDO) is a thermal conversion process that converts organic acids in cellulose hydrolyzates to an essentially oxygen-free organic oil containing a large number of substituted naphthalene compounds.

The current study identifies catalysts and upgrading strategies for the TDO oil using 2-methylnaphthalene as a model compound. Although there have been many studies performed on monocyclic compounds, there have been fewer studies focused on the hydrogenation of naphthalene derivatives. We have focused on nickel as a catalyst because of its low cost and high tolerance to potential impurities in the oil. Furthermore, we have demonstrated that naphthols, which are the predominant oxygen species in TDO oil, are easily deoxygenated over high-surface-area nickel. However, this catalyst is not effective for opening rings. Therefore, a fixed-bed reactor operating in differential mode was used to test nickel supported on oxides with varying acidity to determine the activity and selectivity of the bi-functional catalysts for hydrogenation and ring opening.

563487 Kinetics Study of the Simultaneous Hydrodeoxygenation of Xylitol over a ReO_x-Pd/CeO_2 Catalyst

Xylitol is a sugar alcohol derived from the hemicellulose portion of lignocellulosic biomass, and is mass produced and commonly used as a sugar substitute. However, xylitol can be chemically upgraded into value added platform chemicals such as 1,2-dideoxypentitol and 1,2,5-pentanetriol via a simultaneous hydrodeoxygenation (S-HDO). 1,2-dideoxypentitol and 1,2,5-pentanetriol are 300-5000 times more valuable than the initial cost of xylitol. The S-HDO removes two vicinal hydroxyl groups by undergoing a deoxydehydration to form a double bond and then is hydrogenated to a single bond[1]. The ReO_x-Pd/CeO_2 catalyst was determined to be the best catalyst tested in literature for the S-HDO of sugar alcohols[2]. The S-HDO of xylitol has been previously studied at 160°C and 80bar H_2[1] and our previous works have found that the pressure of this reaction has little effect on the yield down to 10bar H_2. However, the kinetics of this reaction have not been previously studied or reported. A kinetic study of the xylitol S-HDO was conducted at 10bar H_2 with temperatures ranging from 120-170°C. The reaction order, rate constant, and activation energy were elucidated. The effects of varying the xylitol concentration within the solvent were studied to determine the optimum conditions in which to conduct the xylitol S-HDO. Optimal conditions for the xylitol S-HDO with respect to temperature and concentration were elucidated.

References

(1) Ota, N.; Tamura, M.; Nakagawa, Y.; Okumura, K.; Tomishige, K. Hydrodeoxygenation of Vicinal OH Groups over Heterogeneous Rhenium Catalyst Promoted by Palladium and Ceria Support. *Angew. Chemie - Int. Ed.* **2015**, *54* (6), 1897–1900.

(2) Ota, N.; Tamura, M.; Nakagawa, Y.; Okumura, K.; Tomishige, K. Performance, Structure, and Mechanism of ReO X –Pd/CeO 2 Catalyst for Simultaneous Removal of Vicinal OH Groups with H 2. *ACS Catal.* **2016**, *6* (5), 3213–3226.

578409 Novel Bio-Derived Solvents for Extraction of Natural Products

Sustainability is one of the global grand challenges that our modern society faces. Production of biofuels and biochemicals from renewable lignocellulosic biomass has a great potential to sustain our society with minimal greenhouse gas emissions. The major challenge in the conversion of lignocellulose is its low surface accessibility to catalysts, causing the low sugar yield and/or the use of a high amount of catalysts. Here we show the synthesis of bio-based solvents with high biomass dissolution capacity. We hypothesized that: (1) the composition of these bio-based solvents strongly affected their dissolution capacity, and (2) dissolution capacity led to an enhanced surface accessibility to catalysts. We conducted this study by evaluating 20 bio-based solvents on rice straw as our model lignocellulose. The resulting rice straw samples were enzymatically hydrolyzed using 15 mg protein/g glucan. We obtained 97% glucose yield after 72h. Our findings reveal the relationship between solvent composition, dissolution capacity, and sugar yield. This knowledge can be used to design the efficient bio-based solvents for other applications, including organic synthesis and drug formulation.

560588 Understanding and Improving the Recyclability of Supported Nickel and Cobalt Catalysts for the Deoxygenation of Bio-Derived Fatty Acids

Biomass is an abundant feedstock ready for immediate conversion to transportation fuels. Non-edible plants, algae oil, and tallow are examples of feedstocks that exhibit high oil contents, making them optimal fuel candidates. If realized at an industrial scale, second-generation biofuels have been estimated to offset CO_2 emissions as much as 70% when compared to fossil-derived diesel fuels. Unfortunately, the techno-economic outlook of biofuels is currently unfavorable due to a number of costs related to harvesting, extraction, and upgrading.

For over 30 years, the thermochemical catalytic conversion of oily plants and algae has been studied, mostly applying traditional hydrotreating sulfided catalysts and noble-metal platinum and palladium catalysts. More recently, earth-abundant transition metals have been identified as active candidates for the deoxygenation reaction. In our recent work we have shown that zeolite supported transition metals including cobalt and nickel have high activity for the decarboxylation of fatty acids, but still suffer from considerable losses on recycling.

Herein, we report strong catalytic activity under inert atmospheres in our freshly synthesized Ni and Co catalysts. Finally, we provide recommendations for the improvement of non-noble metal catalysts in the deoxygenation of bio-derived feedstocks.

564841 Study on Catalytic Oligomerization of Glycerol over Alumina Supported Ca/La Mixed Oxides

In this work, alumina supported Ca/La mixed oxides as efficient catalysts in oligomerization of glycerol were successfully prepared by the wetness impregnation method using $La(NO_3)$ and $Ca(NO_3)$ solutions as well as boehmite as support, and followed by calcination at 600 °C. Various instruments including BET, TPD, XRD, SEM, and ICP-OES analysis was employed for characterization of these catalysts. The catalytic etherification reaction of glycerol was typically conducted at 240 °C - 280 °C. In a typical run, the reaction was performed at 250 ¢XC with a catalyst loading of 3 wt%, based on the mass of glycerol as the sole reactant and the atomic ratio of Ca/La = 2.7 on the catalyst. After 8 hours of catalytic etherification, nearly 75% of the glycerol could be converted to glycerol oligomers, while the selectivity of di-glycerol and tri-glycerol over all products was near 87%. The reusability and stability of the catalyst were tested. It was found that the recovered catalyst still showed good stability and conversion in two consecutive reactions. After 8 hours, the reaction still achieves 78% for glycerol conversion and 75% as the selectivity.

Keywords: etherification; alumina; glycerol; di-glycerol; tri-glycerol; catalysts

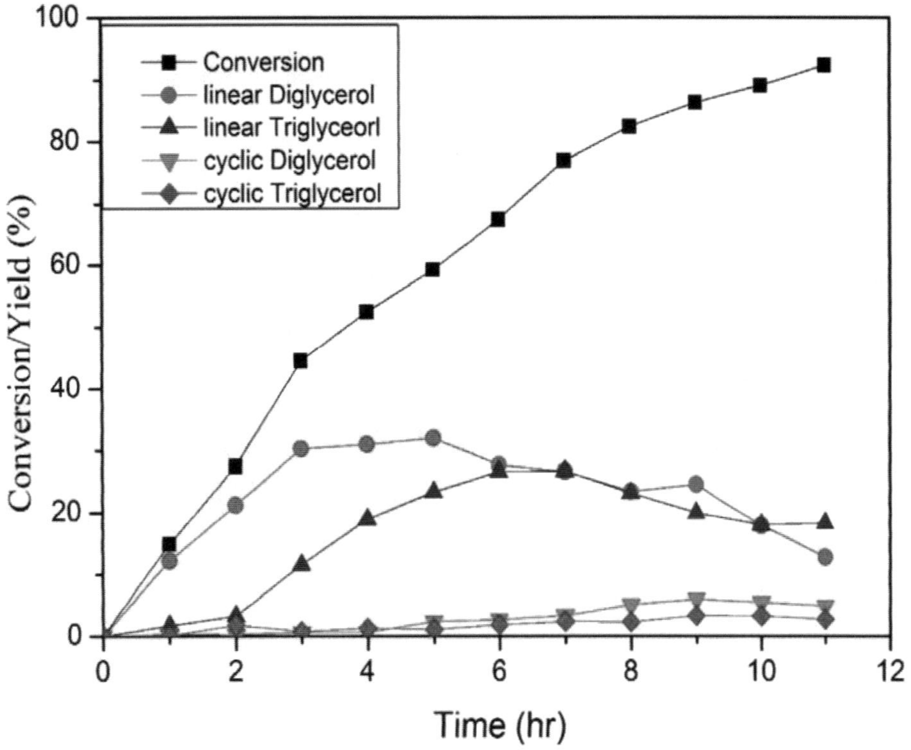

Reaction conditions:

- Reaction temperature = 250 °C,
- Reaction time = 12 h

- Catalyst loadings = 3 wt% of glycerol initially present

572096 Temperature Dependence of pH Dynamic Behavior in the Glucose Oxidase-Ferricyanide-Glucose-NaOH Reaction System in a CSTR

Our work is focused on detailed experimental observations of bistability in the glucose oxidase-ferricyanide-glucose-NaOH system in a continuous stirred tank reactor (CSTR) closed to atmosphere under systematically varied temperature and input NaOH concentrations.

The continuous stirred tank reactor has four inlet streams delivering solution of glucose oxidase, glucose, ferricyanide and NaOH and one output stream. The pH electrode screwed into the reactor keeps the system closed to atmosphere. Temperature is measured by a thermometer placed in the output stream from the reactor. The temperature of stock solutions and reaction mixture is controlled and maintained by a dry thermostat and by a magnetic stirrer and heater. Input concentrations of the enzyme (14.7U/ml), glucose (12mM) and ferricyanide (234mM) are maintained constant, while temperature, which significantly affects the activity of the enzyme [1], and NaOH input concentration are systematically varied. From previous studies it is well known that the system displays multiple steady states and a bistable hysteretic switch [2], but no oscillations were found, unless an external computer controlled loop is added. Here we report pH hysteresis curves for a fixed NaOH input concentration (5mM) and for temperatures in the range from 25°C to 40°C. We found the pH of the two coexisting steady states to range from pH~3 to pH~12. For an optimal temperature ~30°C [1] we also measured pH hysteresis curves with varying NaOH input concentration from 1mM to 10mM. These measurements provide information about the extent of the hysteretic behavior in the system. Interestingly, we also report pH oscillatory behavior with an amplitude of 0.4 pH units and periods between 8 to 15 minutes. Such dynamics has not yet been observed in the system without any external negative feedback loop.

As the simple Michaelis-Menten type of models including pH sensitivity do not describe the observed dynamics (especially the oscillatory dynamics), we formulate a detailed kinetic mechanism of the reaction to account for both the observed bistable and oscillatory behaviors as follows: glucose oxidase (GOX) is a two-substrate enzyme reacting with glucose and ferricyanide ; β-D-glucose binds to the active site of the unprotonated oxidized enzyme and its coenzyme flavin adenine dinucleotide (FAD). The active site of the enzyme is protonated due to glucose creating δ-gluconolactone simultaneously with reduction of the enzyme. The δ-gluconolactone reacts with water forming gluconic acid, which decreases pH. Solution of D-glucose contains both anomers α (36%) and β (64%)[3]. The enzyme consumes β-D-glucose immediately after it mutarotates from α-D-glucose to β-D-glucose. In the next reaction step, the coenzyme FAD forms $FADH^-$ and further reacts with one ferricyanide ion taking first hydrogen from $FADH^-$ to form ferrocyanide, H^+ and FAD. The second molecule of ferricyanide deprotonates FAD forming second ferrocyanide, H^+ and deprotonated FAD [4]. Deprotonated FAD is reprotonated from active site of the enzyme forming unprotonated oxidized form of the enzyme, which can again react with another molecule of β-D-glucose[5]. The unprotonated oxidized form of the enzyme can further be protonated, which reversibly decreases affinity of the enzyme towards the substrate. According to this mechanism, we formulated a model, which accounts for both bistability and oscillatory behavior. To analyze this model we used stoichiometric network analysis [6] and numerical bifurcation theory [7].

[1] J. Singh and N. Verma, Advances in Applied Science Research 4 (3), 250-257, (2013).

[2] V. K. Vanag, D. G. Míguez and I. R. Epstein, The Journal of Chemical Physics 125, 194515:1-12, (2006).

[3] H. S. Lee and J. Hong, Journal of Biotechnology 84, 145-153, (2000).

[4] A. A. Shul'ga, M. Koudelka-Hep, N. F. de Rooij, Sensors and Actuators B 26-27, 432-435, (1995).

[5] M. Meyer, G. Wohlfahrt, J. Knäblein and D. Schoburg, Journal of computer-Aided Molecular Design

12, 425-440, (1998).

[6] V. Radojković and I. Schreiber, Phys. Chem. Chem. Phys. 20, 9910-9921, (2018).

[7] M. Kubíček and M. Marek: Computational methods in bifurcation theory and dissipative structures. Springer Verlag, New York 1983; M. Kohout, I. Schreiber and M. Marek, Compt. and Chem. Eng. **26**, 517-527, (2002).

562640 Describing the Catalytic Role of Alkaline Earth Metals on Initiation during Cellulose Pyrolysis

Rapid thermal decomposition of lignocellulosic biomass yields a complex array of organic volatile compounds that form bio-oil when condensed and can be further upgraded to liquid fuels and chemicals. Lignocellulosic biomass contains inorganic metals such as alkaline earth metals that can catalyze and alter pyrolysis chemistry[1]. However, the role these alkaline earth metals have on the fundamental chemical mechanisms involved in this chemistry remains unknown. This study utilizes experimental and computational techniques to detail the fundamental chemical mechanism on how alkaline earth metals assist in activation of cellulose. Time-resolved kinetic data is obtained via rapid thermal pulsing of a-cyclodextrin, a chemical surrogate of cellulose[2], with the PHASR (Pulsed Heated Analysis of Solid Reactions) system[3-5]. Thin-films of a-cyclodextrin were prepared with varying concentrations of $CaNO_3$ and $MgNO_3$ (up to 2 wt%) to model typical calcium and magnesium concentrations in lignocellulosic biomass. Cellulose conversion exhibits transitory kinetic behavior among all temperatures (370 - 430 °C) and alkaline earth metal concentrations. Catalytic kinetic rate constants were extracted and compared with values during non-catalyzed pyrolysis[4]. Additionally, density functional theory calculations were performed on calcium/magnesium-catalyzed glycosidic bond cleavage and agree with experimentally obtained activation energies, allowing for the identification of an atomistic mechanism with good agreement between experiment and theory.

References:

1. Zhu, C.; Maduskar, S.; Paulsen A. D.; Dauenhauer, P. J., Alkaline-earth-metal-catalyzed thin film pyrolysis of cellulose. *ChemCatChem*, **2016**, 8, 818-829.
2. Mettler, M. S.; Mushrif, S. H.; Paulsen, A. D.; Javadekar, A. D.; Vlachos, D. G.; Dauenhauer, P. J., Revealing pyrolysis chemistry for biofuels production: Conversion of cellulose to furans and small oxygenates. *Energy & Environmental Science* **2012**,5, (1), 5414-5424.
3. Krumm, C.; Pfaendtner, J.; Dauenhauer, P.J., Millisecond pulsed films unify the mechanisms of cellulose fragmentation. *Chemistry of Materials* **2016**, 28, 3108-3114.
4. Zhu, C.; Krumm, C.; Facas, G. G.; Neurock, M.; Dauenhauer, P.J., Energetics of cellulose glycosidic bond cleavage. *Reaction Chemistry and Engineering* **2017**, 2, 201-214.
5. Maduskar, S.; Facas, G. G.; Papageorgiou, C.; Williams, C.L.; Dauenhauer, P.J., Five rules for Measuring Biomass Pyrolysis Kinetics: PHASR Kinetics of Lignocellulosic Biomass. *ACS Sustainable Chemistry & Engineering,* **2018**, 6, 1, 1387-1399.

573263 Towards Pharmaceutical Protein Stabilization: Insights from Theoretical Studies on Peptide Hydrolysis Reactions

Monoclonal antibodies (Mabs) are one of the most lucrative pharmacologics currently on the market. According to Grilo and Mantalaris, the market for Mab production has grown "between 7.2 and 18.3 percent per year".[1] In addition to their primary use as vaccines for diseases such as the flu and HIV, Mabs are also used to treat cancer and autoimmune diseases. One such example is *Avelumab*, a monoclonal antibody created by Pfizer to target the rare type of skin cancer, *Merkel cell carcinoma*.[1] However, formulation efficiency is limited by degradation of the desired products during multiple points of the biomanufacturing process.[2] While the use of additives has greatly reduced degradation, determining the most effective additives is accomplished through a "guess and check" approach, instead of understanding the underlying mechanism of the degradation pathway. One such mechanism of degradation is hydrolysis. At a neutral pH, hydrolysis occurs via the fragmentation of polypeptides into individual amino acids when water molecules are introduced.[3]

Recently, molecular models are being developed for enhanced prescriptive stabilization of antibodies. To better understand the mechanisms, the activation barrier (E_a) and free energy (^oG) of the reaction system must be estimated. One common method of determining the activation barrier of a system is through the use of the Evans-Polanyi correlation, which requires detailed thermochemical understanding of dipeptides and their individual amino acid components. For this purpose, we conducted a computational study of a small subset of dipeptides to predict the thermodynamic properties to begin generating a library of thermodynamic and kinetic values. The optimized geometries of the dipeptides were investigated using the Gaussian n-method at the B3LYP level of theory, as well as different solvation methods to analyze the impact of solvation on thermodynamic and kinetic properties. To validate our methodology, we compared our optimized frequencies to infrared spectra and experimental rate coefficients from current literature. Our studies have established trends in thermodynamic properties (Enthalpy of reaction (^oH_R), Entropy (^oS), and Gibbs free energy of reaction (^oG_R), as a function of peptide functional groups. These learnings, which will promote more efficient additive formulation, are discussed in terms of dipeptides degraded via hydrolysis reaction, at neutral pH.

1. Grilo, A. L. and A. Mantalaris (2019). "The Increasingly Human and Profitable Monoclonal Antibody Market." Trends Biotechnol **37**(1): 9-16.
2. Vazquez-Rey, M. and D. A. Lang (2011). "Aggregates in monoclonal antibody manufacturing processes." Biotechnol Bioeng **108**(7): 1494-1508.
3. Pan, B., et al. (2011). "A molecular mechanism of hydrolysis of peptide bonds at neutral pH using a model compound." J Phys Chem B **115**(19): 5958-5970.

577353 Computational Study on Biomass Fast Pyrolysis: Design Considerations for a Laboratory-Scale Fluidized Bed

Fast pyrolysis is a leading candidate process for converting biomass to liquid fuels. During fast pyrolysis in bubbling bed or circulating bed reactors, biomass particles are rapidly heated through contacting with hot gases and solids, and their constituent components decompose into volatiles, ash, and char. The product vapor/gas composition, which determines the yield of fuel-compatible molecules, is highly dependent on the bubbling intensity which promotes mixing and heat and mass transfer within the biomass particles and at the particle surfaces as they transit through the reactor. Fluidized bed hydrodynamic characterization at smaller scales is a vital first step in reactor scale-up.

In this study, we simulate a 3D bubbling fluidized bed biomass fast pyrolysis reactor from a prior study [149]. This study explores operating effects on hydrodynamics and biomass conversion as the gas flow is increased through the bubbling-to-slugging transition and turbulent regime, with all the other operating variables held constant. We employ MFiX, an open-source software package supported by DOE, which utilizes a continuum (two-fluid) approach for modeling the reactor hydrodynamics. Bubbling intensity and dynamic characteristics were evaluated utilizing pressure-based measurements [http://dx.doi.org/10.1016/j.cej.2016.08.113]. A novel entropy approach based on time irreversibility is introduced to evaluate hydrodynamics.

Mixing, hydrodynamics, and pyrolysis yields are compared which show the effect of fluidizing gas and fluidization regime on biomass fast pyrolysis in bubbling bed reactors of Geldart B particles. This work highlights the importance of initial reactor design for optimizing yield. We will discuss implications on future numerical simulations and experiments based on our observations.

Formatting Issue – Best Available Paper/Graphic

578128 Ultrasound-Assisted Production of Ethyl Butyrate VIA a Lipase Cocktail

ULTRASOUND-ASSISTED PRODUCTION OF ETHYL BUTYRATE VIA A LIPASE COCKTAIL

Fernando C. de Sousa Filho [1], Roberto M. F. de Freitas [1], Bárbara K. N. Gadelha [1], Ellefson E. S. de Oliveira [1], Rodolpho R. C. Monteiro [2], Paula J. M. Lima [2], Maria C. M de Souza [1] and José C. S dos Santos [1,*].

[1] Instituto de Engenharias e Desenvolvimento Sustentável, Universidade da Integração Internacional da Lusofonia Afro-Brasileira, CEP 62790-970, Redenção, CE, Brazil.

[2] Departamento de Engenharia Química, Universidade Federal do Ceará, Campus do Pici, Bloco 709, CEP 60455-760 Fortaleza, CE, Brazil.

*jcs@unilab.edu.br

1. INTRODUCTION

Enzymes are highly specific and efficient biocatalysts that catalyze various chemical and biochemical reactions under certain conditions of temperature, pH, reaction medium, etc (BILAL, et al., 2018). Lipases are enzymes used in biocatalysis, as they have a broad range of compatibility with several substrate species and high stability under multiple conditions. In addition, lipase properties, including selectivity, specificity and activity are easily modulated (SANTOS et al., 2015). The concept of combi-lipases have an increasing interest for heterogeneous substrates, such as oils, once the use of an individual lipase may hardly be "optimal" for all likely components of the oils (ORTIZ et al., 2019).

Ultrasound is the sound energy at frequencies above the range that is audible to humans (> 20 kHz) and which needs a medium to propagate. The positive effects of this technology on enzyme activity and can its potential to accelerate enzymatic reactions, show great potential for industrial applications. However, high frequencies and potencies can cause disruption of the structure of the enzyme, causing denaturation of this macromolecule, while low frequencies and potencies will not result in the desirable effect of cavitation (BODESODE & RATHOD, 2014).

Flavors are composed of different organic chemicals, such as hydrocarbons, alcohols, aldehydes, ketones, acids, esters or lactones. The low volatility and low molecular weight are responsible for a series of sensorial sensations. Flavor esters are widely found in nature and confer pleasant attributes with fruity, floral, spicy and creamy aromas. These properties enable a wide variety of food applications in many beverages, sweets, jellies, compounds, wines and dairy products.

Ethyl butyrate is an flavor ester obtained by esterifying a fatty acid or monocarboxylic acid (butyric acid) with an alcohol (ethyl alcohol). This ester is an important component of aromas of many fruits, for example: pineapple, passion fruit, strawberry and apple (GUMEL; ANNUAR, 2016).

Hereupon, this communication has as main objective to determine the best cocktail of Lipase B from Candida antarctica (CALB), Thermomyces lanuginosus (TLL) and Rhizomucor miehei (RML) for the ultrasound-assisted production of ethyl butyrate.

2. MATERIALS

Lipases from Thermomyces lanuginosus (immobilized on a silicate support, Lipozyme TL-IM), Rhizomucor miehei (immobilized on an anion-exchange resin, Lipozyme RM-IM) and lipase B from Candida antarctica (immobilized on a macroporous resin, Novozym 435) were kindly donated by Novozymes (Madrid, Spain). All other chemicals were of analytical grade and they have been used without any further purification.

The equipment used in all experiments was an ultrasonic bath (Unique Inc., model USC 2800A, Brazil). The equipment presents the capacity volume of 9.5 L with the following dimensions: 300 × 240 × 150 mm (length × width × height). Two disc transducers were placed at the bottom of the reactor. The ultrasonic frequency was 37 kHz and the total ultrasonic power 300 W. Additionally, the equipment has temperature control.

3. METHODOLOGY

In order to determine the best combination of lipases (TLL, RML and CALB) for the production of ethyl butyrate, a 3-factor mixture design and triangular surface analysis was performed using Statistica® 10 (Statsoft, USA). For the esterification, it was considered a 1:1 molar ratio (butyric acid/ethyl alcohol) and a biocatalyst mass of 10% of the butyric acid mass. The reactions were conducted at ultrasonic radiation at 37 kHz and 300 W for 3 hours and 35 °C. The reactions were performed in triplicate and the results expressed as means and standard deviations. Conversion was monitored by determining the acid index through the Ca 5-40 AOCS method.

4. RESULTS AND DISCUSSIONS

The combinations of lipases (CAL-B, RML and TLL) resulting from the simplex-centroid experimental design with interior points and centroid and their respective conversion rates are presented in Table 1. As can be seen in Table 1, the best cocktail of lipases for the ultrasound-assisted production of ethyl butyrate was that proposed by run number 10, which has the same amount of each lipase in study, achieving a conversion of 80.41%.

Table 1 – Experiments performed in the mixture design. Reaction medium: 1:1 molar ratio (butyric acid/ethyl alcohol), biocatalyst mass of 10% of the butyric acid mass. The reactions were conducted at ultrasonic radiation at 37 kHz and 300 W for 3 hours and 35 °C. Further information are given on Section 3.

RUN	CALB	RML	TLL	CONVERSION (%)
1	1.000	0.000	0.000	76.81 ± 2.32
2	0.000	1.000	0.000	35.80 ± 4.14
3	0.000	0.000	1.000	46.55 ± 1.56
4	0.500	0.500	0.000	77.76 ± 2.57
5	0.500	0.000	0.500	38.96 ± 1.83
6	0.000	0.500		

0.500 ***Formatting Issue – Best Available Paper/Graphic***
37.84 ± 3.66
7
0.667
0.167
0.167
79.25 ± 0.89
8
0.167
0.667
0.167
46.97 ± 3.81
9
0.167
0.167
0.667
37.74 ± 2.45
10
0.333
0.333
0.333
80.41 ± 0.94

Analyzing the behavior of each lipase in individual, CAL-B (76.81%) has presented the best conversion when compared to TLL (46.55%) and RML (35.80%), respectively. Also, for the combinations of only two lipases, CALB-RML and CALB-TLL have shown the better values than that for RML-TLL (37.84%). By these results and by analyzing the Pareto chart (Figure 1), for the ultrasound assisted production of ethyl butyrate, it can be stated that CALB, among the lipases studied, has the major effect on the conversion. TLL and RML are both 1,3-regio specific lipases, whereas CALB is a non-specific lipase. The combination of the 1,3-specific with non-specific lipases may have increased the reaction rate by attacking the different parts of the substrates, showing the specificity of RML and TLL for the substrate and the high activity of CALB, even being a non-specific lipase.

Figure 1 – Pareto chart. Reaction medium: 1:1 molar ratio (butyric acid/ethyl alcohol), biocatalyst mass of 10% of the butyric acid mass. The reactions were conducted at ultrasonic radiation at 37 kHz and 300 W for 3 hours and 35 °C. Further information are given on Section 3.

Besides, the results shown above point out the efficiency of the ultrasonic technology for the production of ethyl butyrate when compared to traditional techniques. Souza et al., (2017), using CALB obtained conversions higher than 90% for the production of ethyl butyrate under 25 °C, 150 rpm and a molar ratio of 1:1 after 8 hours of reaction, while in this study conversions of around 80% were obtained after only 3 hours of reaction. Badgujar and Bhanage (2015) produced benzyl butyrate, anisil butyrate and o-cresyl butyrate using the ultrasound technique and obtained a conversion yield of 99% after 3 hours of reaction at 52 °C, while in this study conversions of around 80% were obtained under 35 °C.

5. CONCLUSIONS

Here, the use of a cocktail of lipases for the assisted-production of ethyl butyrate was proposed. The optimum combination was found to be 33,3% of RML, 33,33% of TLL and 33% of CALB, showing the specificity of RML and TLL for the substrate and the high activity of CALB, even being a non-specific lipase. Besides, the results shown above point out the efficiency of the ultrasonic technology for the production of ethyl butyrate when compared to traditional techniques.

REFERENCES

AOCS – American Oil Chemists Society; Official and Tentative Methods, 3ª ed., Chicago, 1985, vol.
BADGUJAR, K.C.; BHANAGE, B.M. The combine use of ultrasound and lipase immobilized on co-polymer matrix for efficient biocatalytic application studies. Journal of Molecular Catalysis B: Enzymatic, v. 122, p.255-264, 2015. Elsevier BV. http://dx.doi.org/10.1016/j.molcatb.2015.09.012.
BANSODE, S.R.; RATHOD, V.K. Ultrasound assisted lipase catalysed synthesis of isoamyl butyrate. Process Biochemistry, v. 49, n. 8, p.1297-1303, 2014. Elsevier BV. http://dx.doi.org/10.1016/j.procbio.2014.04.018.
BILAL, M. et al. Magnetic nanoparticles as versatile carriers for enzymes immobilization: A review. International Journal of Biological Macromolecules, v. 120, p.2530-2544, 2018. Elsevier BV. http://dx.doi.org/10.1016/j.ijbiomac.2018.09.025.
GUMEL, A. M.; ANNUAR, M.S.M. Thermomyces lanuginosus lipase-catalyzed synthesis of natural flavor esters in a continuous flow microreactor. 3 Biotech, [s.l.], v. 6, n. 1, p.6-24, 11 jan. 2016. Springer Nature. http://dx.doi.org/10.1007/s13205-015-0355-9.
ORTIZ, C. et al. Novozym 435: THE "PERFECT" LIPASE IMMOBILIZED BIOCATALYST?. Catalysis Science & Technology, p.1-126, 2019. Royal Society of Chemistry (RSC). http://dx.doi.org/10.1039/c9cy00415g.
SANTOS, J.C.S. dos et al. Versatility of divinylsulfone supports permits the tuning of CALB properties during its immobilization. Rsc Advances, v. 5, n. 45, p.35801-35810, 2015. Royal Society of Chemistry (RSC). http://dx.doi.org/10.1039/c5ra03798k.
SOUZA, M.C.M. de et al. Production of flavor esters catalyzed by lipase B from Candida antarctica immobilized on magnetic nanoparticles. Brazilian Journal of Chemical Engineering, v. 34, n. 3, p.681-690, 2017. FapUNIFESP (SciELO). http://dx.doi.org/10.1590/0104-6632.20170343s20150575.

565792 Food Waste Repurposing to an Integrated Biorefinery

Food waste repurposing provides an alternative waste management strategy towards meeting goal 12 of the United Nations sustainable development goals. We have successfully repurposed potato peel waste (PPW) for manufacturing value added products. We report high antioxidant activity of potato peel extractives (PPE), providing a natural alternative to chemically synthesized antioxidants for human health. Our optimized dehydration system resulted in the highest reported yields (54%) of hydroxymethylfurfural (HMF), directly from potato peels, for manufacturing biobased chemical precursors. Our response surface methodology (RSM) studies reveal a synergistic contribution of $AlCl_3$ concentration-time and $AlCl_3$ concentration-temperature on glucose dehydration. Also, our synthesized biochar is effective in removing pesticides from contaminated water, improving water quality. Converting PPW into antioxidants, HMF, and biochar, we obtain revenues of about $8500 per MT of dry PPW. This provides the opportunity for successful translation of our technology to an economically profitable process using zero value food waste. This study provides a sustainable valorization framework that can be extended to other types of FW and offers a template for assessing the economics of biomass based biorefineries for manufacturing multiple renewable products.

571804 Mixed Al/B Catalysts: A Bifunctional Groups Modified Solid Acid Catalyst with Nano-Size Mesoporous for Efficient Conversion of Glucose to 5-Hydroxymethylfurfural in DMSO

With increasing energy and resource consumption correlates with increasing greenhouse gas emissions on climate change, lignocellulosic biomass as a renewable non-food resources, has been intensively used to develop bio-based platform chemicals and bio-fuels, which could partially or even fully replace the currently utilized fossil-based chemicals in well-established chemical processes. Amongst, 5-Hydroxymethylfurfural (5-HMF) was identified as one of high-potential C6-platform chemicals and can be served as a bio-refinery building block for the production of agrochemicals, polymers, pharmacetuticals, furanic polyesters, solvent, as well as fuel components.

To date, 5-HMF could be prepared from glucose, fructose and directly biomass by various catalysts in terms of homogeneous acids, ionic liquids, metal chlorides, zeolites and acidic resins. Compared with glucose, fructose as feedstock to produce 5-HMF is more efficient, however it is not cost economical. By comparison, the cost of the more abundant resource glucose is about a quarter that of fructose, making it more attractive to prepare 5-HMF through glucose.

Herein, a mixed Al/B catalyst with bifunctional groups modified solid acid with nano-size mesoporous and used as environmentally friendly bifunctional catalysts for the conversion of glucose to 5-hydroxymethylfurfural (5-HMF). The physicochemical properties of the Al/B catalysts were characterized by SEM-EDX, XRD, XPS, NH3-TPD, FTIR, TG-DTA and nitrogen adsorption-desorption isotherms. Optimization of reaction parameters including solvent, substrate type, initial glucose concentration, temperature, time, catalyst dosage, Al/B ratio was systematically investigated. The results showed that Al/B catalyst has the ability to synergistically catalyze the isomerization of glucose and the dehydration of fructose in a "one-pot" reaction, which could afford an impressive 5-HMF yield of 58% in the DMSO system. More importantly, the Al/B catalyst still retained the satisfied catalytic activity and stability after seven recycling cycles.

574154 Hydrothermal Carbonization of Microalgae for Hydrochar Production

Microalgae are viewed as a potentially sustainable source of chemicals, biologicals, and fuels that can help chemical manufacturing lower its greenhouse gas emissions and thus combat climate change. The fast growth rate of photosynthetic microalgae and their ability to utilize low-grade water and nutrients could help improve the cost competitiveness of renewable products compared to their fossil counterparts. Using hydrothermal carbonization (HTC), algal biomass can be thermochemically converted to hydrochar, a solid product that can be used in a variety of environmental and energy applications, such as solid nutrient amendment, water retention, solid fuel, capacitors, low cost adsorbents, and conversion into industrial chemicals. The process takes place at temperatures lower than those of fast pyrolysis and hence is less energy intensive, while yielding a product highly enriched in carbon. In this project, we focus our attention on the production of high-value hydrochar and testing it for application in industry. The microalga *Nannochloropsis oculata* was cultivated in photobioreactors, harvested, and dried to a certain moisture content. The algal biomass was then heated in batch mode in a stainless steel Parr reactor to a final temperature in the 180-240°C range over various residence times in the absence and presence of catalysts to determine the effect of conditions and catalyst on the yield and quality of the hydrochar. This product was analyzed for carbon content, porosity, surface area, and structure and its properties were compared to those reported in the literature. Activation steps to further improve key properties of the hydrochar are under investigation in an effort to meet the carbon black specifications required for high-value industrial applications.

557928 Hydrogen Donation from Bio-Acids over Carbon Nano Tube (CNT) Based Bi-Functional Catalysts: Density Functional Theory Study

Bio-acids (e.g. acetic acid and formic acid) have drawn considerable interest as alternative hydrogen source for in-situ hydrodeoxygenation of bio-oil to produce biofuels. However, their potential and decomposition mechanism as hydrogen donors have not been adequately explored. In this study, the decomposition pathways of bio-acids were investigated by DFT modelling with the emphasis on their performance on in-situ hydrogen donation. As promising green materials, metal supported on nitrogen doped carbon nanotubes (CNTs) was evaluated as catalysts for the decomposition of bio-acids. Adsorption modelling of acetic acid (AcOH) indicated Mo loading led to the highest adsorption energy compared to nine other transition metals, and N_p (pyridinic nitrogen doped)-CNT enabled the strongest binding to AcOH among a series of CNTs based supports, followed by intact CNT and N_g (graphitic nitrogen doped)-CNT. It was found that different doping types of nitrogen on CNT not only affected the adsorption of bio-acids, but also determined the catalysis performance and consequent reaction pathways of bio-acids decomposition. Modelling results indicated Mo/N-CNTs could significantly facilitated the decomposition of both bio-acids by lowering the reaction energy barriers, in comparison with the radical reactions without catalysts. The electron acceptor property of Mo/N_p-CNT gave rise to dissociative adsorption of both acids molecules, which significantly facilitated the decomposition of bio-acids, directly contributing to the hydrogen donation of formic acid (FA). Mo/N_p-CNT also enabled electron enrichment on both sides of C-CH_3 bond so that facilitated homolytic cleavage of in AcOH. The electron donation property of Mo/N_g-CNT made the hydrogen donation from lower fragments of methyl and hydroxyl much more facile, by arousing more electron enrichment to the hydrogen atoms, and resulted in eventual homolytic cleavage of the hydrogen atom. It can be concluded that FA performed better in hydrogen donation than that of AcOH, and Mo/N_g-CNT was supposed to be the most effective catalyst to facilitate the hydrogen donation from FA compared with Mo/Np-CNT and sole Mo catalyst.

574475 Engineering Lignin Transformation Mechanisms to Create Value-Added Products Using Atomistic Modeling

Engineering lignin transformation mechanisms to create value-added products using atomistic modeling

Tanzina Azad[1], Jonathan D. Schuler[1], Maria L. Auad[1,2], Thomas Elder[3], Andrew J. Adamczyk[1,*]

[1]Department of Chemical Engineering, Auburn University, Auburn, AL, USA.

[2]Center for Polymer and Advanced Composites, Auburn University, Auburn, AL, USA.

[3]United States Department of Agriculture (USDA) Forest Service, Southern Research Station, Auburn, AL, USA.

[*]Corresponding author aja0056@auburn.edu

Among the three polymers that form plant cells (i.e., cellulose, lignin, and hemicellulose), lignin is the second most abundant natural polymer and one of the least utilized biomass sources. Lignin's high natural abundance, carbon content, and functionalized nature make it a promising candidate for targeted valorization to fuels and polymer composites, as well as serving as a building block for various value-added products of renewable origin.[1] Due to the nature of its structure, lignin has the potential to be one of the largest renewable sources for aromatic compounds to be used in the chemical industry with an approximate market of \$130 billion USD.[2] Despite this demonstrated benefit, lignin is still largely utilized as a low-grade fuel because its degradation reaction mechanisms and energetics under various reactor process conditions, as well as basic atomic-level structure, are still not fully understood. Therefore, computational modeling of lignin at the atomistic scale has the potential to provide new insights to help guide experimental reaction design and process optimization towards a tailored valorization process with high selectivity and yield.

In this work, we are analyzing the transformation of lignin macromolecules by cleaving the b-O-4 bond homolytically, which simulates pyrolysis reactor conditions. Among the three monolignols that are precursors for the biosynthesis of lignin (i.e., p-hydroxyphenyl (H), Guaiacyl (G), and Syringyl (S) units), only G- and S-units are analyzed in our study. Though the concept of lignin modeling is not recent, very few studies are available that explored lignin molecules with a large number of monolignol units. In this study, we have modeled lignin structures comprised of 20 repeating monolignol units. We have focused upon conformer analysis and Bond Dissociation Energetics (BDEs) calculations for various lignin macromolecules. Our methods include both classical molecular mechanics and quantum chemical methods for advanced conformational sampling and Density Functional Theory (DFT) calculations, respectively. We have developed a novel and robust method for lignin macromolecule conformational sampling, which maps the conformational energy landscape efficiently and provides multiple low energy structures that are then used to determine the corresponding DFT reaction energetics of degradation.

We have also constructed a lignin conformer library which is a product of our developed conformational sampling method. This structural library is very useful for further computational studies and cross validation with experiments. This presentation will discuss the advanced conformation sampling method in the context of various commonly used force fields parameters, and provide specific guidance on force field parameter selection for lowest energy conformer identification. Our computational models correlate very well with available experimental data on structure and energetics. Lastly, the presented results demonstrate how theoretical investigations are currently impacting new materials discovery and synthesis understanding in the development of novel polymers and advanced composites.

1. Roberto Rinaldi, Robin Jastrzebski, Matthew T. Clough, John Ralph, Marco Kennema, Pieter C. A.

Bruijnincx, Bert M. Weckhuysen, Paving the way for Lignin valorization: Recent advances in Bioengineering, Biorefining and Catalysis, Chem. Int. Ed. 2016, 55, 8164 – 8215.
2. Frost & Sullivan. (2014, 7 May) *Full Speed Ahead for the Lignin Market with High-Value Opportunities as early as 2017* [Press Release]. Retrieved from http://www.frost.com/

574524 Evaluation of Al, Ti, and Nb Oxides As Support for $H_3PMo_{12}O_{40}$ to be Used As Catalyst in Biodiesel Production from Low-Grade Oils

Heteropolyacids (HPA) with Keggin structures, such as $H_3PW_{12}O_{40}$, $H_4SiW_{12}O_{40}$, $H_3PMo_{12}O_{40}$, and $H_4SiMo_{12}O_{40}$, have been described as efficient catalysts in trans/esterification reactions due to their tolerance to water and free fatty acids contents, with particularly well-suited characteristics of high proton mobility and stability due to their strong Brønsted-profile. Despite the numerous advances over the past years, a few challenges remain as focus of study, particularly in increasing the thermal stability, specific area, and solubility in polar media. The versatility array of HPA is considerably increased when such catalysts are supported onto solid matrices, such as activated carbon, and on Zr-, Ti-, Nb-, and Al-oxides. Nonetheless, the characteristics of the support have some direct implications on the performance of the HPA catalytic properties. While zirconia has been demonstrated as one of the best supports for HPA, the use of a catalyst prepared as HPA supported on Zr-oxides is hindered by the costs associated with the matrices, which often affect the economics of bulk-production chemicals, as biodiesel. In this sense, this study evaluated the synthesis of a 12-phosphomolybdic acid catalyst supported on Al_2O_3, TiO_2, and Nb_2O_5 and its application on simultaneous esterification and transesterification reaction to produce biodiesel using low-grade lipid feedstock, i.e., acid-rich macaw palm oil (*Acrocomia aculeata*). Results show that under optimized conditions, the Al-oxide supported-catalyst presented the highest surface area (31.2 $m^2\ g^{-1}$), and a considerable acidity value of 7.83 mmol $H^+\ g^{-1}$. While the catalysts supported on Ti-oxide presented the highest acidity (8.21 mmol $H^+\ g^{-1}$), the surface area was the lowest observed on all three matrices (13.2 $m^2\ g^{-1}$). Considering the reaction conditions tested on a pressurized reactor to produce standard-grade ethyl esters, the HPA catalyst supported on Ti-oxide required the least amount on a weight basis in regards to the oil used, i.e., 5 wt.% of HPA/TiO_2 was enough to produce a mixture with 98.22% of ethyl esters and viscosity of 4.81 $mm^2\ s^{-1}$, which fits within the standard values for using as fuel. The HPA/Al_2O_3 and HPA/Nb_2O_5 catalysts, on the other hand, were used at higher concentrations (13 and 20 wt.%, respectively) to produce ethyl esters of biodiesel grade. Nonetheless, the use of the catalysts supported on Al-oxide required lower reaction temperature (190 °C, compared to 210 °C for HPA/Nb_2O_5 and 200 °C for HPA/TiO_2) and also a lower ethanol-to-oil molar ratio (50:1 for HPA/Al_2O_3 and 90:1 for HPA/Nb_2O_5 and 60:1 for HPA/TiO_2) when compared to the others. Thus, the Al_2O_3 support demonstrated to be feasible under the reaction conditions to convert low-grade oil to ethyl esters of fuel grade. These results may, for instance, provide a novel alternative to the utilization of HPA with a special look on the production of bulk chemicals, as biodiesel, since the costs associated with the use of Al_2O_3 as support can potentially lower the costs associated with the production of the catalyst.

578511 Effect of Process Parameters and Heterogeneous Catalysts on the Upgrading of Micro-Algal Bio-Crude Oil

Hydrothermal liquefaction of microalgae has been proven to be a viable technique to produce biofuels. In spite of the high oil yields and conversion efficiencies, algae biocrude oil have high heteroatom, Total Acid Number (TAN) content which makes it unsuitable for the current petrochemical industry. This was addressed by upgrading the biocrude oil (microalgae *micractinium sp.)* using three different heterogeneous catalysts (MO_2C, HZSM-5, and Pt/C) and various process parameters. The upgraded biocrude yields were higher at 350 °C compared to 400 °C. All the upgraded biocrudes have higher carbon and hydrogen content values relative to the original biocrude. The heteroatoms (N, S, O) were considerably reduced with the use of the catalysts mentioned earlier. Among the catalysts under study, HZSM-5 catalyst was effective in removing the oxygen the best. The TAN values decreased considerably compared to the original biocrude owing to the removal of acidic constituents and the HHV have been improved significantly suggesting energy densification.

571994 Catalytic Transfer Hydrogenolysis of Xylitol over Bifunctional Pd-Based Catalysts

Xylitol is known as one of the key platform chemicals, which can be converted into renewable products, including ethylene glycol (EG), 1,2-propanediol (1,2-PDO) and other derivatives. They are widely used as antifreezes, solvents and pharmaceuticals.[1] Up to date, it has been reported that conversion of xylitol to those renewables have been carried out in the aqueous phase under harsh conditions with high temperature ($T > 200$ °C) and high H_2 pressure (4 ~ 8 MPa), which cause several technological issues such as significant formation of gaseous by-products, fast catalyst deactivation as well as high capital cost and energy consumption.[2,3] It is still a great challenge to achieve high glycol yields at mild conditions, due to the complexity of the parallel and consecutive reactions in xylitol hydrogenolysis.[4] In this work, we proposed a unique process of catalytic transfer hydrogenolysis (CTH) in the absence of externally added H_2 for facile conversion of xylitol to value-added glycols.

Taking xylitol as a model compound, in this work, we have designed a series of bimetallic Pd-Cu, Pd-Ni, Pd-Pt and Ni-Cu catalysts and confirmed their remarkable performances in CTH reactions to 1,2-PDO, EG and lactic acid. Based on experimental results, it is found that bimetallic Pd-Pt catalysts supported on TiO_2 has obvious superior CTH performances by comparing the conversion and products distribution to monometallic Pd and Pt under identical conditions. In particular, at nearly 100% xylitol conversion, a selectivity of 41.1% to EG and 1,2-PDO was obtained on Pd-Pt/TiO_2 at 220 °C, 1.0 MPa nitrogen atmosphere in absence of external H_2. In contrast, the main product of hydrogenolysis of xylitol on monometallic Pd/TiO_2 was lactic acid with the low conversion. Similarly, monometallic Pt/TiO_2 catalyst shows strong tendency for C-C breaking. In our presentation, the structural properties including effect of particle size on catalytic activity and selectivity will be systematically studied, with the aim to establish structure-activity correlations for C-C and C-O cleavage of xylitol during CTH process.

Reference

[1] Sun, J. Y.; Liu, H. C. *Green Chemistry* 2011, 13 (1), 135-142.

[2] Riviere, M.; Perret, N.; Cabiac, A.; Delcroix, D.; Pinel, C.; Besson, M. *Chemcatchem* 2017, 9 (12), 2145-2159.

[3] Jin, X.; Shen, J.; Yan, W.; Zhao, M.; Thapa, P. S.; Subramaniam, B.; Chaudhari, R. V. *ACS Catalysis* 2015, 5 (11), 6545-6558.

[4] Liu, H. L.; Huang, Z. W.; Kang, H. X.; Li, X. M.; Xia, C. G.; Chen, J.; Liu, H. C. *Applied Catalysis B-Environmental* 2018, 220, 251-263.

571954 Catalytic Conversion of Glucose to Tartaric Acid over Au-Pt/TiO_2 Catalysts

Replacing fossil-based feedstocks with renewable resources is considered to be key for reducing carbon emission and environmental pollution. Catalytic conversion of biomass to fuels and chemicals is very important for establishing environment-friendly chemical processes. Aqueous phase oxidation of bio-derived sugars is crucial for sustainable conversion of biomass to value-added carboxylic acids and derivatives.[1] Tartaric acid (TA) has been widely used for additives of food, dyes and pharmaceuticals. Traditionally, TA is mainly synthesized via oxidation of Maleic anhydride, a two-step reaction generating significant amounts of toxic substances and waste products.[2] Up to date, there is no detailed study being reported for selective oxidation of sugars to TA using heterogeneous catalysts. Therefore, we proposed AuPt/TiO_2 catalyst for facile TA synthesis (50%) from glucose in base-free conditional 100 °C 10 bar O_2.

In this context, we systematically studied the transformation of TiO_2 crystalline phases to explore the strong Metal-Support Interaction for enhanced oxidation performances. It is found that AuPt nanoparticles supported pure anatase TiO_2 and a mixture of anatase and rutile phases have shown relatively better activity (conversion:100%) for glucose oxidation compared with pure rutile TiO_2(conversion:<70%). As revealed by literature survey, it is believed that O vacancy and the phase junctions between anatase and rutile provide the activity sites for oxidation reactions.[3] The influence of particle size, as well as metal-metal and metal-support interaction on catalyst performances will be presented in this work. The outcome of this study will provide insights into rational design of effective metal catalysts for oxidation of sugars to value-added chemicals.

Reference

1. Derrien, E. *et al.* Aerobic Oxidation of Glucose to Glucaric Acid under Alkaline-Free Conditions: Au-Based Bimetallic Catalysts and the Effect of Residues in a Hemicellulose Hydrolysate. *Industrial & Engineering Chemistry Research* **56**, 13176-13190, doi:10.1021/acs.iecr.7b01571 (2017).
2. Xuan, J. & Feng, Y. Enantiomeric Tartaric Acid Production Using cis-Epoxysuccinate Hydrolase: History and Perspectives. *Molecules (Basel, Switzerland)* **24**, doi:10.3390/molecules24050903 (2019).
3. Zhang, J., Xu, Q., Feng, Z., Li, M. & Li, C. Importance of the relationship between surface phases and photocatalytic activity of TiO2. *Angewandte Chemie-International Edition* **47**, 1766-1769, doi:10.1002/anie.200704788 (2008).

564808 Biofuel Production from Hydrotreatment of Fatty Acids Using Supported Ni/Mo Bimetallic Catalysts

Ni/Mo bimetallic catalysts supported on zeolite ZSM-5, *aka* Ni-Mo/ZSM5, were successfully prepared by the wetness impregnation method to hydrotreat fatty acids, mainly palmitic acid, with an aim for the production of liquid alkane fuels. The contents of Ni and Mo on Ni-Mo/ZSM5 were found from 8 wt% to 10 wt%, based on the mass of zeolite ZSM-5 support. Various instruments, such as XRD, SEM, BET, XPS, NH_3-TPD and H_2-TPR, were employed to characterize these catalysts. Typically, the catalyzed hydrodeoxygenation (HDO) reaction of palmitic acid was carried out in a Parr reactor at 573 K with an initial pressure of hydrogen at 35 bar at this temperature. With Ni-Mo/ZSM5 catalysts, the conversion of palmitic acid could reach near 99% at 573K after HDO reaction for 4h. Furthermore, a significant portion of isomerized alkanes were obtained in the product as revealed by GC-MS. That is, both hydrodeoxygenation and hydroisoermization reactions took place conjointly with Ni-Mo/ZSM5 catalysts.

Keywords: Biofuel, Palmitic acid, Ni-Mo/ZSM5 catalyst, Hydrodeoxygenation (HDO).

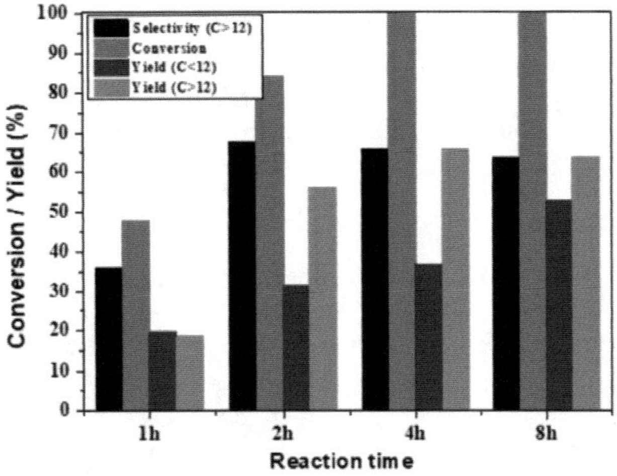

Reaction conditions: T = 300 °C; P_{H2} = 35 bar

Reactants: 1 g Palmitic acid; Solvent: 80 mL Dodecane; Catalyst: 0.25 g Ni-Mo/ZSM5

567565 Large-Scale Exploitation of Glucose Reductive Aminolysis: Comparison of Jet-Loop and Trickle-Bed Reactor Performance

Introduction

The reductive aminolysis of glucose with dimethylamine (DMA) has been presented as a renewable route towards dimethylaminoethanol (DMAE) and tetramethylethylenediamine (TMEDA), which can serve as precursors for surfactants [1]. In our previous work we have elaborated on the elucidation of the reaction mechanism and construction of a corresponding kinetic model for glucose reductive aminolysis [2]. In the current work we aim at optimizing the product yields in two very different reactor types, i.e., a jet-loop and a trickle-bed reactor [3, 4]. To achieve this goal 1D reactor simulations were performed, using the previously constructed kinetic model, and a comparison was made with pilot tests.

Results and Discussion

A graphical representation of the reactors, along with the reactor mass balances, has been shown elsewhere for the jet-loop reactor with a reactor vessel and loop volume of 4.75 m^3 and 0.63 m^3 respectively and for the trickle-bed reactor with a length of 1.17 m and a diameter of 0.015 m [5]. Verification experiments have been performed for the latter at the pilot scale. Removing the generated heat in the jet-loop reactor is achieved using a heat exchanger in the recycling loop while a water cooled jacket serves this purpose in the trickle-bed reactor.

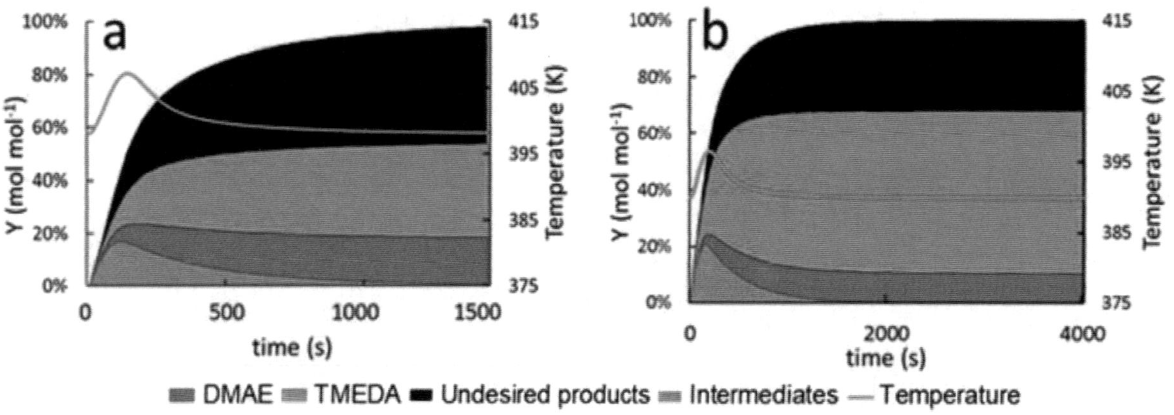

Figure 1: Molar product yields of the reductive aminolysis of glucose with dimethylamine as a function of time in the jet loop reactor. a) ($n_{glucose}^0$ = 2250 mol, $n_{DMA}/n_{glucose}^0$ = 12 mol mol^{-1}, $n_{H2}/n_{glucose}^0$ = 5 mol mol^{-1}, p_{tot} = 7.5 MPa, T = 398 K, W_{cat} = 35 kg). b) ($n_{glucose}^0$ = 850 mol, $n_{DMA}/n_{glucose}^0$ = 40 mol mol^{-1}, $n_{H2}/n_{glucose}^0$ = 10 mol mol^{-1}, p_{tot} = 7.5 MPa, T = 390 K, W_{cat} = 35 kg).

The composition of the mixture in the jet-loop reactor is presented in Figure 1 (a). The selectivity towards the TMEDA amounts to 37%, while that towards DMAE is limited to 16%. A major fraction of the feed is converted into degradation products, such as dimethylformamide. The temperature exhibits a maximum as a function of the time, simultaneously with the intermediates concentration and, hence, when hydrogenation reactions, which generate the most pronounced exothermicity, reach their highest rates. This maximal temperature increase is limited to 13 K and the heat is efficiently removed as evidenced from the fast temperature decrease after the maximum. Yet, given the high activation energy of the degradation reactions, the latter occur to a significant extent during this (limited) temperature excursion.

When varying the temperature while keeping all other operating conditions fixed TMEDA exhibits a maximum yield of 42% at 386 K (not shown). At higher temperatures degradation significantly increases and, hence, the formation of both DMAE and TMEDA is reduced. Considering higher amounts of DMA in the feed, while keeping the other operating conditions fixed, leads to higher TMEDA yields (not shown). The latter is a direct result of the lower glucose concentration resulting in less pronounced degradation.

Figure 1 (b) shows the optimized product spectrum in the jet-loop reactor. Simulation results critically depend on the catalyst mass, a total catalyst mass of 35 kg seeming to be the most interesting one. Lower catalyst masses result in product spectra with more degradation while higher catalyst masses result in more DMAE formation and, hence, a lower TMEDA yield. The initial amount of glucose is 850 mol and a molar ratio of DMA to glucose of 40 mol mol^{-1} is selected, while the temperature is 390 K and the total pressure is 7.5 MPa. Ultimately, the TMEDA yield amounts to 57%, while the DMAE yield amounts to 12%. Thus, the overall product spectrum is very similar to the one obtained on the lab scale [2].

While efficient heat removal was achieved in the jet-loop reactor, this was not the case for the trickle-bed reactor, see also Figure 3 (a), and, hence, isothermal simulations were performed first to assess the effect of the individual operating conditions, see Figure 2.

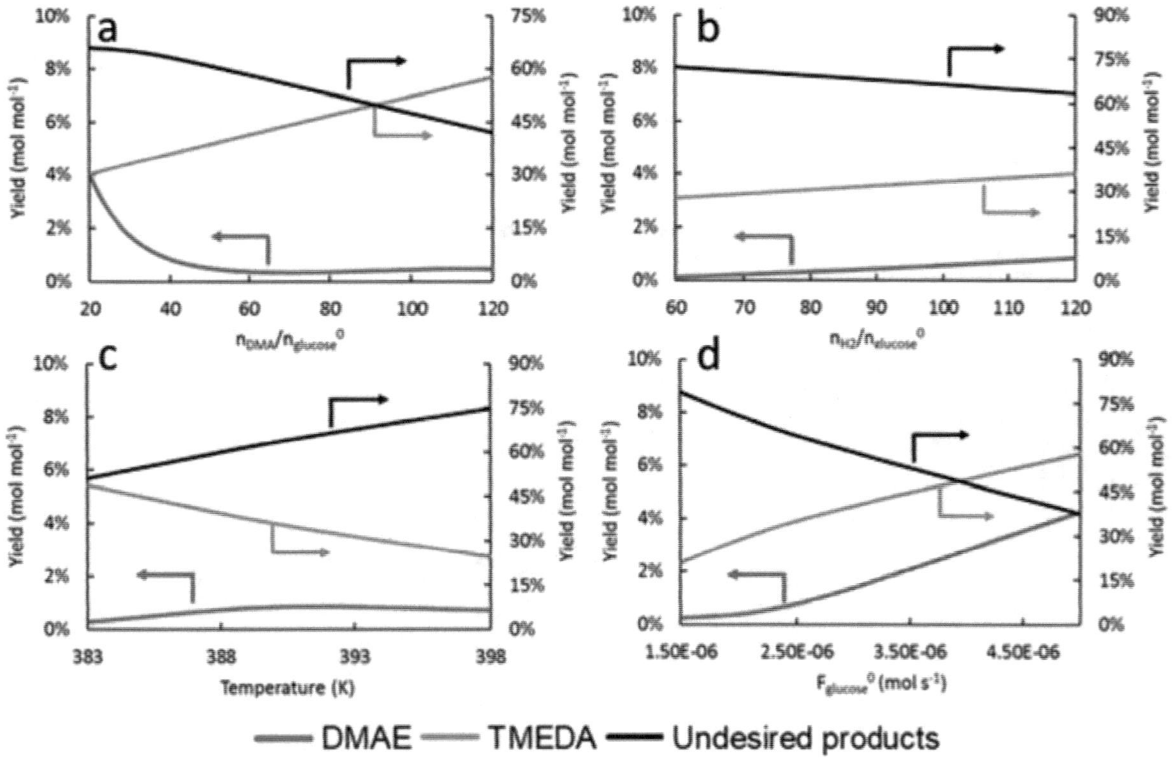

Figure 2: Molar DMAE and TMEDA yields as a function of varying operating conditions in the isothermally operated trickle bed reactor. (a) Variation of $n_{DMA}/n_{glucose}^0$ while $F_{glucose}^0 = 2.57 \cdot 10^{-6}$ mol s^{-1}, $n_{H2}/n_{glucose}^0 = 120$ mol mol^{-1}, T = 390 K, p_{tot} = 7.5 MPa, (b) Variation of $n_{H2}/n_{glucose}^0$ while $F_{glucose}^0 = 2.57 \cdot 10^{-6}$ mol s^{-1}, $n_{DMA}/n_{glucose}^0 = 40$ mol mol^{-1}, T = 390 K, p_{tot} = 7.5 MPa, (c) Variation of temperature while $F_{glucose}^0 = 2.57 \cdot 10^{-6}$ mol s^{-1}, $n_{H2}/n_{glucose}^0 = 120$ mol mol^{-1}, T = 390 K, p_{tot} = 7.5 MPa, (d) Variation of $F_{glucose}^0$ while $n_{H2}/n_{glucose}^0 = 120$ mol mol^{-1}, $n_{DMA}/n_{glucose}^0 = 40$ mol mol^{-1}, T = 390 K, p = 7.5 MPa.

A higher $n_{DMA}/n_{glucose}^0$ ratio results in a significant TMEDA yield enhancement, from 36% at a $n_{DMA}/n_{glucose}^0$ ratio of 40 mol mol^{-1} to 57% at a $n_{DMA}/n_{glucose}^0$ ratio of 120 mol mol^{-1}, as can be seen from Figure 2 (a). Several phenomena contribute to the enhanced TMEDA yield, firstly the higher $n_{DMA}/n_{glucose}^0$ ratio leads to lower glucose concentrations and, hence, leads to suppressing the degradation reactions. Secondly, more DMA molecules are required during TMEDA formation as compared to DMAE formation and, hence, a higher $n_{DMA}/n_{glucose}^0$ ratio yields more TMEDA. Thirdly, the keto- enol tautomerism, which is a key step in tuning the selectivity between DMAE and TMEDA, is base-catalyzed by DMA. Finally, higher mass transfer coefficients are achieved with a higher DMA feed rate.

Figure 2 (b) shows that a higher $n_{H2}/n_{glucose}^0$ ratio results in higher desired product yields, mainly TMEDA, and a reduced amount of degradation.

A higher temperature results in more pronounced degradation and a corresponding decrease in the TMEDA yield as can be seen from Figure 2 (c). The DMAE yield exhibits a weak optimum but overall the DMAE yield remains below 1%.

Higher total feed flow rates again result in enhanced mass transfer, see Figure 2 (d) for the variation of $F_{glucose}^0$ and, hence, result in enhanced hydrogenation rates. This ultimately leads to a significant increase of the TMEDA yield from 18% to 59% when $F_{glucose}^0$ is increased from $1.29 \cdot 10^{-6}$ mol s^{-1} to $5.14 \cdot 10^{-6}$ mol s^{-1}, nevertheless the DMAE yield remains below 5%.

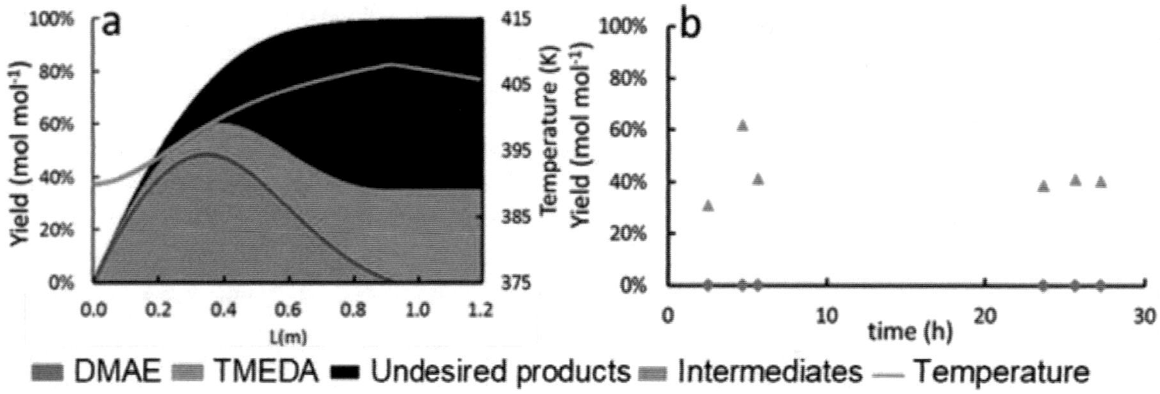

Figure 3: Molar product yields of the reductive aminolysis of glucose with dimethylamine in the trickle bed reactor. ($F_{glucose}^0 = 2.57 \cdot 10^{-6}$ mol s^{-1}, $n_{DMA}/n_{glucose}^0 = 120$ mol mol^{-1}, $n_{H2}/n_{glucose}^0 = 120$ mol mol^{-1}, $p_{tot} = 7.5$ MPa, T = 390 K, $\rho_{bed} = 1000$ kg$_{cat}$ m$_r^{-1}$) (a) Calculated product yields. (b) Experimentally measured product yields as a function of time.

Figure 3 (a) shows the simulated glucose aminolysis product spectrum when the reaction exothermicity and heat exchange are taken into account. It can be seen from this figure that temperature control is rather poor leading to an unavoidable temperature increase of 18 K. Correspondingly, a maximum TMEDA yield amounting to 37% is simulated, which is about 20% lower than the isothermally simulated TMEDA yield as a consequence of more pronounced degradation at the higher temperatures (not shown). The experimental validation in the considered reactor is shown in Figure 3 (b). The steady state TMEDA yield amounts to 40%, while almost no DMAE is formed. Thus, the model succeeds in predicting the trickle bed reactor yields accurately when the most important phenomena are accounted for.

Conclusions

Industrial-scale glucose aminolysis simulations have shown that TMEDA is the most abundantly obtained product in a jet-loop reactor with yields amounting to 57%, while a non-negligible amount of DMAE is formed. The product spectrum, thus, corresponds closely to the one obtained in (fed-)batch mode on the lab scale, thanks to an acceptable temperature control. This temperature control is, on the contrary, the main issue in the trickle-bed reactor where TMEDA yield losses up to 20% are observed as a consequence of poor temperature control and, hence, more pronounced degradation at higher temperatures. The model predicted and experimentally observed TMEDA yield of 40%, hence, indicate that mass transfer and temperature control should be enhanced to exploit this complex chemistry to its full potential.

Acknowledgement

This work was supported by Flanders Innovation & Entrepreneurship VLAIO (IWT) via the intermediary of FISCH/CATALISTI, contract 145020–Carboleum.

References

1. Pelckmans, M., et al., ACS Catalysis, 8 (2018): p. 4201-4212.

2. Poissonnier, J., et al., Applied Catalysis B: Environmental, 227 (2018): p. 161-169.

3. Stitt, E., Chemical Engineering Journal, 90 (2002): p. 47-60.

4. Ranade, V.V., R. Chaudhari, and P.R. Gunjal, *Trickle bed reactors: Reactor engineering and applications.* 2011: Elsevier.

5. Poissonnier, J., J.W. Thybaut, and G.B. Marin, Industrial & Engineering Chemistry Research, 56 (2017): p. 14192-14199.

573716 Characterization of Physical and Chemical Properties of Bio-Crude Oil from Hydrothermal Liquefaction of Food Waste

Sustainable management of food waste is one of the most urgent challenges facing the society with 1.3-billion-ton food waste being generated annually in the worldwide. The traditional managements (e.g. landfilling) raise economic and environmental concerns including low processing efficiency, greenhouse gas emission, and water/soil pollution. Considering the abilities of recycling value-added products, stabilizing organic matter, and reducing waste volume, hydrothermal liquefaction (HTL) is deemed as a an alternative remediation technique, which can convert moisture-rich food waste into bio-crude oil, that can be further upgraded into liquid fuels. In this study, two types of food wastes (pre-consumption and post-consumption) were collected from a New Mexico State University- campus dining facility. HTL of organic fractions of the above food wastes were conducted using a 100 mL batch reactor at 200-290°C, 30 min reaction time and ~15% solids load. The highest bio-crude oil yield (28.26 wt.%, dry basis) was obtained at a temperature of 240 °C and overall product distribution was investigated. Feedstock and the product bio-crude oil, aqueous phase and char were characterized by various analytical methods. Detailed characterization of bio-crude oil was performed by fatty acid methyl ester (FAME) analysis, gas chromatography-mass spectroscopy (GC-MS), high-resolution Fourier transform-mass spectroscopy (FT-MS), Fourier transform infrared spectroscopy (FT-IR), and thermogravimetric analysis (TGA) in order to comprehensively evaluate bio-crude oil make up and chemistry.

559591 Bifunctional Co-Mn Oxide Catalysts for Catalytic Transfer Hydrogenation of Levulinic Acid to γ-Valerolactone

The rapidly diminishing fossil fuel reserves and significant exhaust emission to environment have promoted the transition from fossil energy to renewable alternatives. Biomass, known as the most abundant and renewable carbon resource, can be used to product value-added chemicals and transportation fuels [1]. Therefore, conversion of biomass is of great significance to solve current energetic and environmental issues. Levulinic acid (LA) was considered as one of the most promising platform molecules and can be formed as a as by-product from 5-hydroxymethylfurfural (HMF) [2-4]. Hydrogenation of LA can produce γ-valerolactone (GVL), which is a versatile building blocks agent for green solvent, food additive and energy storage intermediates [5]. Up to date, LA conversion to GVL is primarily achieved under elevated temperature and H_2 pressure. Poor activity of catalysts and significant generation of various by-products are major issues. Catalytic transfer hydrogenation (CTH) display several economic and environmental advantages over traditional hydrogenation. In this work, we studied inexpensive CuMn oxide catalysts for CTH of LA to GVL under mild condition.

Formic acid (FA) is a co-product in production of LA, using FA as H-donor can make the conversion process of LA more economical [6]. In our work, we first propose a cheap bifunctional CoMn oxide catalyst to convert the mixture of LA and FA. It is found that conversion of LA and selectivity of GVL could reach to 45% and 50% separately without external H_2. To study the effect of catalyst's structure on reaction, we tune valence state of Co or Mn active site, interaction of metal support and component of the bifunctional catalyst. The catalysts are characterized by XRD, XPS, BET, TEM and SEM technique. In addition, we will also study the reaction pathway.

Reference

[1] Osatiashtiani, A.; Lee, A. F.; Wilson, K. *Journal of Chemical Technology and Biotechnology* 2017, 92 (6), 1125-1135.

[2] Besson, M.; Gallezot, P.; Pinel, C. *Chemical Reviews* 2014, 114 (3), 1827-1870.

[3] Luterbacher, J. S.; Rand, J. M.; Alonso, D. M.; Han, J.; Youngquist, J. T.; Maravelias, C. T.; Pfleger, B. F.; Dumesic, J. A. *Science* 2014, 343 (6168), 277-280.

[4] Mika, L. T.; Csefalvay, E.; Nemeth, A. *Chemical Reviews* 2018, 118 (2), 505-613.

[5] Yan, L.; Yao, Q.; Fu, Y. *Green Chemistry* 2017, 19 (23), 5527-5547.

[6] Upare, P. P.; Jeong, M. G.; Hwang, Y. K.; Kim, D. H.; Kim, Y. D.; Hwang, D. W.; Lee, U. H.; Chang, J. S. *Applied Catalysis a-General* 2015, 491, 127-135.

567436 Ionic Liquid-Mediated Rapid Catalytic Conversion of Lignocellulosic Grass to Biofuel Precursors in Microwave Reactors

The Biofuel Policy of India (2018) recommends the use of non-edible lignocelluloses for producing second generation bioethanol to meet the country's blending targets for petrol set by the Paris Climate Accord (2011), and solve the connected problems of climate change and India's dependence on foreign oil. Lignocelluloses are composed of three different polymers – cellulose, hemicelluloses, and lignin – bonded to each other by hydrogen and covalent bonds that make the lignocellulose structure recalcitrant. Switchgrass, Napier grass, Miscanthus, etc., are among the non-food lignocelluloses commonly used for producing 2G ethanol through the process of pretreatment (that breaks the inter-polymeric bonds), hydrolysis (that ruptures the β,1-4 glycosidic bonds linking the monomeric units in cellulose and hemicelluloses), and microbial fermentation (that converts the monomers to bioethanol).

This study uses a tall, thin, locally available, cellulose-rich grass – which qualifies neither as a food source nor as cattle-feed – as feedstock for biofuel production. The June grass used in this work is extremely rich in cellulose (82.3%), and has small amounts of hemicelluloses (1.3%), lignin (9.5%), ash (1.9%) and, moisture (4.5%), with an overall crystallinity index of 48%. We use a microwave reactor that provides an ionic liquid-mediated one-pot synthesis system for converting lignocelluloses to biofuel precursors [1]. The raw untreated grass is converted to biofuel precursors such as glucose, fructose, 5-Hydroxymethylfurfural (HMF), Levulinic acid (LA) and, Formic acid (FA), in only 36 minutes, when catalyzed by copper chloride in a BmimCl (ionic liquid) medium at temperatures ranging from 120-200ºC. When both the reactor temperature and the water concentration are used as reactor-scale parameters to control the product distribution, the maximum product yields are obtained as 88%, 25%, 18% and, 7%, for glucose, HMF, LA and FA, respectively.

The aprotic ionic liquid (BmimCl) rapidly ionizes to form a Lewis acid catalyst with copper chloride and water, and creates a supramolecular complex with cellulose. This polar supramolecule aligns itself along the microwave's rapidly alternating electromagnetic field, resulting in frequent dipole rotations that lead to molecular collisions, and dissipation of electromagnetic energy as friction and heat. This increases the reaction temperature and accelerates the reaction rate, thus rapidly rupturing the glycosidic bonds in the cellulose to produce glucose monomers that are further converted to HMF, LA and FA [1].

Reference: [1] Paul, S. K., Chakraborty, S. (2018). Microwave-assisted ionic liquid-mediated rapid catalytic conversion of non-edible lignocellulosic Sunn hemp fibers to biofuels. Bioresource Technology, 253, 85-93.

576431 Main Effects Observed in the Conversion of Fructose in HMF

5-hydroxymethyl-2-furfural (HMF) is a furanic compound considered as an important chemical platform once it is obtained from renewable raw materials and it has wide application in the industry, acting as an intermediate in the production of fine chemicals, polymer materials, biofuels and pharmaceuticals. This work aimed an experimental study of the reaction of HMF formation by heterogeneous catalysis, varying catalysts and reagents, with determination of yield and selectivity for optimization, as well as kinetic data. Experiments in aqueous medium were carried out in a Parr reactor using a combination of a substrate (glucose or fructose) and a catalyst (macroporous or supergel ion exchange resins). The maximum HMF yield was 20 % in 3.5 h and the conversion was 84 % after 7.5 h in the reaction of fructose catalyzed by the supergel resin (named SGC650H) at 110 °C, which indicated a poor performance towards HMF using water as solvent. This result is attributed to the HMF rehydration under high temperature conditions to produce levulinic and formic acids, so the best alternative to increase HMF selectivity is to use anhydrous solvents. Using dimethylsulfoxide (DMSO) as solvent, the yields significantly increased to almost 100 % in 3 h reacting at 110 °C. Besides conducted in batch reactions, experiments in continuous packed bed reactors were also performed with reaction systems containing fructose, DMSO and SGC650H resin. Analyzing the reactions performed with this system in both reactors, the increase in temperature and the reduction of the particle size of the catalyst favors the HMF formation, in contrast to the fructose feed concentration. Considering a pseudo first order kinetic model, the available data allowed to calculate rate constants and Arrhenius parameters, the latter varied with the concentration of reagent fed. This variation was linear in the Cremer-Constable diagram, which means that the model was insufficient to represent the reaction. Langmuir-Hinshelwood rate equations (LHHW) models were also tested by the rate determining step method, but none had adjusted to the data, and thus a rate law was not determined by the methods applied.

576553 Computer-Aided Analysis of Protease Catalytic Activity to Enhance Process and Product Development

With a promise of being more efficient and environment friendly, the industrial application of enzymes has become one of the rapidly growing sectors. However, studies on the modification of enzymes and their respective catalytic processes for the non-biomedical field is comparatively small with respect to biomedical applications, such as drug development. An approach of significant importance to contribute to these studies on catalytic activity and to reduce chemical footprint from experimental setups is the molecular modeling of enzyme function. There have been a variety of atomistic modeling tools that have shown promise for the study of enzymatic catalysis, including quantum mechanical (QM) cluster, hybrid quantum mechanical(self-consistent field)/molecular mechanics (QM(SCF)/MM), and hybrid QM(EVB)/MM methodologies. The empirical valence bond (EVB) method is a hybrid QM/MM method that describes reactions by mixing diabatic states that correspond to classical valence bond structures, which represent the reactant, intermediate (or intermediates), and product states[1]. Here we will focus upon the QM(EVB)/MM methodology, where key force field parameters are estimated from QM cluster (DFT) calculations.

Our results will focus upon the computational study of serine protease activity. In this study, the Gibbs free energy of activation ($\Delta g^{\ddagger}_{cat}$), rate coefficient (k), and the Gibbs free energy of reaction (ΔG_{rxn}) were calculated using QM(EVB)/MM and free-energy perturbation (FEP)/umbrella sampling methods. These key thermochemical and kinetic parameters were utilized for detailed reaction path analysis on peptide bond cleavage with more than hundred numbers of samplings; Gibbs free energy surfaces of reaction were developed for reaction in water and in the native enzyme, also in noncatalyzed condition. Furthermore, influence of key environmental variables affecting catalysis such as temperature and pH were also analyzed, which can present paramount importance in process optimization for commercialized enzymes or for the scale-up of promising enzyme candidates. Moreover, key amino acid residues in the native enzyme active site were mutated to determine overall effects on catalytic activity. These learnings will be discussed in the context of enzyme design.

1. Adamczyk, A. J.; Warshel, A., Converting structural information into an allosteric-energy-based picture for elongation factor Tu activation by the ribosome. *Proc. Natl. Acad. Sci. U. S. A.* **2011,** *108* (24), 9827-9832.

559613 Optimization of the Upgrading of Ethanol to n-Butanol Using in-Situ IR Spectroscopy in the Guerbet Reaction Network

As the crude oil reserves are finite, the intensified use of renewable sources for fuels is necessary. The most common renewable fuels used today are ethanol from fermentation processes and fatty acid methyl esters (biodiesel). In comparison with ethanol, higher alcohols are advantageous fuels because of their higher energy density. A suitable reaction for the upgrading of ethanol to higher alcohols is the Guerbet-reaction which is formally a condensation of two alcohols under elimination of water. [1] Either pure ethanol or a blend of ethanol and methanol are preferably used as substrates, resulting in either n-butanol or iso-butanol as desirable Guerbet-alcohols. In homogeneously catalyzed systems, the reaction proceeds over three steps (see Scheme 1): (1) dehydrogenation of two alcohol molecules to aldehydes, (2) aldol condensation, (3) hydrogenation of the a,ß-unsaturated aldehyde to the higher alcohol. The redox steps are catalyzed by a transition metal catalyst, while the aldol condensation is catalyzed by a strong base. The most important side-reaction is the over-oxidation of acetaldehyde. The resulting ethyl acetate or acetic acid neutralizes the base and thereby inhibits the aldol condensation. [2]

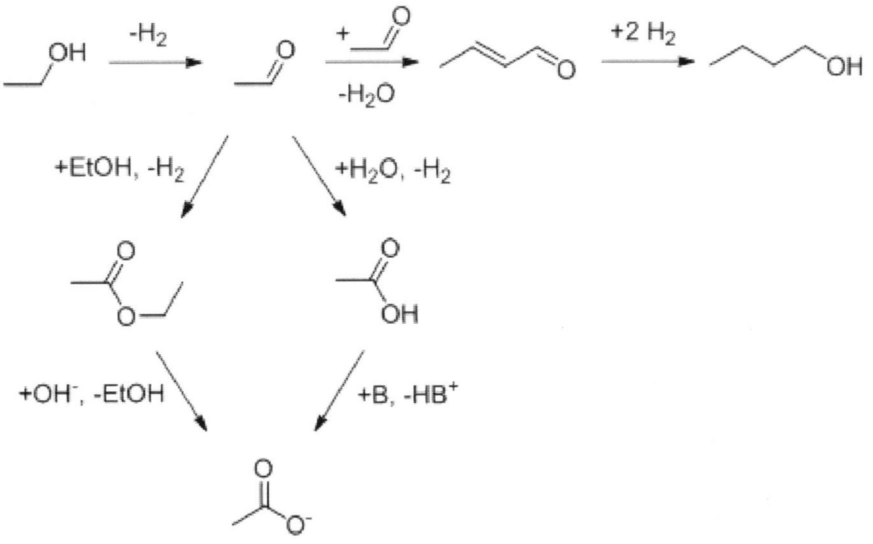

Scheme 1: Reaction network of the Guerbet condensation from ethanol to n-butanol (upper row) including the side reaction to acetate.

In this work, we present an in-situ IR-spectroscopic study on the Guerbet-reaction of ethanol, catalyzed by the redox catalyst Ru-MACHO® and KOtBu. The aim is to prevent the over-oxidation of acetaldehyde in the reaction system to reach higher yield and selectivity of n-butanol. Ru-MACHO® was first employed for the hydrogenation of esters to alcohols and is proven to be a suitable catalyst for the acceptorless dehydrogenation of alcohols for hydrogen production. The variable parameters of the reaction are the catalyst loading (<1 mol-%), the base concentration (1-10 mol-%) and the temperature (120-150 °C). Furthermore, this work focuses on the effect of hydrogen pressure on the reaction, since the equilibrium between hydrogenated and dehydrogenated catalyst has a great impact on the reaction rates of the two redox steps of the Guerbet reaction (see Scheme 2).

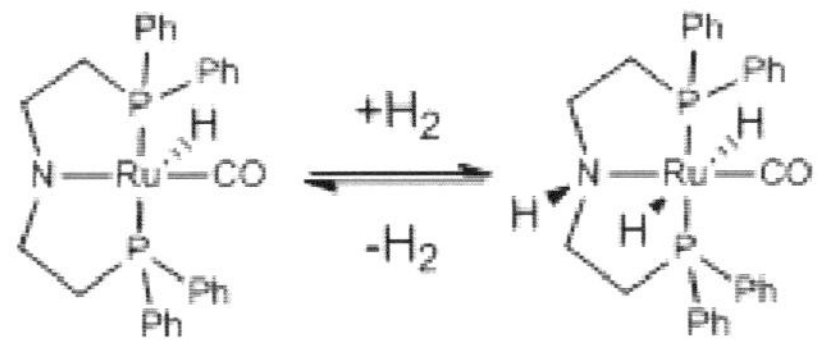

Scheme 2: Hydrogenation/dehydrogenation equilibrium of the Ru-MACHO-catalyst.

The use of in-situ IR-spectroscopy enables to analyze the kinetic behavior of the reaction by resolving the time-dependent concentrations of the present species. In particular, the high time resolution of the method makes details visible that are often overlooked, e.g. accumulating intermediates or non-ideal kinetic behavior. [3]

[1] H. Aitchison, R. L. Wingad, D. F. Wass, ACS Catal. 6 (2016) 7125-7132.

[2] R. Mazzoni, C. Cesari, V. Zanotti, C. Lucarelli, T. Tabanelli, F. Puzzo, F. Passarini, E. Neri, G. Marani, R. Prati, F. Vigano, A. Conservano, F. Cavani, ACS Sustainable Chem. Eng. 7 (2019) 224-237.

[3] A. Ohligschläger, C. Gertig, D. Coenen, S. Brosch, D. Firaha, K. Leonhard, M. A. Liauw, Chem. Eng. J. 368 (2019) 649-658.

565015 HMF Synthesis from Glucose By Reaction-Extraction System Using a Microreactor

1. Introduction

5-Hydroxymethylfurfural (HMF) is one of the valuable chemicals obtainable via biomass thermochemical conversion. Basically glucose, one of the main components of lignocellulosic biomass, converts into HMF under an acidic aqueous condition. However, under such a condition, HMF further hydrolyze to levulinic acid. Suppression of this hydrolysis is essential for an effective HMF production. As one of the suppression techniques, biphasic reaction system was proposed by Román-Leshkov *et al.*[1] The system consists of an acidic aqueous phase as the reaction phase and an organic phase as the extraction phase. Because HMF is lipophilic, the product HMF via reaction in an acidic aqueous phase immediately transfers into the organic phase, which prevents further conversion. After the biphasic system was introduced, the average yield of HMF from glucose improved to about 60 mol% [2,3]. Our previous study employed a segmented flow with the biphasic system, and succeeded in producing HMF at the yield of 76 mol% [4]. A segmented flow is a unique flow which is realizable using a microreactor. When more than two kinds of immiscible fluids flow in a thin tube, each fluid flows alternately as segment. In each segment an internal circulation occurs, which promotes mass transfers between phases by reducing boundary layers' thicknesses and also by preventing accumulations of transferred components near boundaries. In this report, the improvement of HMF yield and recovery of HMF from the extraction phase were examined.

2. Experimental method

As a material, glucose (FUJIFILM Wako Pure Chemical, Japan) or fructose (KISHIDA CHEMICAL, Japan) was solubilized into phosphate buffer saline (PB). PB was prepared by solubilizing NaH_2PO_4 $2H_2O$ (Nacalai tesque, Japan) into dilute phosphoric acid. $AlCl_3$ $6H_2O$ (Nacalai tesque) was also dissolved in PB when catalytic conversion was examined. As an extraction phase, methyisobutyl ketone (MIBK) (FUJIFILM Wako Pure Chemical) or 2-sec butylphenol (SBP) (Tokyo Chemical Industry, Japan) was used. All the chemicals were used as purchased without any further purification.

2.1. HMF synthesis using a microreactor

The experimental apparatus consisted of two high pressure pumps, SUS tubes, a PTFE tube, hastelloy tubes, a union tee, and a back pressure regulator. The sugar solution in PB and the organic phase (MIBK or SBP) were fed by the high pressure pump, respectively, and collided at the T-mixer. After the T-mixer, the fluids flowed through the PTFE tube, the hastelloy tube, and the SUS tube. The PTFE tube was installed to visualize the flow state. The hastelloy tube with the inner diameter of 2 mm and the length of 1.6 m or 19.7 m was for the reactor section, which was soaked in the oil bath. The SUS tube with the inner diameter of 2 mm and the length of 1 m was for the cooling section, which was soaked in the ice bath. All the channels from the high pressure pumps to the cooling section were pressurized at 3.45 MPa by the back pressure regulator installed after the cooling section. The depressurized product was collected at the ambient temperature and pressure in the glass bottle. The reaction temperature, residence time, sugar type, sugar concentration, pH of PB, and organic phase/aqueous phase volumetric ratio (O/A) were changed to evaluate the effects on the HMF yield. In addition, the effect of Lewis acidic catalyst was examined using the sugar solution prepared with $AlCl_3$ $6H_2O$ (5 wt% of sugar). The product was analyzed by the high performance liquid chromatography (HPLC).

2.2 HMF back extraction from organic phase to water phase

Double pipe extractor was constructed for the HMF extraction from product solution. The outer tube was PFA tube with the outer diameter of 1.6 mm and the inner diameter of 1.0 mm, whereas the inner tube was the porous PP tube with the outer diameter of 0.63 mm, the inner diameter of 0.33 mm, and the pore diameter of 0.2 μm. The product extraction phase of MIBK from synthesis part (1 wt% glucose, 16.8 min,

180°C, O/A = 2) flowed in the outer channel, whereas deionized water flowed in the inner channel at the same volumetric flow rates. The obtained extract and raffinate were analyzed by HPLC.

3. Results and discussion

3.1 HMF synthesis using a microreactor

Fructose was converted into HMF by the prepared system. The concentration of fructose, reaction temperature, residence time, and O/A were 1 wt%, 180 °C, 5 min, and 2, respectively. The HMF yield was about 1.8 fold when SBP was used as an extraction phase (SBP: 76.8 %, MIBK: 43.4 %). After the reactions, 97 % of HMF was in the extraction phase when SBP was used while only 75 % was in the extraction phase when MIBK was used. This result indicated that the extraction ability (or capacity) had significant effect on HMF yield. This was confirmed by increasing O/A, which resulted in the HMF yield increase.

Because Lewis acid is known to promote an isomerization of glucose to fructose, the use of Lewis acidic catalyst of $AlCl_3$ was examined. The reaction condition was determined according to our previous study [4], where the concentration of glucose, reaction temperature, residence time, and O/A were 1 wt%, 180 °C, 47 min, and 3, respectively, with the use of SBP as the extraction phase. The HMF yield reached 84.9 mol% as the result, which was much higher comparing with the one without the catalyst (76 mol%). Thus, the benefit of catalyst use on HMF conversion was confirmed.

3.2 HMF extraction from organic phase

The synthesis experiments revealed the extraction ability of MIBK was inferior to that of SBP. However, the inferior extraction ability might be the advantage when whole the process was taken into account. Because HMF is not tolerant to heat, the distillation is not suitable for the purified HMF recovery. This means the recovery requires the operations such as back extraction, crystallization, or drying. The inferior extraction ability works as the advantage in these processes. Therefore, the back extraction of the product HMF in MIBK after the synthesis reaction was examined using a constructed microextractor. After the examination, the chromatogram of the obtained MIBK decreased the intensity of HMF only. In addition, the obtained water phase only delivered the peak of HMF. Thus, using the constructed microextraction apparatus, HMF was selectively extracted from product MIBK to deionized water. When the extraction was conducted by counter-current, 19 min of residence time enabled the extraction to the equilibrium, which extracted 74 % of HMF into the water phase.

4. Conclusions

HMF synthesis by biphasic system was examined using a microreactor. Two kinds of extraction agents were examined, and the extraction ability was clarified as the important factor for high HMF yield. However, HMF was selectively recovered from the extraction phase when MIBK was used. When SBP was used, by adding the Lewis acidic catalyst of $AlCl_3$ for the isomerization promotion, the HMF yield successfully reached as high as 84.9 mol%.

References

[1] Yuriy Román-Leshkov *et al.*, Science, 312 (2006)

[2] Vinit Choudhary *et al.*, JACS, 135 (2013)

[3] Yomaira J. Pagan-Torres *et al.*, ACS Catal., 2 (2012)

[4] Yosuke Muranaka *et al.*, I&EC Res., 56 (2017)

578369 Preparation of a New Enzymatic Biocatalyst Via the Taguchi Method: Application to Lipase a from Candida Artarctica Immobilized Onto Cashew Apple Bagasse

Due to their high stability, high substrate specificity, high region, chemo and stereoselectivity, lipases are the most commonly used enzymes in biocatalysis. Among lipases, lipase A from Candida antarctica (CALA) is a very attractive biocatalyst, once it presents high thermostability, selectivity for trans fatty acids, stability in the acid pH range and high chemo-selectivity in relation to the amine groups. However, once lipases in their free form are soluble in water, it is necessary to immobilize them to improve their biocatalytic properties and enable their recovery and reuse easier. Among the methods for enzyme immobilization, the chemical ones allow to form covalent bonds are between the lipases and the support, resulting in rigidity and improved enzymatic stability. An alternative support for enzyme immobilization is cashew bagasse in natura, which is rich in hydroxyl groups, allowing multipoint covalent bonds. Activating the support with divinylsulfone (DVS) allows such types of bonds to be obtained on supports containing on their surface a very broad range of amino, thiol or hydroxyl groups, such as cashew bagasse in natura. In this study, CALA was immobilized onto cashew bagasse in natura activated with DVS and optimal conditions (DVS concentration: 1, 3, 5 and 7.5; buffer molarity: 5, 25, 100 and 350 mM; pH: 3, 5, 7 and 12.5; temperature: 4, 15, 25 and 30 °C and time of activation: 0,5, 3, 12 and 24 hours) for the immobilization procedure was determined. For this communication, an advanced experimental design by the Taguchi method with a standard orthogonal matrix L16' (the 'L" and "16" represent the Latin square and the number of experiments, respectively) was used to examine five factors at four levels in order to maximize the derivative activity. For the preparation of the support, cashew bagasse in natura was washed three times with water, dried at 60 °C for 24h and milled in a hammer mill to obtain an average particle size of less than 0.177mm. The support was then activated as follows:1.0g of cashew bagasse in natura was added to 20 mL of 350 mM sodium carbonate solution, at neutral pH, containing 4.5 M divinyl sulfone. The resulting solution was kept under constant stirring for 30 minutes at 15 °C; the support was then washed with distilled water and stored at 6 °C. For the immobilization procedure, 1.0 mg of enzyme per g of support was used in the presence of 0.01% Triton-X and 5 mM sodium phosphate buffer and pH 7.0, for 24 h and 25 °C, under continuous agitation. After immobilization, the biocatalysts were incubated in 100 mM bicarbonate buffer at pH 10.0 (1:10 w/v) at 25 °C for 24 h, then in 1 M EDA pH 10.0 also for 24 h at 4 °C. Finally, the biocatalysts were washed with distilled water, vacuum dried and stored at 4 °C. Then, the hydrolytic activity of soluble and immobilized CALA was performed, as well as the protein concentration was determined. Once the experimental design was finished, the optimal conditions was found to be: DVS concentration: 3, buffer molarity: 3 mM, pH: 3; temperature: 30 °C and time of activation: 12 hours, obtaining a derivative activity of 4.31 U/g.

563530 Reaction of Acetic Acid and Chicken Eggshells to Form Calcium Acetate

In this work, we analyze the effect of the particle size, acetic acid concentration and excess of acetic acid on the reaction of acetic acid and chicken eggshells to produce calcium acetate. Inductively coupled plasma mass spectroscopy (ICP-MS) and X-ray diffraction (XRD) were used to quantify the amount of calcium carbonate in chicken eggshells. Calcium acetate was qualitatively and quantitatively characterized by X-ray diffraction (XRD) and thermogravimetric analysis (TGA), respectively. The reaction was investigated using a face-centered central composite design (FCCD) and evaluating the optimal conditions to produce calcium acetate. A simplified shrinking-core model was fitted to the experimental data suggesting that the reaction is mass transfer limited. The highest experimental conversion achieved was 95.82% with an excess of moles of acid about 50%, an acid concentration of 5% w/w and a particle size of 175 μm. The predicted conversion obtained by the statistical model was 98.78 %, implying an error of 3%.

577853 Enhanced Glucose Dehydration to 5-Hydroxymethyl Furfural through Encapsulated Phosphotungstic Acid in Metal Organic Frameworks

5-hydroxymethy furfural (HMF) is a versatile bio-based renewable chemical platform because it can serve as a building block for the productions of liquid fuels and fine chemicals. HMF can be produced by glucose dehydration. Typically, both Lewis and Brønsted acid sites are needed for glucose isomerization to fructose and fructose dehydration to HMF, respectively. This study explores the synergy between phosphotungstic acid (PTA) and MIL-101(M) (M= Cr, Al, and Fe) for cascade reaction of glucose isomerization to fructose and fructose dehydration to 5-hydroxymethylfurfural (HMF). We hypothesize that the synergy between Brønsted acid from PTA and Lewis acid/Brønsted base from MIL-101(M) enhances the selectivity toward HMF in ionic liquid. We conducted the experiments by encapsulating PTA in MIL-101(M). This paper reveals the synergy between the BrØnsted and Lewis acid sites for glucose dehydration to HMF. The aluminum nodes provide Lewis acidity and the presence of the $-NH_2$ functional groups from linkers gives Brønsted bases. Together, they enhance the glucose isomerization to fructose. The encapsulated PTA dehydrated the resulting fructose to HMF. We obtained 54% HMF selectivity at 38% conversion using 120°C for 2h. The encapsulated MOFs exploit the high selectivity from single-site metal nodes and encapsulated species for additional catalytic sites for cascade reactions. This strategy allows the design of the multi-functional catalysts for one-pot reaction.

567594 Coverage-Dependent First Principles Microkinetic Models for Copper-Catalyzed Hydrogenation of Carbonyl Groups: Role of Water and Sacrificial Alcohols

Promoted copper catalysts are emerging as catalysts for the hydrogenation of aldehyde and ketone compounds, a key step in biomass conversion. Their superior selectivity and low material cost make them an attractive alternative to the conventional platinum and palladium catalysts in the hydrogenation of carbonyl functionalities. However, the reaction mechanism and nature of the active site(s) remain debated. Both a single site mechanism in which only metallic sites contributes, and a multiple-site mechanism in which multiple copper species and/or support active sites are involved, have been proposed. Catalyst design and optimization is severely hampered by this lack of fundamental insight. We applied coverage-dependent first principles microkinetics to elucidate the hydrogenation mechanism under different process conditions.

In the hydrogenation of carbonyl functionalities, two pathways are generally available, depending on which atom is hydrogenated in the first step (Figure 1, blue). When the carbon atom is hydrogenated first, a stable alkoxy intermediate is formed, i.e., the alkoxy pathway. Alternatively, if the oxygen atom is hydrogenated first, an unstable alkyl intermediate is formed, i.e., the alkyl pathway. In a first study,[1] we showed that the alkoxy pathway is dominant over copper under typical process conditions. Hydrogenation of alkoxy intermediates is often rate-limiting in the hydrogenation of carbonyl groups over copper catalysts, leading to high alkoxy coverages. We showed, however, that adding a small amount of water facilitates hydrogenation of these stable alkoxy intermediates via proton transfer from adsorbed water molecules and surface hydroxyl species. As a result, the rate-limiting step shifts from hydrogenation of the alkoxy intermediate, to water regeneration. These findings also provide an alternative interpretation for the role of promoters in Cu-catalysis. Adding ZnO as a promoter drastically reduces the barriers for hydrogenation of surface oxygen and surface hydroxyl species, from 119 to 37 $kJmol^{-1}$ and from 131 to 35 $kJmol^{-1}$ respectively, and, as a result, greatly increases the global reaction rate.

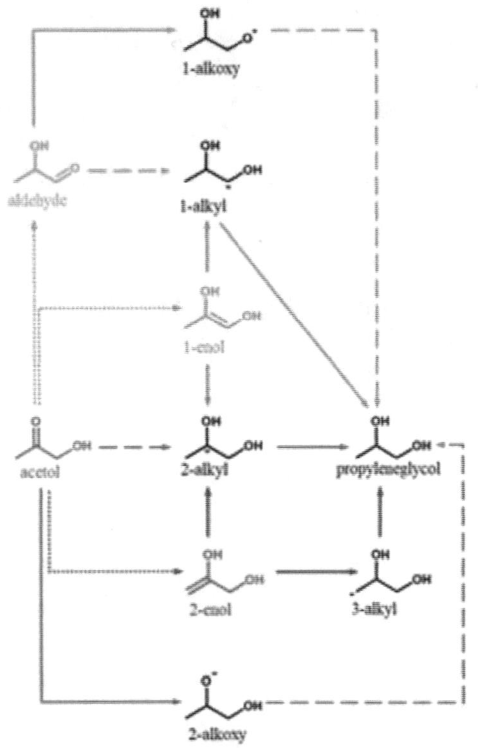

Figure 1. Pathways for the hydrogenation of acetol: direct pathways (blue) and via keto-enol tautomerism (dotted lines): 2-hydroxypropanaldehyde (orange), 1,2-dihydroxypropylene (green) and 2,3-hydroxypropylene (red). Hydrogenation of a carbon atom (solid lines) occurs via direct hydrogenation, hydrogenation of an oxygen atom (dashed lines) can occur via direct hydrogenation or via proton transfer from surface hydroxyl species or from adsorbed water.[2]

Alternatively, when keto-enol tautomerism is considered, 3 additional hydrogenation pathways open up (Figure 1). We showed that for alpha-hydroxy ketones, such as the industrially important acetol, the difference in stability between the adsorbed ketone and enol tautomer is sufficiently low (<30 $kJmol^{-1}$) for the enol pathways to become kinetically relevant.[2] Under dry conditions (Figure 2A), the majority (84%) of acetol is hydrogenated via the 1-enol pathway, consistent with recent isotope labeling studies, while the remaining 16% is hydrogenated through the acetol pathway. Both 1-enol and alkoxy hydrogenation are rate-controlling steps under these conditions. The superior activity for C=C over C=O hydrogenation is rather surprising due to the high selectivity of copper in the hydrogenation of unsaturated and aromatic ketones.

The presence of water again drastically impacts the hydrogenation reaction mechanism and three regimes can be identified as a function of the inlet H_2O:ketone molar ratio (Figure 2B). At low water partial pressures, close to dry conditions, the 1-enol pathway dominates. By adding water however, proton transfer pathways open up to hydrogenate the stable alkoxy intermediates. As a result, the contribution of both the acetol and aldehyde pathways increase, up to a H_2O:ketone molar ratio of 0.5. When more water is added, the 1-enol becomes kinetically irrelevant and the contribution of the aldehyde pathway starts to go down again, and the mechanism is dominated by the acetol pathway. Under those conditions, the rate is again controlled by regeneration of the hydroxyl species and the adsorbed water molecules.

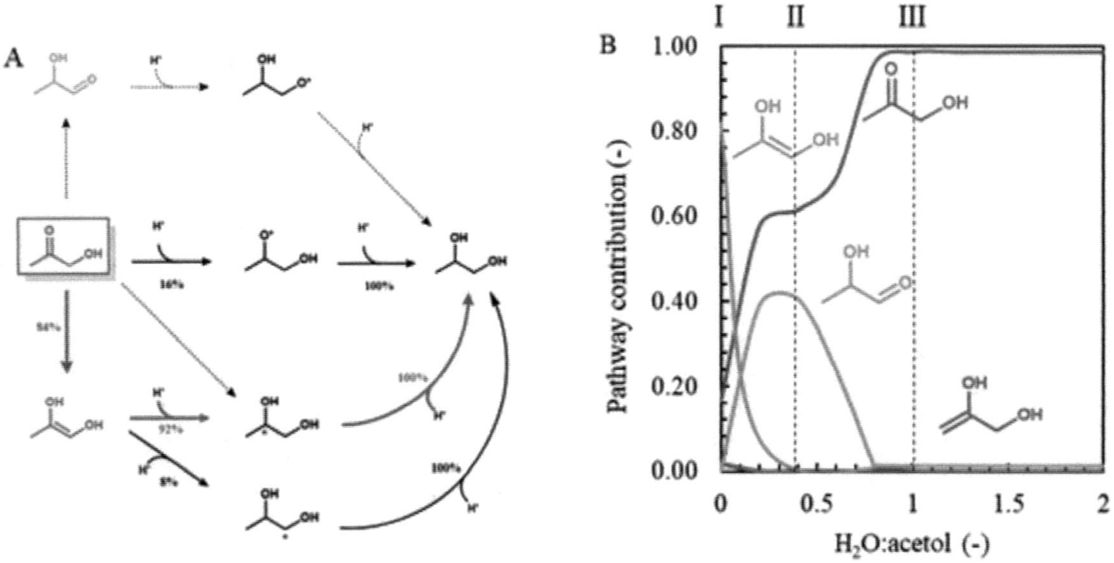

Figure 2. A. Reaction path analysis for the hydrogenation of acetol to propylene glycol under dry conditions. The numbers indicate the fraction of each components consumed in a particular reaction step. Dotted arrows indicate reactions with a fraction below 0.5% while green arrows indicate reaction steps with a fraction above 50%. **B.** Effect of the water to acetol feed flow ratio on the dominant reaction pathway for the hydrogenation of acetol. The contribution indicates the fraction of acetol consumed in each particular pathway. Conditions: 500 kg_{cat} s mol_{acetol}^{-1}, 200 °C , p_{H2} = 5 MPa, p_{acetol} = 2.5 MPa.

Finally, the hydrogenation of the alkoxy intermediate can be facilitated by the introduction of a sacrificial alcohol.[3] Rather than following a dehydrogenation-hydrogenation mechanism, which was generally hypothesized to be the dominant pathway, the proton of the alcohol is directly transferred to the ketone. This proton transfer between the sacrificial alcohol and the alkoxy intermediate enhances the reaction rate, similar to water. Proton transfer from a sacrificial alcohol also results in an alkoxy intermediate, (*sac*-alkoxy), which rather than being regenerated back to the alcohol, is dehydrogenated to its corresponding ketone. As a result, the rate-limiting step shifts from hydrogenation of the alkoxy intermediate, under dry conditions or regeneration of water, under wet conditions, to hydrogenation of the ketone to its alkoxide and dehydrogenation of the *sac*-alkoxy species. Because of the higher degree of rate control for ketone hydrogenation, we find that the addition of gas phase hydrogen has a surprising catalytic effect – it speeds up the overall reaction without being consumed.

The detailed microkinetic simulations illustrate the complexity of this seemingly simple reaction, and show the importance of self-consistent, coverage-dependent microkinetic simulations to discover the different kinetic regimes.

(1) De Vrieze, J. E.; Thybaut, J. W.; Saeys, M. Role of Surface Hydroxyl Species in Copper-Catalyzed Hydrogenation of Ketones. *ACS Catal.* **2018**, *8* (8), 7539–7548.

(2) De Vrieze, J. E.; Thybaut, J. W.; Saeys, M. Role of Keto-Enol Tautomerization in the Copper-Catalyzed Hydrogenation of Ketones. *ACS Catal.* **2019**, *9*, 3831–3839.

(3) De Vrieze, J. E.; Urbina Blanco, C. A.; Thybaut, J.; Saeys, M. Autocatalytic Role of Molecular Hydrogen in Copper-Catalyzed Transfer Hydrogenation of Ketones. *Submit*.

571561 Catalytic Deoxydehydration of Glycerol to Allyl Alcohol with 2-Hexanol As H-Donor – a Detailed Study of the Reaction Mechanism

Allyl alcohol is a very promising starting material for a variety of molecules such as acrylonitrile, acrolein and acrylic acid – all, important intermediates for the chemical industry: The ammoxidation of allyl alcohol to acrylonitrile was reported with a high yield (more than 80%) by Guillon et al.[i],[ii]. The selective oxidation of allyl alcohol to produce acrylic acid yields up to 85 % was submitted at rather low temperatures (300°C) over MoVOx catalyst.[iii] However, the main bottleneck for these applications lies in the sourcing of allyl alcohol. Nowadays, allyl alcohol is produced by the selective catalytic hydrogenation of acrolein, which is obtained by the selective catalytic oxidation of propylene. Recently we reported the synthesis of allyl alcohol with a high yield (>86%) coming from deoxydehydration of glycerol (DODH) over Rhenium based catalyst with 2-hexanol as sacrificial hydrogen-donor and the corresponding reaction mechanism explaining the key role of water.

Experimental

For the DODH of glycerol, a rhenium based catalyst was prepared by an incipient wetness impregnation method using a 75 wt% perrhenic acid aqueous solution (Aldrich). After one hour, the impregnated catalyst was dried at 110 °C for 12 h and then calcined under static air from 110 °C to 500 °C (5 °C / min) for 3 h. The catalytic performance was evaluated using a pressure-resistant glass tube equipped with a magnetic stirring bar, it was loaded with 100mg of catalyst, glycerol (92 mg, 1 mmol), 100 mg, and 2-hexanol (3.3 mL). The reaction temperature inside the tube kept at 148 °C for 2.5h. The conversion and selectivity were determined by GC analysis using benzene(20 mg, 0.25 mmol) as an internal standard.

Results and discussion

The DODH of glycerol over rhenium based catalyst exhibited full conversion after 2.5 h, yielding up to 86% of allyl alcohol. From the reaction curve, it becomes visible that the selectivity in allyl alcohol increased during the 2 first hours of the reaction, suggesting the presence of an intermediate species. With respect to these results, the by-products were deeply analyzed using GC-MS, allowing notably the identification of hexenes, ethers, and acetals. The identification of acetals was most surprising. The latter is formed by condensation of glycerol with acrolein, indicating that acrolein is the intermediate species in the DODH reaction. The latter is formed by dehydration of glycerol and then further hydrogenated to allyl alcohol. The significant formation of acetal at the beginning of the reaction can be explained by the role of water in the equilibrium of this reaction. in other words, the acetalization between acrolein and glycerol is an equilibrium reaction, giving water as a co-product. When the amount of water in the reaction mixture is low, the equilibrium is on the side of the acetals. Here, other identified side reactions come into the game: 2-hexanol was found to be dehydrated to various hexenes and condensated to an ether during the reaction. The as-formed water helps to shift the equilibrium of the acetalization to the reactant side (formation of acrolein), whereby the selectivity of the final product (allyl alcohol) increased with the reaction time. The corresponding mechanism was further validated by using individual reactions with acrolein, water, and 2-hexanol as reactants.

Conclusions

The DODH reaction of glycerol over a heterogeneous rhenium based catalyst using 2-hexanol as H-donor showed increasing selectivity in allyl alcohol over reaction time, suggesting the formation of an intermediate. The deep analysis of the reaction mixture revealed the presence of acetals formed from the reaction between acrolein and glycerol. With respect to the equilibrium of this reaction, water was found to play a key-role.

[i] C. Liebig, S. Paul, B. Katryniok, C. Guillon, J-L. Coururier, J-L. Dubois, F. Dumeignil, W.F. Hoelderich, Appl. Catal. B: Env. 132-133 (2013) 170.

[ii] C. Guillon, C. Liebig, S. Paul, A-S. Mamede, W.F. Hoelderich, F. Dumeignil, B. Katryniok, Green Chem. 11 (2013) 3015.

[iii] T. Murayama, B. Katryniok, S. Heyte, M. Araque, S. Ishakawa, F. Dumeignil, S. Paul, W. Ueda, ChemCatChem 8 (2016) 1

570527 Esterification of Isobutyl Alcohol and Acetic Acid - Kinetic Study

Currently there is an increasing pressure at the chemical industry to generate safer and environmentally friendly products. As a results the industry is experiencing a change in the production paradigms, following a transition from fossil-based to biobased products. In particular, this is a major trend in the segment of organic solvents. Several green solvents, which have been identified according to sustainability criteria, are gaining pace at the industrial scale for their good performance and reduced impacts. Among these, esters of fermentation-derived alcohols and acids are of particular interest because their broad range of applications. From this group, isobutyl acetate standouts as an excellent solvent for coatings, inks, lacquers, and cosmetic products (e.g. nail polish). Also, it is a valuable ingredient in the perfumery, flavors and fragrances industry for its fruity aroma.

Entirely biobased Isobutyl acetate can be produced by esterification of isobutyl alcohol (obtained from fusel oil) and acetic acid (from vinegar fermentation). The process involves an acid catalyst as reaction promoter, and because the need for suitable sensory properties, most traditional homogeneous catalysts cannot be used because they affect the organoleptic profile (odor and color). Thus, focusing on the required organoleptic profile for the fine chemicals industry, the use of heterogeneous catalysts (e.g. ion exchange resin) would be preferred. Taking into account that the reaction conversion is limited by the chemical equilibrium, the implementation of a reactive distillation process would enable higher conversions and savings in product purification. However, in order to explore the usefulness of a reactive distillation process for isobutyl acetate production, there is need to develop the corresponding kinetic model.

Thus, this work focused on the kinetic study of the Fisher esterification between isobutanol and acetic acid, using Amberlyst 15 and sulfuric acid as catalyst. Preliminary experiments helped to rule out mass transfer limitations during experiments, and the Weiss-prater criteria was used as surrogate for ensuring kinetic regime. Experiments were carried out based upon a suitable experimental design including temperature, mole ration of reactants and catalyst loading as main variables. Additional experiments were carried out to evaluate chemical equilibrium constant at different temperatures. The whole set of experimental data were correlated with mole fraction and activity-based kinetic expressions; the last to account for non-idealities of the reactive mixture. Both models fitted well with experiments, and can be used for further design of a reactive distillation process.

572022 Sustainable Solutions for Selective Glycerol Oxidation into Dihydroxyacetone Using Commercial Catalysts

The increasing world's energy demand and the need for greener solutions for its production and utilization enhanced renewable energy industries, including the biodiesel industry as a potential direct alternative to fossil diesel, for which the world's production increased from 20 Mm^3 in 2010 to 36 Mm^3 in 2017, and it is expected to keep increasing. Biodiesel growing production is leading to the accumulation of crude glycerol, the main reaction by-product with almost no economic value, and its valorization is crucial to increase biodiesel production sustainability and competitiveness. Crude glycerol purification is only economically feasible for big biodiesel industries; however, several valorization routes into high added value products have been proposed, although very few reached the commercial phase. Selective glycerol oxidation may be a solution, as the high added value products ensure the process robustness against raw material price fluctuations. The state-of-the art of dihydroxyacetone (DHA) production by glycerol aerobic catalytic oxidation already suggests some routes to reach a solid commercial application, as nowadays DHA industrial bio fermentation production is not satisfactory and a more efficient production process is needed to overcome the market deficit.

DHA production by catalytic glycerol aerobic oxidation was firstly reported in 1993 using carbon supported Pt/C and Pt-Bi/C catalysts and, since then, the scientific community focus has been mainly focused on the development of catalyst to increase the process selectivity towards DHA and glycerol conversion. Simultaneously, two main operating conditions were established for glycerol oxidation: basic or base free conditions. Although working under strong basic conditions (typically by adding NaOH) greatly increases reaction rate, Pt group catalysts are unstable and metal leaching may occur, DHA is unstable, leading to the formation of glyceric acid (GCA) and organic acids salts, which may precipitate, leading to process complications at industrial scale.

In the Lab-scale tests, three main catalysts groups were constituted to efficiently convert glycerol into DHA: Pt based, Au based and noble bimetallic catalysts. For Pt based catalysts, base free conditions were usually applied and the highest DHA yields were obtained with Bi and Sb promoted Pt catalysts supported in carbon materials, namely carbon nanotubes, and mesoporous silicas, due to its well defined structure and medium sized porous, favoring glycerol's secondary hydroxyl group oxidation and avoiding DHA overoxidation. For Au catalyzed glycerol oxidation, the mandatory presence of strong bases, which resulted in poor or null DHA selectivity, was overcome in 2010, by using specific support materials, which increased catalysts activity under base free conditions. In 2014 high DHA yields were firstly reported using Au catalysts supported in late transition metal oxides as CuO, and later using Au/ZnO, which are so far the most promising results, opening new possibilities for reaction process design as Au catalysts are far more resistant to overoxidation compared with Pt group metals. Bimetallic catalysts, such as AuPt, showed enhanced performance compared with the monometallic catalysts, namely in terms of stability and activity, but only activated carbon (AC) supported AuPt-Bi catalysts achieved moderate DHA yields.

It is generally difficult to achieve very high DHA yields in batch reactors, due to catalysts deactivation and because with increasing reaction times, glycerol conversion increases but DHA overoxidation into compounds like hydroxypyruvic acid occurs, decreasing DHA selectivity. Concerning catalysts deactivation, although poisoning by surface overoxidation, metal leaching and sintering processes may occurs under certain operating conditions, the major cause of catalysts deactivation is the strong adsorption of reaction products onto catalysts surface, namely GCA, as most of the produced organic acids are well-known chelating agents. Catalyst washing with water revealed to be insufficient to restore the performance of the majority catalysts tested, whereas thermal treatment resulted for some catalysts.

To the best of our knowledge, about twenty-five years after the first work about selective glycerol aerobic oxidation into DHA, the process is still in laboratorial scale and far from reach commercial application in industry, mainly due to the unsatisfactory DHA yields achieved, the use of expensive noble metal catalysts and catalyst deactivation problems and also due to the fact that most of the catalyst tested have been synthesized by research groups and were not available at commercial scale.

Based on the above-identified problems, some of the most promising commercial catalysts available for selective glycerol aerobic oxidation into DHA were tested, to evaluate their performance and reactions kinetics. The commercial catalysts Pt5%/AC (Aldrich®), Pt5%-Bi1.5%/AC (Johnson Matthey®) and Pt5%-Bi5%/AC (Evonik®) were purchased and tested in a semi-batch reactor under base free conditions. The main operating conditions as temperature, initial glycerol concentration, oxygen pressure and oxygen inlet flow rate were screened. High glycerol conversions were obtained, namely Pt5%/AC and Pt5%-Bi5%/AC catalysts achieved glycerol conversions above 90% within the first 6h of reaction. It was possible to verify that the performance of Pt-Bi catalysts outcomes the performance of Pt catalysts in terms of DHA selectivity. The best results within the platinum catalysts tested in terms of DHA yield was attained with the Pt5%-Bi1.5%/AC from Johnson Matthey®.

Hence, this work demonstrates that commercially available Pt-Bi/AC catalysts may represent a potential solution for the scale up of DHA synthesis through glycerol's aerobic oxidation.

569233 Catalytic Conversion of Cellulose into Levoglucosenone Using Homogenous Brønsted Acid Catalysts in Polar Aprotic Solvents in a Flow Reactor

Levoglucosenone (LGO) is a high value anhydro – sugar used in the synthesis of pharmaceutics, renewable solvents, polymers; and other specialty and high value chemicals. LGO is commonly produced from the pyrolysis of cellulose in low yields (12-20%). Huber and coworkers reported yields of LGO of up to 51% from cellulose dehydration using 20mM sulfuric acid (H_2SO_4) as catalyst in polar aprotic solvents such as tetrahydrofuran (THF) in a batch reactor.[1] Recently, our research group in collaboration with Huber's at the University of Wisconsin – Madison studied the production of LGO from cellulose in a batch reactor using lower concentrations of sulfuric acid combined with solid acid catalysts in (THF). The study found similar LGO yields to those obtained with higher concentrations of sulfuric acid. Itwas also found that the accumulation of water during the reaction of cellulose to LGO in a batch reaction system promotes the formation of other products over LGO. Here we show the results of using a flow reactor instead of a batch one for the dehydration of cellulose to LGO using different Brønsted acid catalyst such as H_2SO_4, 1-propanesulfonic acid and phosphoric acid in polar aprotic solvents such as γ-valerolactone, THF and acetone. We show the effect of the reaction temperature and the catalyst concentration on the LGO yield. We demonstrate that using different sources of cellulose has an effect on the yield of LGO. Also, we show that the crystallinity of the cellulose has a significant effect on the production of LGO. We demonstrate that LGO can be effectively produced from cellulose using Brønsted acid catalysts in polar aprotic solvents using a flow reactor.

1. Cao, F.; Schwartz, T. J.; McClelland, D. J.; Krishna, S. H.; Dumesic, J. A.; Huber, G. W. Dehydration of cellulose to levoglucosenone using polar aprotic solvents. Energy & Environmental Science 2015, 8 (6), 1808-1815.

577095 The Use of Aqueous Ethanol As a Solvent for the Catalytic Hydrogenation of Lactose to Lactitol

The selection of a solvent can be particularly influential for performing three-phase hydrogenation reactions, where gaseous hydrogen and liquid phase reactant react over a solid catalyst. Descriptions in the literature indicate that the hydrogenation of sugar compounds such as glucose and xylose, via the reduction of the sugar compound's carbonyl group using hydrogen, can be promoted by the selection of aqueous alcohol rather than water as a solvent. Thus far, this line of research has not been explored regarding the hydrogenation of lactose to lactitol.

As a major component of waste streams from the dairy industry, upgrading lactose to valuable commodities by chemical processing would help incentivize sustainable handling of this material. At Ohio University's Sustainable Energy and Advanced Materials Laboratory (SEAM Lab), an experimental process system has been designed and constructed to perform batch-wise hydrogenation reactions, and work is underway to perform lactose hydrogenation in pure water and aqueous ethanol solutions. This research will contribute to the study of the solvent effect in the hydrogenation of sugar compounds, and will evaluate the feasibility for potential enhancements to dairy waste processing.

577808 Aqueous Phase Reforming of Ethanol over Nickel Supported on Oxygen Ion Conducting Supports

Aqueous phase reforming of ethanol was carried out over nickel catalysts supported on yttrium-stabilized zirconium (YSZ), gadolinium-doped ceria (GDC), and samarium-doped ceria (SDC), which are materials which show high oxygen ion conductivity (OIC) and contrasted to an alumina support. Those supports that with a high OIC demonstrate superior ethanol conversion and hydrogen production compared to the non-conducting support. In previous work, it was found that ceria was superior to alumina as a support. Ceria has a higher OIC than alumina though other factors will affect the activity. In other experiments, OIC supports have been shown to enhance the water-gas shift reaction as well as carbon burnoff. This work will determine how the OIC supports enhance catalytic activity.

Preliminary results from thermogravimetric analysis of used (12 hours) catalysts shows that the GDC support actually leads to the most carbon formed despite the high activity. It is known that the type of carbon formed plays an important role in deactivation. Further results with longer time-on-stream experiments that determine the amount and type of carbon formed will be carried out. Furthermore, the catalyst particle size is known to have an effect on carbon formation, so results from TEM and chemisorption will be presented. The overall goal is to unravel the relationship between support material, carbon formation, and nickel particle size to explain the observed reactor results and determine how important the role of OIC is in the aqueous phase reforming of ethanol.

559009 Polyelectrolytes Enabled Co-Localization of Enzyme Cascades for One-Pot Synthesis

Co-localization of enzyme cascades for one-pot synthesis hold great promise for the green production of fine chemicals because it circumvents the intermediate purifications, and meanwhile, facilitates the substrate transport to the subsequent enzyme, mimicking that occurs a living cell in nature. However enzymes from different origin may differ dramatically in their pH window of stability and/or activity, and thus hinder the operation and the performance of the enzyme cascade.

Here we present a method of using polyelectrolyte to co-localize enzyme cascades, in which the polyelectrolyte generates a favorable microenvironment in terms of pH in the vicinity region of the conjugated enzyme and thus enables a high activity and stability.

To demonstrate the effectiveness of this method, we chose chloroperoxidase (CPO) from *Caldariomyces fumago* and xanthine oxidase (XO) from bovine milk as a model cascade and tested its performance in the oxidization of p-xylene to p-toluic acid. After the conjugation of CPO to polymethacrylic acid (PMAA) whereas XO to chitosan quaternary ammonium salt, respectively, CPO-PMAA@XO-chitosan was obtained by electrostatic interactions between oppositely charged polyelectrolytes. The co-localization of enzymes was confirmed by confocal laser scanning microscopy, which showed the average distance between CPO and XO was 6.2 nm. The oxidization of p-xylene to p-toluic acid operated at various pH showed that, compared to the native counterpart, the CPO-PMAA@XO-chitosan received a 7.6-fold increase in throughput and moreover a wider pH window of activity and stability. The effectiveness of this method was also demonstrated in GOx-PMAA@HRP-chitosan cascade. The enhanced activity and expanded operation window suggests that this assembly method is promising for one-pot enzyme cascade reaction.

568595 Converting Biogas to Liquid Fuels By Low Energy Electrical Corona Discharge Processes

Considered as a renewable resource, biogas has abundantly available sources worldwide, such as dairies, sewage, landfills, etc. However, it is usually disposed of by flaring because of its low economic value. Therefore, the development of technologies to use biogas as feedstock to produce liquid fuels like gasoline or diesel would have a positive economic and environmental impact and stimulate new approaches in industrial chemistry.

In this work, a new process that enables the conversion of biogas to liquid transportation fuels with high energy efficiency has been developed. The approach is to employ low energy electron impact ionization through the use of a non-thermal plasma generated by an atmospheric pressure electric corona discharge. Additionally, microtechnology has been combined with this corona discharge technology because micro-scale geometry provides strong gradients with respect to temperature, concentration, pressure, and reactive species and the efficient heat and energy management [1].

Multi-discharge reactors with 10-discharge and 100-discharge have been built, and their performance proves the concept of chemical conversion of biogas and demonstrates the ability to produce long-chain hydrocarbons. In order to identify which parameters are most important in the optimization of the conversion process, the experimental design has been developed as a guide for experimental work. Major parameters under investigation include the power level per discharge, the discharge gap, the reactive gas flow rate, the synthetic biogas composition, the concentration of inert, and the gas pressure. A high LHV energy efficiency up to 95% can be achieved, and methane conversion can reach up to 65% with high selectivity (45~55%) towards C2+ hydrocarbons. Additionally, methods to prevent the production of coking and wax have been developed so the reactors can be adapted for continuous operations.

Reference:

[1] D. Mariotti, R. M. Sankaran (2010), "Microplasmas for nanomaterials synthesis," *Journal of Physics D: Applied Physics*, *43* (32), pp. 323001.

578244 Mechanisms of C-O and C-C Bond Activation of Acetic Acid Hydrodeoxygenation over Pt-Mo Catalysts

Promising new technologies for biomass conversion involve production of bio-oils, which need to be upgraded to remove oxygen-containing hydrocarbons. Reactions of carboxylic acids are some of the slowest steps in the hydrodeoxygenation of biomass-derived feedstocks. Therefore, acetic acid, the simplest carboxylic acid, can serve as a useful model compound in the development of improved hydrodeoxygenation (HDO) catalysts. Furthermore, acetic acid itself is a significant component of bio-oils. Although supported Pt catalysts are known to be efficient in HDO, they exhibit insufficient activities and low selectivities with undesirable C-C bond breaking reactions.

In this study, more active and selective HDO catalysts were developed by adding Mo to Pt. While Mo catalysts do not have any activity, Pt-Mo bimetallic catalysts exhibit significantly improved activities and selectivities even at a small concentration of added Mo. The nature of the promotion effects was studied by characterizing a series of Pt, Mo and Pt-Mo catalysts with multiple techniques, including H_2 temperature programmed desorption, transmission electron microscopy (TEM), scanning transmission electron microscopy with energy-dispersive X-ray spectroscopy (STEM/EDS) and in situ X-ray absorption near edge structure/extended X-ray absorption fine structure (XANES/EXAFS). In addition, reaction kinetic results were obtained in a flow reactor, and the reaction mechanism was analyzed with dispersion-corrected density functional theory (DFT-D) calculations.

Addition of Mo to Pt improved the reaction activity by more than an order of magnitude. More importantly, the selectivities were also improved. The formation of C_1 products (CO and CH_4) went down to less than 10 mol % for the Pt-Mo bimetallics. The formation of C_2 products, mostly CH_3CH_3, increased to more than 90 mol %. Furthermore, Pt-Mo catalysts were stable under the reaction conditions for at least 10 hours. The STEM/EDS and XAS results demonstrate that Mo formed subnanometer clusters that were fairly uniformly distributed on the support, including the surface of Pt nanoparticles. The dispersion of Pt nanoparticles was not affected by the Mo addition. With increasing Mo concentration, the reaction activity of the Pt-Mo bimetallics increased, went through a maximum and then declined. The DFT-D calculations show that although Mo atoms themselves are catalytically inactive, they promote neighboring Pt atoms by changing the stability of adsorbates and surface reaction energies. Mo atoms are oxophilic and, therefore, serve as preferential adsorption sites for acetic acid and other oxygenates. Due to the presence of Mo on Pt, C-O bond splitting reactions become more thermodynamically preferable, and C-C bond splitting reactions become suppressed.

While Mo catalysts are inactive and Pt catalysts have low activities and selectivities, Pt-Mo catalysts exhibit high activities and selectivities in acetic acid HDO. In addition, the Pt-Mo catalysts are stable. Although Mo atoms are catalytically inactive, they serve as preferential adsorption sites for oxygenates and, thus, promote neighboring Pt atoms by changing adsorbate properties and reaction energies. Beyond biomass processing, Pt-Mo formulations are promising catalysts for a wide variety of reactions that require a transformation of molecules with an oxygen atom due to oxophilicity of Mo. Beyond catalysis, Pt-Mo nanoparticles offer an exciting potential in sensing and biomedical applications as well as in numerous other science and technology fields that require tuning of surface-oxygen interactions.

564667 $La_xBa_{1-x}FeO_3$ Perovskite Oxide for Low Temperature Thermochemical Conversion of Carbon Dioxide

Reverse water-gas shift chemical looping (RWGS-CL) is a thermochemical process to convert carbon dioxide to carbon monoxide by a lower temperature and high selectivity. The resulting carbon monoxide can be subsequently used for chemical energy storage in the form of hydrocarbon-based fuels and chemicals. Based on the catalyst requirements of RWGS-CL, this research focused on the synthesis and characterization of the $La_xBa_{1-x}FeO_3$ perovskite-type catalysts, which can form oxygen vacancies for CO_2 conversion. Temperature-programmed reduction (TPR), temperature-programmed oxidation (TPO) and long-term chemical looping experiments were performed to identify reaction temperatures, redox properties and CO yield. X-ray diffraction was utilized to identify the crystal structure. Comparing with other perovskite-type catalysts, $La_xBa_{1-x}FeO_3$ has the advantage of converting CO_2 to CO at a low temperature of 500 °C. $La_{0.25}Ba_{0.75}FeO_3$ shown a high CO yield of 1660 umol/g-catalyst and $La_{0.5}Ba_{0.5}FeO_3$ shown a high stability in the conversion process which is an important catalyst property using for industrial scale.

578628 Hydroxyl-Mediated Ethanol Selectivity of CO_2 hydrogenation

Oxide-supported Rh nanoparticles have been widely used for CO_2 hydrogenation, especially for ethanol synthesis. However, this reaction operates under high pressure, up to 8 MPa, and suffers from low CO_2conversion and alcohol selectivity. This paper describes the crucial role of hydroxyl groups bound on Rh-based catalysts supported on TiO_2 nanorods (NRs). The RhFeLi/TiO_2 NR catalyst shows superior reactivity (˜15% conversion) and ethanol selectivity (32%) for CO_2 hydrogenation. The promoting effect can be attributed to the synergism of high Rh dispersion and high-density hydroxyl groups on TiO_2 NRs. Hydroxyls are proven to stabilize formate species and protonate methanol, which is easily dissociated into $*CH_x$, and then CO obtained from the reverse water–gas shift reaction (RWGS) is inserted into $*CH_x$ to form CH_3CO*, followed by CH_3CO* hydrogenation to ethanol.

559849 H^+ Conductivity and O_2 Impermeability of Ultrathin Oxide Layers with Built-in SiO_2 Separation Membrane for Scalable Artificial Photosynthesis

Replacing fossil fuels with renewable resources to meet the global need requires a technology that is scalable to the unprecedented level of several terawatts. Natural photosynthesis is the sole existing technology that produces energy dense chemicals on the terawatt scale (> 100 TW). Its key design feature is the closed cycle of H_2O oxidation and formation of the primary reduction products on the shortest possible length scale, nanoscale, while separating the incompatible redox environments by an ultrathin membrane. This offers the advantage of minimizing efficiency-degrading mass transport processes and unwanted side reactions.

To incorporate the key feature into artificial photosystems, we assembled ultrathin (2 nm), pinhole-free, molecular wires-embedded SiO_2 membrane on planar and nanotube constructs. This membrane system spatially separates the H2O oxidation CO_2 reduction, but enables (photo-)electrochemical communication between the incompatible redox reactions by transmitting protons and electrons in a precisely controlled manner, while preventing O_2 diffusion causing unwanted reverse reactions. This unique mass-transfer behavior on planar and nanotube configurations was systematically studied via cyclic voltammetry, electrochemical impedance spectroscopy, and visible light-sensitized short circuit current experiments. The embedded molecular wires' integrity before and after the mass (proton) transfer process was confirmed by time-resolved optical spectroscopy and grazing angle ATR-FT-IR or IRRAS characterization.

572624 3D Nitrogen-Doped Graphene Aerogel-Supported MnO Nanoparticles As Efficient Electrocatalyst for CO_2 Conversion to CO

3D Nitrogen-Doped Graphene Aerogel-Supported MnO Nanoparticles as Efficient Electrocatalyst for CO_2 Conversion to CO

Bike Zhang, Mengchu Wang, Yi He, Yao Shi

Institute of Industrial Ecology and Environment, College of Chemical and Biological Engineering, Zhejiang University, Hangzhou 310027, China

Abstract: Electrochemical CO_2 reduction reaction (CO_2RR) is a promising option for converting CO_2 waste streams into useful chemicals, which has attracted wide attention. The viability of this process, however, is still facing several crucial limitations such as the prohibitively high electrochemical overpotential and poor product selectivity due to the competitive process of the hydrogen evolution reaction (HER). Herein, for the first time three-dimensional (3D) N-doped graphene aerogel (NGA) support MnO nanoparticles (MnO/NGA) as efficient CO_2RR catalyst synthesized via a facile and scalable hydrogel route is applied for the CO_2RR. This novel catalyst exhibits an extremely low overpotential of 0.16 V (vs RHE), high Faradaic efficiency of 86% for CO at potential of -0.82 V (vs RHE), which is much better than most of reported Mn-based materials to our knowledge with a current density of 7.6 mA cm^{-2}, as well as good long-term stability. The unique 3D crumpled porous nanostructure of NGA with large specific area facilitate reactant diffusion in pores, maximize the electrode/electrolyte contact area and promote electron transport. Well crystallized MnO nanoparticles are believed to be the metallic sites active sites towards CO_2, which have a clear positive correlation with the catalyst activity. Evidence is provided by the significant differences of CO_2 electrocatalytic performance between MnO/NGA-2 (MnO with good crystallinity) and MnO/NGA-1 (Mn in ionic chemical state). In addition, doped nitrogen can effectively suppress the HER. Our results highlight on a promising new class of low-cost, abundant electrocatalysts for CO_2RR.

Key words: MnO, 3D N-doped graphene aerogel, electrocatalysts, CO_2 reduction

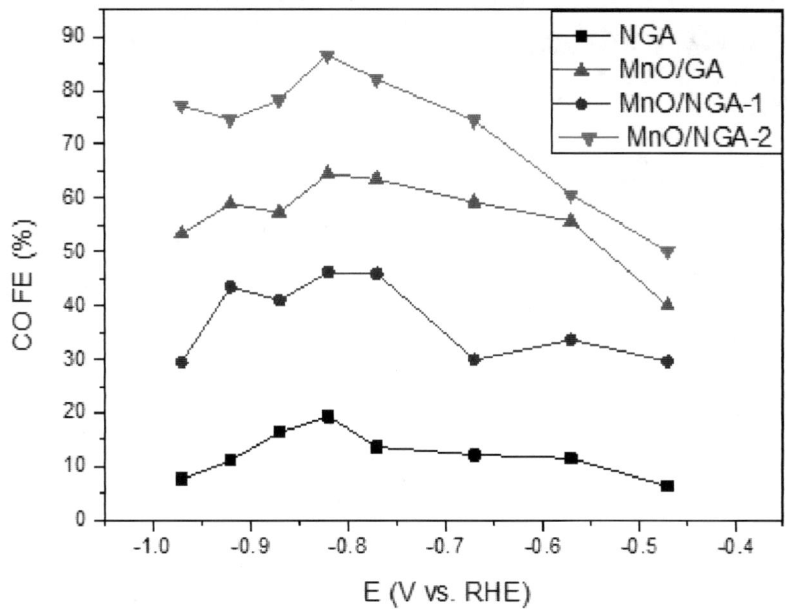

Fig.1. Faradaic efficiency (FE) over MnO/NGA catalysts. The data were obtained by chronoamperometry (CA) for 1 h, in 0.5 M $KHCO_3$ saturated with CO_2 (pH=7.2) at room temperature.

562309 Non-Noble Metals Doped Bi-Nanoparticles for Electrochemical CO_2 Reduction

Abstract:

The electrochemical reduction of CO_2 by using ionic liquid (ILs) and metal nanoparticles have been attracting research efforts. Our previous work has shown that the electrochemical reduction of CO_2 to CO can be realized over surface-activated bismuth nanoparticles (NPs) with 1-butyl-3-methylimidazolium ([bimi][OTf]) as the electrolyte. Particularly, by using the activated 36-nm Bi NPs on glassy carbon as the electrode, the Faradic efficiency of CO could reach 96.1%. In this work, non-noble metals (Cu, Sn) doped Bi NPs are prepared and then adopted as the cathode to achieve C_2 products in MeCN/H_2O electrolyte with the assistance of [bimi][OTf]. By virtue of controlling the reaction conditions, e.g., the pH of the solution, the doping concentration of the metal and reaction temperature, the optimal Cu, Sn-doped Bi NPs capable of effectively making the CO* intermediate continue to react can be obtained.

570507 Enhancing the Electrochemical Production of Syngas from Captured CO_2 by Using Silver Nano-Structured Electrocatalysts

Global warming has become one of the biggest challenges of the modern era. The most critical of all factors that have exacerbated this problem is the rapid increase in CO_2 pollution. Looking for solutions to this issue, the present work aims to develop a breakthrough technology for the capture and electrochemical reduction of CO_2. The innovation of this project relies on the fact that for the first time, post combustion capture solutions are being incorporated into an electrochemical cell, as electrolyte and CO_2 source, for the conversion of CO_2 into syngas. Despite the promising results obtained through the use of this concept, one of the priorities identified for future work, is the need to develop a better electrocatalyst to improve the invention. Therefore, the goal of the present study is to enhance the electrochemical production of syngas from captured CO_2 by using novel nanostructured Ag electrocatalysts. To reach this goal, the first objective is to establish a procedure to fabricate Ag nanostructures with specific sizes and shapes. The second objective is to produce syngas from CO_2 captured in switchable polarity solvents (SPS) by using the fabricated Ag electrocatalysts. Syngas with H_2:CO rations between 2 and 4, conversions and yields at over 60% are targeted.

576585 Electrochemical Reduction of CO_2 on Transition Metal-P Block Catalyst Compositions

Among all the pollutants in the atmosphere, CO_2 has the highest impact on global warming and with the rising levels of this pollutant, studies on developing various technologies to convert CO_2 into carbon neutral fuels and chemicals have become more valuable. Electrochemical reduction is one of the solutions to convert CO_2 to value added hydrocarbon fuels using non-precious, earth-abundant nanocatalysts making this process cost-effective. To understand the activity of catalysts for a particular reaction, we should be able to tailor the catalyst atom by atom. With the advances in computing power and quantum modeling tools, researchers are able to design and study different types of "*in silico*" materials. Previous experimental results indicate transition metal-p block catalysts such as oxides show improved catalyst activity and desired product selectivity. However, the design principle and reaction mechanism are poorly explored.

In this work, we present a detailed computational study of electrochemical reduction of CO_2 (CO_2RR) to methane and methanol over different transition metal-oxide and sulfide catalysts using Density Functional Theory calculations. In addition to the catalyst structure, we studied reaction mechanisms using free energy diagrams that explain the product selectivity with respect to the competing hydrogen evolution reaction. From these diagrams, we hypothesized that transition metal oxides and sulfides favor methanol over methane formation at lower overpotentials.

Furthermore, we developed scaling relations to find the key intermediate species for CO_2RR on these catalyst materials. We have found CO* as the descriptor (key species) from these relations and modifying the binding free energy of this species would modify the catalyst activity. We developed thermodynamic volcano plots for each product relating descriptor (CO*) binding energy to all other intermediate species binding energy to characterize and rank the activity of catalysts studied so far and determine the optimal binding energy region of the descriptor. This volcano plot will provide guidance to our future work on improving the activity of current transition metal-p block family of catalysts such as metal oxides and sulfides and develop new catalysts for this important reaction.

Table 1 below shows the reducing potentials for each product on different catalyst compositions.

Catalyst	Reducing potential (0V vs. SHE)			
	CO	HCOOH	CH_3OH	CH_4
MoO_3	-2.18	-3.44	-2.09	-2.09
WO_3	-2.17	-3.11	-0.84	-0.65
ZnO-Wurtzite	-1.65	-2.63	-0.69	-0.69
ZnO-Zincblende	-1.66	-2.67	-1.28	-1.28
ZnS-Wurtzite	-1.35	-2.21	-0.44	-0.47

577555 Synthesis and Characterization of Bimetallic Core@Shell Structured Nanoparticles for Electrochemical Reduction of CO_2 into Formic Acid

Synthesis and Characterization of Bimetallic Core@Shell Structured Nanoparticles for Electrochemical Reduction of CO_2 into Formic Acid

Wei Jyun Wang[a], Su Ha[a], Louis Scudiero[b]

[a] *The Gene and Linda Voiland School of Chemical Engineering and Bioengineering, Washington State University, WA 99164*

[b] *Chemistry Department and Material Science and Engineering Program, Washington State University, WA 99164*

The electrochemical reduction of CO_2 is considered as an attractive strategy to convert CO_2 into formic acid that can be directly used as the energy carriers in a direct formic acid fuel. However, there is still a need to develop novel cathode materials with high catalytic activity and stability for this technology to be viable. Currently, Palladium (Pd) is one of the most efficient electrocatalysts that is able to convert CO_2 into formic acid at low overpotential (< 0.2 V), but it is very expensive and its surface is susceptible to CO poisoning. One strategy to improve the electrocatalytic performance, reduce CO poisoning, and reduce the amount of Pd needed for catalyst synthesis is to produce new bimetallic core@shell nanocatalysts. Those novel bimetallic materials can decrease the bonding strength of key intermediate species on the Pd surface improving its catalytic activity and long-term stability. This was achieved by modifying the electronic properties of Pd shell of Cu@Pd core@shell structured nanoparticle. Synthesized samples of nanoparticles in this work were characterized by UV-Vis spectroscopy, X-ray photoelectron spectroscopy (XPS), Transmission electron microscopy (TEM), cyclic voltammetry (CV) and controlled potential electrolysis (CPE). The core@shell structure of the bimetallic nanoparticles was verified by UV-Vis and TEM results. XPS and CV data suggest that the transfer of charges from the Pd shell to the Cu core decreases the bonding strength of key intermediate species on the surface of Pd. This electronic perturbation of the catalyst surface lowers the energy barrier for CO_2 hydrogenation into formic acid and make the formation of CO on the surface of Pd less favorable. The CV shows that the CO_2 reduction current density is 0.21 mA/cm^2 on the surface of Cu@Pd, which is about 6.2 times higher than the current density on commercial Pd when overpotential is 0.15 V. The CPE data shows the long-term stability of the Cu@Pd catalyst is higher than commercial Pd. The production of formate/formic acid will be quantified by a titration method.

563330 Copper-Nitrogen-Doped Carbon Nanostructures for an Efficient CO_2 Reduction Reaction

Copper-Nitrogen-Doped Carbon Nanostructures for an Efficient CO_2 Reduction Reaction

Huiyuan Cheng, Xuemei Wu*, Gaohong He*

State Key Laboratory of Fine Chemicals, Research and Development Center of Membrane Science and Technology, School of Chemical Engineering, Dalian University of Technology, Dalian, P. R. China

*Corresponding author: Email: xuemeiw@dlut.edu.cn

hgaohong@dlut.edu.cn

Abstract

Electrochemical reduction of carbon dioxide (CO2RR) to value-added carbon products is an effective approach to fuels generation and CO_2 remediation[1,2]. However, this desired technology is confronted many challenges as the inert property of CO_2 causes extremely high overpotential and the competitive hydrogen evolution (HER) side reaction[1,3,4]. Among studied catalyst formulations, transition metal and nitrogen doped carbon (M-N-C) nanomaterials have been emerging as promising electrocatalysts for its superiority in low cost and excellent electrocatalytic activity[5-7]. Herein, we propose a facile but efficient strategy to facilitate the CO2RR through rationally tailoring the coordination environment between copper and nitrogen atoms embedded in porous carbon matrix, which derived from pyrolyzing Cu-doped zeolitic imidazolate frameworks-8(ZIF-8). Figure 1 shows the preparation scheme of Cu-N-C CO2RR electrocatalysts. Cu^{2+} ions were controllable doping into ZIF-8 frameworks, followed by calcination in inert atmosphere. The obtained porous carbon nanomaterials inherited rhombic dodecahedron shape and used directly as CO2RR electrocatalysts without any acid etching. The Cu-N-C catalyst with optimal structural and chemical configuration was achieved through investigating pyrolysis temperature and Cu doping contents. As shown in Figure 2, the optimized Cu-N-C catalyst exhibits particularly high selectivity with CO faradaic efficiency over 90% in the potential range from -0.6 to -1.1V (*vs.* the RHE) and reaches the maximum value of 98.3% at -0.9V(*vs.* the RHE) with a current density of 7.3 $mAcm^{-2}$. Both experimental and theoretical calculations demonstrate that a hybrid Cu coordination site of Cu0/Cu(II)-N is more beneficial to CO_2 activation than hydrogen evolution reaction. The result provides a significant guideline to the design of high efficiency CO_2 reduction electrocatalysts.

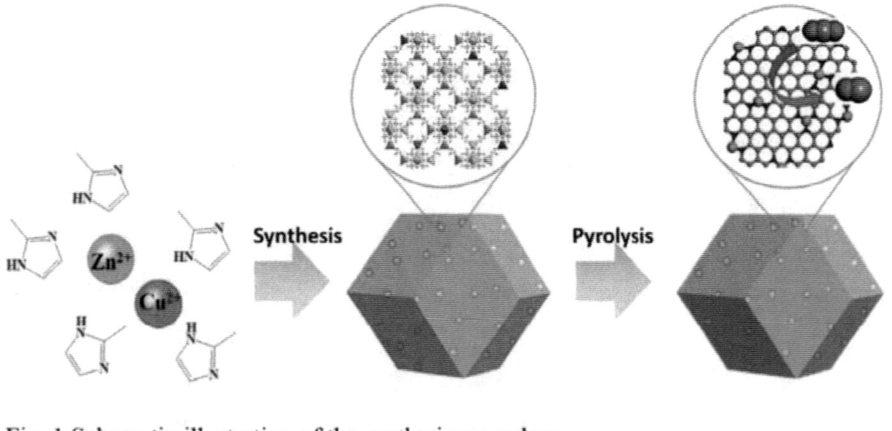

Fig. 1 Schematic illustration of the synthesis procedure.

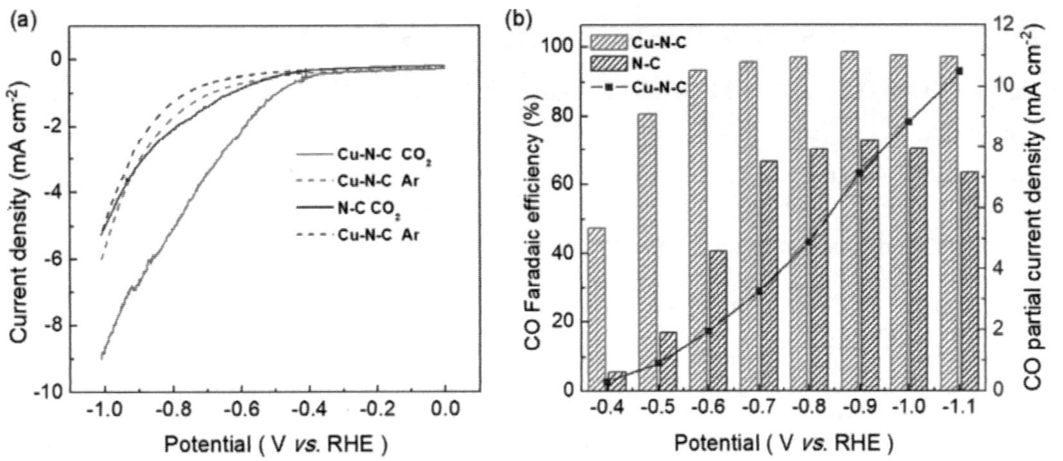

Fig. 2 Performance of Cu-N-C electrocatalyst and N-C (pyrolyzed ZIF8). (a) Linear sweep voltammetric curves in the Ar (dotted line) or CO_2-saturated (solid line) 0.1M $KHCO_3$ aqueous solutions with a 5mv s^{-1} scan rate. (b) CO Faradaic efficiencies and particial current density on applied potencial.

Acknowledgment

We acknowledge financial contribution from National Natural Science Foundation of China (Grant No. U1663223, 21776034), Changjiang Scholars Program (T2012049)

References

1. Pan Y, Lin R, Chen Y, et al. Design of Single-Atom Co-N-5 Catalytic Site: A Robust Electrocatalyst for CO2 Reduction with Nearly 100% CO Selectivity and Remarkable Stability. *Journal of the American Chemical Society*. Mar 28 2018;140(12):4218-4221.

2. Yan C, Li H, Ye Y, et al. Coordinatively unsaturated nickel-nitrogen sites towards selective and high-rate CO2 electroreduction. *Energy & Environmental Science*. May 1 2018;11(5):1204-1210.

3. Pan F, Zhang H, Liu K, et al. Unveiling Active Sites of CO2 Reduction on Nitrogen-Coordinated and Atomically Dispersed Iron and Cobalt Catalysts. *Acs Catalysis*. Apr 2018;8(4):3116-3122.

4. Li X, Bi W, Chen M, et al. Exclusive Ni-N-4 Sites Realize Near-Unity CO Selectivity for Electrochemical CO2 Reduction. *Journal of the American Chemical Society*. Oct 25 2017;139(42):14889-14892.

5. Lu P, Yang Y, Yao J, et al. Facile synthesis of single-nickel-atomic dispersed N-doped carbon framework for efficient electrochemical CO2 reduction. *Applied Catalysis B-Environmental*. Feb 2019;241:113-119.

6. Yin P, Yao T, Wu Y, et al. Single Cobalt Atoms with Precise N-Coordination as Superior Oxygen Reduction Reaction Catalysts. *Angewandte Chemie-International Edition*. Aug 26 2016;55(36):10800-10805.

7. Lai Q, Zheng L, Liang Y, He J, Zhao J, Chen J. Meta-Organic-Framework-Derived Fe-N/C Electrocatalyst with Five-Coordinated Fe-N-x), Sites for Advanced Oxygen Reduction in Acid Media. *Acs Catalysis*. Mar 2017;7(3):1655-1663.

573667 Doped Transition Metal Nitrides As Efficient Electrocatalysts for Electrochemical Reduction of CO_2

Doped transition metal nitrides as efficient electrocatalysts for electrochemical reduction of CO_2

Mohammadreza Karamad, Samira Siahrostami and Ian Gates

Department of Chemical and Petroleum Engineering, University of Calgary, 2500 University Drive NW, Calgary, AB T2N 1N4, Canada.

Electrochemical reduction of CO_2 to hydrocarbons and alcohols using intermittent renewable energy sources is a promising approach to store energy into chemical bonds for automotive and industrial applications.[1] Among many known catalysts, very few, such as copper, are capable of reducing CO_2 to significant amount of hydrocarbons and oxygenates, albeit at high overpotentials.[2]

In the past, numerous experimental and theoretical efforts have been dedicated to understand CO_2 reduction reaction mechanism as well as identify efficient catalysts.[3–6]Transition metal nitrides have attracted a lot of attention for electrochemical reactions such as oxygen reduction reaction, ammonia synthesis and hydrogen evolution in recent years.[7–9] In the present study, we employ Density Functional Theory (DFT) calculations to study a wide range of transition metal nitrides with and without heteroatom dopants for CO_2 reduction reaction. We identify several stable and active catalysts. This study suggests that transition metal nitrides are promising catalysts for electrochemical reduction of CO_2.

References:

1. Verma, S., Kim, B., Jhong, H.-R. "Molly", Ma, S. & Kenis, P. J. A. A Gross-Margin Model for Defining Technoeconomic Benchmarks in the Electroreduction of CO_2. *ChemSusChem* **9**, 1972–1979 (2016).
2. Hori, Y. Electrochemical CO 2 Reduction on Metal Electrodes. in *CO2 Reduction on Metal Electrodes. Modern Aspects of Electrochemistry* (eds. Vayenas, C. G., White, R. E. & Gamboa-Aldeco, M. E.) 89–189 (Springer New York, 2008).
3. Kuhl, K. P. *et al.* Electrocatalytic conversion of carbon dioxide to methane and methanol on transition metal surfaces. *J. Am. Chem. Soc.* **136**, 14107–14113 (2014).
4. Peterson, A. A., Abild-Pedersen, F., Studt, F., Rossmeisl, J. & Nørskov, J. K. How copper catalyzes the electroreduction of carbon dioxide into hydrocarbon fuels. *Energy Environ. Sci.* **3**, 1311 (2010).
5. Karamad, M., Hansen, H. A., Rossmeisl, J. & Nørskov, J. K. Mechanistic Pathway in the Electrochemical Reduction of CO_2 on RuO_2. *ACS Catal.* **5**, 4075–4081 (2015).
6. Li, Y. & Sun, Q. Recent Advances in Breaking Scaling Relations for Effective Electrochemical Conversion of CO_2. *Adv. Energy Mater.* **6**, 1600463 (2016).
7. Abroshan, H. *et al.* Ultrathin Cobalt Oxide Overlayer Promotes Catalytic Activity of Cobalt Nitride for the Oxygen Reduction Reaction. *J. Phys. Chem. C* **122**, 4783–4791 (2018).
8. Abghoui, Y. *et al.* Enabling electrochemical reduction of nitrogen to ammonia at ambient conditions through rational catalyst design. *Phys. Chem. Chem. Phys.* **17**, 4909–4918 (2015).
9. Abghoui, Y. & Skúlason, E. Hydrogen Evolution Reaction Catalyzed by Transition-Metal Nitrides. *J. Phys. Chem. C* **121**, 24036–24045 (2017).

565187 Artificial Thylakoid to Coordinate Photo-Enzyme-Coupled Catalysis for Carbon Dioxide Upgrading

Artificial Thylakoid to Coordinate Photo-Enzyme-Coupled Catalysis for Carbon Dioxide Upgrading

Shaohua Zhang[a,c], Jiafu Shi[b,c], Zhongyi Jiang [b,c]

[a] *Key Laboratory for Green Technology of Ministry of Education, School of Chemical Engineering and Technology, Tianjin University, Tianjin 300072, China*

[b] *School of Environmental Science and Engineering, Tianjin University, Tianjin 300072, China*

[c] *Collaborative Innovation Center of Chemical Science and Engineering (Tianjin), Tianjin 300072, China*

Artificial photosynthesis has become the hotspot in energy and environmental relevant fields. Generally speaking, the well-defined catalyst and the well-organized complex reaction system jointly determine the efficiency of artificial photosynthesis process. Currently, most efforts are devoted to the intensive exploitation of efficient catalysts, while less efforts are devoted to the rational construction of artificial photosynthesis system by coordinately manipulating different catalytic reactions therein.

Inspired by thylakoid in chloroplast, we explore a capsular structure to coordinate photocatalysis and biocatalysis for the construction of artificial photosynthesis system. The capsular structure, termed as artificial thylakoid, has a wall composed of cadmium sulfide quantum dots shielded by an amorphous titania wall. The artificial thylakoid could absorb visible light to activate the single- and multiple-enzymatic CO_2 reduction by oriented transfer of photo-generated electrons through the capsule wall for NADH regeneration. The photo-generated holes and its reactive oxygen species are 'imprisoned' in the lumen of the artificial thylakoid, avoiding its poison effect on the enzymes in solution. The artificial thylakoid affords a compatible design strategy to coordinate photocatalysis and biocatalysis of the photo-enzymatic system. Under visible light illumination, CO_2 is converted to formate and methanol with a production rate of 1500 $\mu M\ h^{-1}$ and 99 $\mu M\ h^{-1}$ that were much higher than previous particulate photo-enzymatic systems. Moreover, CO_2 can be converted to formate over several light-dark cycles with a quantum yield of 0.66±0.13%, comparable to the year-long average determined for natural green plants (~0.2 to 1.6%).

575409 A Novel Approach to Kinetic Modeling of High Temperature CO_2 Adsorbents

Hydrogen is produced from natural gas (or biomass) by the two-step steam methane reforming (SMR) and the water-gas shift (WGS) reactions. This is an ideal opportunity for sorption-enhanced reaction, in which CO_2 is chemically fixed to a suitable substrate, thus driving the equilibrium in the WGS shift step to the right. Lithium-based ceramics, notably Li_4SiO_4, have been evaluated as potential CO_2 sorbents for this purpose. Several different approaches to modify the lithium silicate composition and morphology have been tried to improve the performance as measured by thermogravimetric analysis (TGA). A difficulty in assessing performance has been the lack of adequate models to reduce TGA data to physically significant parameters; the commonly used double -exponential model has no mechanistic basis and yields parameters that have no basis in the physics and chemistry of these materials. In this talk, we show an adaptation of the Ishida and Wen core-shell model that yields physically meaningful reaction rate constants and diffusion coefficients, and show how this information is used to direct our efforts to develop novel high-temperature CO_2 sorbent materials.

572655 Extrusion of Perovskite Oxide/ Silica Composites for Thermochemical Carbon Dioxide Conversion

Perovskite-type oxides have been used to convert CO_2 to CO in a reverse water-gas shift chemical looping (RWGS-CL) process. Currently, the process is at the micro-reactor scale and used catalyst in powder form. However, formed materials are needed if the scale is to be increased. Our research focus is on increasing the conversion yield of perovskite oxides with SiO_2 support and scaling up the powder form perovskite-type oxides to pellet by extrusion. The binder added during the extrusion process must be removed to avoid affecting the performance of the catalyst and to provide porosity, without substantially lowering the mechanical strength. X-ray diffraction was utilized to identify the crystal structure and determine the crystal stability after the conversion. Temperature programmed reduction (TPR) and temperature programmed oxidation (TPO) experiments were performed to identify the conversion temperature and redox properties of the materials. Compared to pure perovskite-type oxides, sample $La_{0.75}Sr_{0.25}FeO_3$ mixed with SiO_2 (mass ratio was 50:50) had better conversion performance and stronger mechanical strength after extrusion.

575421 Understanding Mechanisms of CO_2 Electroreduction on Cu(100) with Kinetic Monte Carlo Simulations

Converting carbon dioxide (CO_2) to value-added chemicals or fuels through electrocatalysis is an appealing approach to tackle both CO_2 emission and energy storage challenges. Cu catalyzes direct electrochemical reduction of CO_2 to hydrocarbons in an aqueous bicarbonate solution with a high current density. However, the underlying mechanism remains elusive due to multistep kinetic pathways and complexities of surface reactions1. Traditionally, the microkinetic model is used with energetics from quantum chemical simulations within a mean-field approximation. Going beyond this approximation and ideally taking into account adsorbate-adsorbate interactions in surface reactions is important for unraveling kinetics of surface reactions. In this poster, we employed kinetic Monte Carlo simulations with DFT-derived neural network potentials to understand the effect of pH and operating potentials on CO_2 reduction on Cu(100) surface2. Particularly, we show that with proper consideration of CO-CO interactions the kinetic Monte Carlo method can better capture observed trends of kinetic current than the traditional microkinetic method.

1. Peterson, A. A., Abild-Pedersen, F., Studt, F., Rossmeisl, J. & Nørskov, J. K. How copper catalyzes the electroreduction of carbon dioxide into hydrocarbon fuels. *Energy Environ. Sci.* **3**, 1311–1315 (2010).
2. Goodpaster, J. D., Bell, A. T. & Head-Gordon, M. Identification of Possible Pathways for C–C Bond Formation during Electrochemical Reduction of CO2: New Theoretical Insights from an Improved Electrochemical Model. *J. Phys. Chem. Lett.* **7**, 1471–1477 (2016).

556955 Promoting CO_2 Hydrogenation Activity and Methanol Selectivity of Cu/CeO_2 Catalyst By W-Doping

The direct hydrogenation to methanol provides a facile CO_2 reactivation pathway for the carbon capture and utilisation scheme. Amongst the various CO_2 hydrogenation catalysis systems, Cu/CeO_2 has received considerable research attention owing to the ability of CeO_2 to efficiently activate CO_2 at its surface oxygen vacancies through the highly active Ce^{4+}/Ce^{3+} redox cycle. However, the application of Cu/CeO_2 for direct methanol synthesis is limited by its mediocre selectivity towards methanol. In the present study, we present a W-doping approach to promote the performance of the Cu/CeO_2-based catalysts. The synthesised W-Cu/CeO_2 achieves an eight-fold increase in turn-over-frequency (TOF) for CO_2 conversion over unmodified Cu/CeO_2, as well as a methanol selectivity of >86% at 230 °C and 3.5 MPa, with stable performance over 72 h time-on-stream. XPS results suggest that the improved TOF is attributed to the enhanced activity of the Ce^{4+}/Ce^{3+} redox pair. Operando DRIFT data reveal that the methanol synthesis proceeds via the CO pathway on both Cu/CeO_2 and W-Cu/CeO_2. The observed (i) strong CO_2 and CO adsorption and (ii) weak methoxy adsorption are responsible for the superior performance of the W-Cu/CeO_2 catalyst.

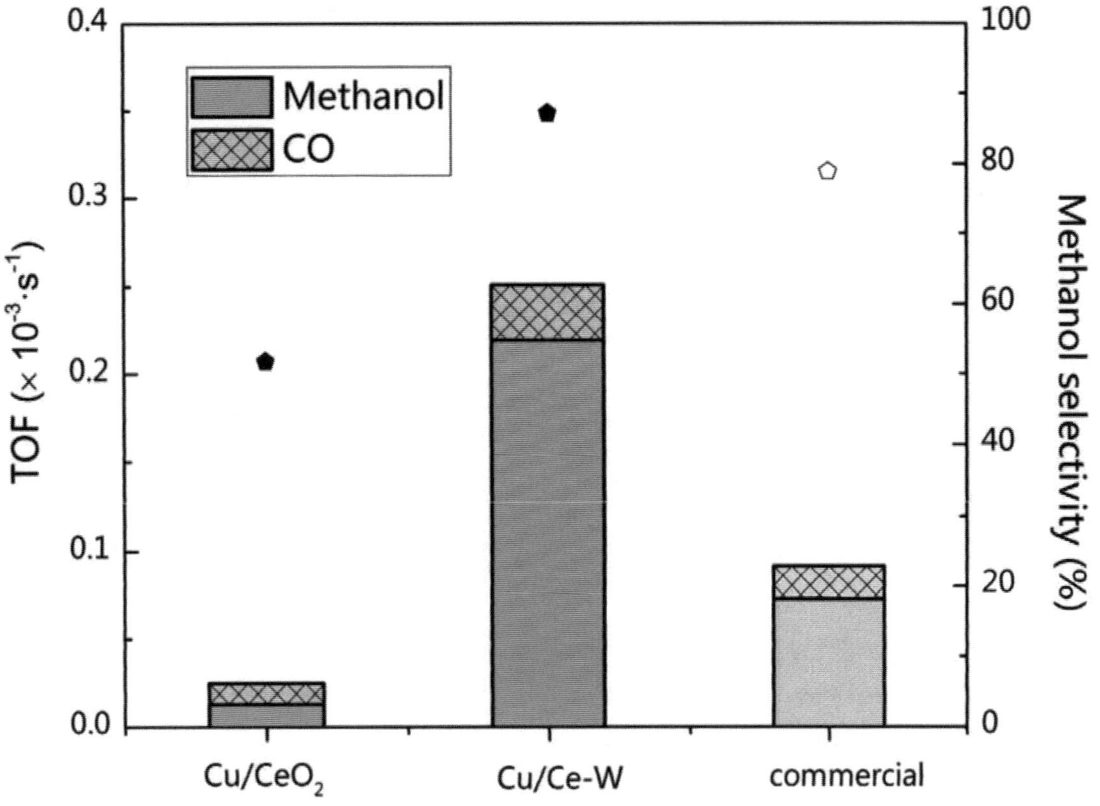

560063 A Computational Investigation of Catalytic Upgrading of CO_2 to Methanol over Indium Oxide

Carbon dioxide (CO_2) valorization has been identified as a great challenge in the face of excess anthropogenic carbon emissions. Current industrial processes to convert CO_2 into useful platform chemicals are limited by low selectivities and middling conversions. Recent attention has been given to oxide catalysts that have demonstrated excitingly high selectivities in experiments. Specifically, indium oxide (In_2O_3) has been shown to produce methanol, an important platform chemical, at comparable conversions to industrial catalysts and at selectivities approaching 100%[1]. However, our understanding of this promising catalyst is currently limited.

In this work, we investigate potential reaction mechanisms using density functional theory (DFT) calculations and microkinetic modeling (MKM). Through methodical active site selection and extensive exploration of different reaction pathways, we generate an explicitly calculated reaction mechanism through DFT energy and vibrational frequency calculations. We compare paths to methanol and CO, i.e., the reverse water gas shift reaction to reveal what controls selectivity. Using the appropriate statistical thermodynamic models, we produce a comprehensive MKM parametrized using these DFT values that enabled us to evaluate the proposed mechanism against experimental data. Comparison to other literature mechanisms will also be presented. Finally, we investigate why this oxide is so selective by performing electronic structure calculation analysis and discussing catalyst and chemistry descriptors.

1. Martin, O., A. J. Martin, et al. (2016). "Indium Oxide as a Superior Catalyst for Methanol Synthesis by CO_2 Hydrogenation." Angewandte Chemie-International Edition 55(21): 6261-6265.

563192 DFT Investigation of the Mechanism and Site Requirements for Reverse Water Gas Shift on Transition Metal Sulfides

Shale and other natural gas sources can be acidic, containing varying amounts of CO_2 and H_2S. The removal of these acidic components requires expensive chemical treatment though the problem of CO_2 disposal remains. Alternatively, a catalytic process based on a sulfur tolerant catalyst, like MoS_2,[1] can be capable of hydrogenating CO_2 in the presence of H_2S. The work[1] suggests the heterogeneity of edge sites as established via a Boltzmann distribution, thereby indicating that active sites evolve as a function of reaction condition, i.e. it could either be coordinative unsaturated sites (CUS), a Mo-O pair, brim sites and/or the undercoordinated Mo-S pair under different regimes of H_2, H_2S and H_2O partial pressures. With the help of electronic structure calculations, we elucidate the mechanism of reverse water gas shift chemistry on the plausible active sites. Specifically, the two catalytically active regions on MoS_2, the Mo-edge and S-edge are considered in this study. Our results suggest the importance of coordinative unsaturation (CUS) on Mo-edge and Mo-S pair on S-edge, which supports the redox over the associative mechanism, consistent with transient kinetic experiments,[1] for reverse water gas shift chemistry in the absence of H_2S. Furthermore, in presence of H_2S, our results suggest a H_2S mediated mechanism through which we explore the feasibility of CO production versus the production of COS, which supports the experimental evidence suggesting that H_2S can either behave like a co-catalyst or co-reactant under the presence of H_2S (>60 ppm).[1]

The details of our calculations and models, the likely reaction mechanism, and supporting experimental evidence will be presented and discussed.

References

[1] "Inhibitor, co-catalyst, or co-reactant? Probing the different roles of H_2S during CO_2 hydrogenation on MoS_2 catalyst", Lohit Sharma, Ronak Upadhyay, Srinivas Rangarajan, Jonas Baltrusaitis*, submitted.

567139 Influence of Coordination Environment Around Anchored Single-Site Cobalt Catalyst for CO_2 Hydrogenation

As the field of carbon dioxide capture and storage is expanding as a means of mitigating CO_2 emissions, transforming carbon dioxide into value added chemicals and fuels represents a more economically viable approach. Recently, various technologies of upgrading CO_2 have been explored, most notably electrochemical conversion and thermo-catalytic conversion. Typical catalysts for CO_2 hydrogenation consist of supported cobalt nanoparticles, which exhibit optimal catalytic activity in the size range of 5-15nm[1]. However, such catalysts have several factors that can affect catalytic activity, such as surface crystal faceting, particle size, or the relative ratio of exposed surface defects[2]. As CO_2 hydrogenation has been shown to be a highly surface sensitive reaction, the surface structures of nanoscale catalyst can be prohibitively complex to establish a clear structure/activity relationship to guide the design of active catalyst.

In this study, cobalt single site catalyst supported on silica were explored due to their highly uniform active sites; allowing for definitive claims as to which surface species are responsible for specific reaction mechanisms. To characterize the structure and dispersion of the single-site catalysts, techniques such as UV-vis, XAFS, XPS, TPR, and Raman were utilized under ambient conditions as well as under reductive environments to simulate reaction conditions. This thorough understanding of the surface moieties under ambient and reductive environments coupled with their corresponding catalytic performance during CO_2 hydrogenation allows us to discern how the transition between isolated atoms to small nanoparticles affects the reaction mechanism. To understand the surface under a reductive environment, we performed *in situ* XAFS, and pretreatment studies via XPS, using hydrogen as our reductant to elucidate the nature of the active site. We have found that under reductive environments below 500°C, the surface remains almost entirely in the Co^{2+} (Td) geometry. Our findings from our CO_2 hydrogenation studies have shown that isolated atoms promote the preferential formation of CO via the RWGS reaction while small ensembles between Co^{2+}(Td) and Co^0 form both methane and CO, suggesting a change in reaction mechanism; which will ultimately be elucidated by this work.

[1] a)V. Iablokov, S. K. Beaumont, S. Alayoglu, V. V. Pushkarev, C. Specht, J. Gao, A. P. Alivisatos, N. Kruse, G. A. Somorjai, *Nano Letters* **2012**, *12*, 3091-3096; b)Y. Zhu, S. Zhang, Y. Ye, X. Zhang, L. Wang, W. Zhu, F. Cheng, F. Tao, *ACS Catalysis* **2012**, *2*, 2403-2408.

[2] a)G. Melaet, W. T. Ralston, C. S. Li, S. Alayoglu, K. An, N. Musselwhite, B. Kalkan, G. A. Somorjai, *Journal of the American Chemical Society* **2014**, *136*, 2260-2263; b)J. Jimenez, A. Bird, M. Santos Santiago, C. Wen, J. Lauterbach, *Energy Technology* **2017**, *5*, 884-891.

571574 Carbon Dioxide Hydrogenation

Hydrogenation of carbon dioxide may on the one hand contribute to reduction of CO_2 emissions, on the other hand we might make use of CO_2 to raise the energy density of hydrogen based energy carriers by orders of magnitude. The hydrogenation products methane and methanol provide the physical properties for improved energy density of hydrogen based energy carriers. Both constituents do also easily release hydrogen by steam reforming, making them highly feasible for fuel cell powering. However, hydrogenation of CO_2 is not free of charge and does need sophisticated stable catalysts, easy to prepare, easy to recycle, robust and stable for long term usage. A suitable catalyst system can improve the reaction conditions concerning temperature, pressure, CO_2 conversion, and product yield. State of the art catalysts in industrial methanol production are $Cu/ZnO/Al_2O_3$, for methane synthesis Ni-based catalysts with Al_2O_3 backbone or SiO_2 backbone have established.

Changing the catalyst backbone from Al_2O_3 to MgO is very promising due to the appreciable CO_2 capture capacity of MgO as well as the low technological demand for recycling of spent catalysts. Due to the caustic properties of MgO based catalyst systems a product yield close to equilibrium conversion can even be obtained at moderate operation pressure. Exemplarily Ni doped MgO catalysts offer great performances in methanation of CO_2 at ambient pressure. For catalyst optimization the Ni-load of MgO has been varied between 11 wt% and 27 wt%. For methanol synthesis the catalyst performance of Cu/MgO and Cu/ZnO/MgO systems has been investigated and compared.

561471 Mesoporous Silica Supported Perovskite Oxide for Low Temperature Thermochemical CO_2 Conversion

Today, carbon dioxide (CO_2) emission levels are higher than at any other point in human history, and chemical reactions can be used to establish a synthetic carbon cycle, which would reduce emissions and produce liquid fuel. The reverse water-gas shift chemical looping process (RWGS-CL) is a reaction of interest because it is effective for converting large amounts of CO_2 and the resulting carbon monoxide (CO) can be used to produce hydrocarbons that are liquid fuel or chemical intermediates. Perovskite oxides are an effective catalyst for the RWGS-CL based on converting CO_2 to CO at efficient conversion temperatures, ability to form oxygen vacancies, and structural stability. The perovskite oxide used is $La_{0.75}Sr_{0.25}FeO_3$ (LSF). Recent studies have shown silica (SiO_2) as the most efficient support material for the perovskite oxide catalyst used in RWGS-CL. The form of SiO_2 that has been experimented with to provide support for LSF is SBA-15, because of its large pore volume and surface area, coupled with potential to support smaller particles of the perovskite increasing the number of active sites available, which would enhance conversion. Here we have analyzed adaptations to the synthesis procedure of perovskite oxides and SBA-15 to develop a single synthesis procedure for a composite material. These materials are characterized by X-ray Diffraction, N_2-physisorption, and temperature-programmed reduction, oxidation, and RWGS-CL experiments.

573260 Promoter and Support Effects on Transition Metal Carbides for CO_2 Hydrogenation

Carbon dioxide hydrogenation is a reaction of interest as an alternative route to produce chemicals and fuels. To make this reaction pathway competitive with existing production methods, heterogeneous catalysts have been employed to manipulate the reaction mechanism and promote high product selectivity.

Transition metal carbides (TMCs) have been shown to outperform traditional precious bimetallic catalysts and are highly active and selective for the reverse water-gas shift (RWGS) reaction, the first step of CO_2 hydrogenation to light olefins.[1-2] Additionally, by modifying TMCs with alkali metal promoters, the CO yield increases significantly and approaches the thermodynamic limit at 300 °C. These experimental results are supported by density functional theory (DFT) calculations, which show enhanced CO_2 adsorption and reduced CO_2 dissociation barriers on the alkali metal-promoted catalyst, compared to the pristine carbidic surfaces.[3]

Although supported TMCs (i.e. WC and Mo_2C) have been shown to be effective RWGS catalysts, little is known about the effects of electronic promotion and the support-metal carbide physicochemical interactions on promoting high CO product selectivity. To investigate these structure-property relationships, Mo_2C and WC are first synthesized on a variety of metal oxide supports to vary both chemical (Lewis and Brønsted acidity) and structural properties (amorphous SiO_2 vs. silicalite-1). Acidic and basic properties of the catalysts are probed via NH
desorption, while bulk and surface-sensitive X-ray characterization techniques are used to describe the metal carbide crystalline structure, phase and oxidation state with varying support composition. By combining the above techniques with reactor studies, *in situ* diffuse reflectance FTIR and synchrotron-based XAFS measurements, structure-property relationships are developed relating physicochemical interactions between the carbidic active phase and metal oxide support to determine adsorption behavior and catalytic performance. Further research in this area will lead to a framework to direct rational design of future catalysts for CO_2 hydrogenation to chemicals and fuels.

1. M. D. Porosoff, J. W. Baldwin, X. Peng, G. Mpourmpakis, and H. D. Willauer, *ChemSusChem*, **10** (11), pp. 2408–2415, (2017).

2. S. Posada-Perez, F. Vines, P. J. Ramirez, A. B. Vidal, J. A. Rodriguez, F. Illas, *Physical Chemistry Chemical Physics*, **16**, pp. 14912-14921, (2014).

3. M. D. Porosoff, J. W. Baldwin, X. Peng, G. Mpourmpakis, H. D. Willauer, *ChemSusChem,* **10**, pp. 2408-2415, (2017).

578465 A Metal- and Solvent-Free Synthesis of aminoalcohols under Continuous Flow Conditions

The use of multifunctional organocatalysts and continuous flow platform are commonplace in modern chemical transformation. Herein, we describe a method for immobilization of trifunctional organocatalysts on porous composite hollow fiber and demonstrate their application as heterogeneous catalyst and continuous-flow microfluidic reactor for chemical transformation. The polyamide imide hollow fibers (PAIHF's) are functionalized with aminosilanes and bromine source to immobilize covalently hydrogen-bond donor groups (-OH and -NH), and nucleophilic [Br^-] species upon the fiber surface and provide trifunctional acid-base-nucleophilic organocatalysts and microfluidic reactors. The cooperative effects of Br/APS/PAIHF trifunctional organocatalysts are elucidated in the CO_2 cycloaddition and hydroxyalkylation of aniline under batch and continuous flow synthesis. Our results indicated that the synergistic cooperative effect of trifunctional organocatalyst on PAIHF leads to a maximum 1-(phenylamino)propan-2-ol selectivity of 97.1% at 60% aniline conversion and 0.02 cm^3/min flow rate. While knowledge of the acid-base-nucleophilic trifunctional cooperativity is still limited, these findings demonstrate useful structure-property trends that can be used to design more efficient organocatalysts for sustainable chemical transformation.

574865 Probing the Oxygen Evolution Reaction Efficacy of Pure and Doped NiOOH (0001) Using Hybrid Density Functional Theory

The splitting of water into oxygen and hydrogen is a pre-requisite for a clean, hydrogen-based fuel economy. Alternatively, the electrons and protons produced during the process can be used to reduce carbon dioxide. While the hydrogen evolution reaction involves two electrons, no water molecule, and exhibits a standard potential of 0.0 V, the oxygen evolution reaction (OER) involves four electrons, two water molecules, and exhibits a standard potential of -1.23 V, thereby involving a greater level of complexity. Nickel oxyhydroxide (NiOOH) is a promising 2D material candidate to replace expensive, conventional noble metal catalysts. However, so far, understanding of the electrocatalytic activity of the most-abundant, basal (0001) plane of this material has been obtained using standard density functional theory (DFT) calculations with a Hubbard-like correction term (DFT+U) for the *d* electrons, wherein the results depend on the on-site Coulomb (U) and exchange (J) parameters. For doped systems, this presents an added level of complexity, because of the use of different values of the U and J parameters for various elements, in order to appropriately avoid the self-interaction error for the localized *d* electrons. In this work, we instead apply hybrid DFT calculations, which include a fraction of exact exchange, to estimate lower bounds for the OER overpotential on both pure and doped versions of NiOOH, for the associative OER mechanism. By contrasting and rationalizing results, including the magnetic moments and the oxidation states of the adsorbed species and active sites, from these two different levels of theory, we are able to provide new insights into the electrocatalytic activity of the NiOOH basal plane.

572047 Defect-Engineered Interface for Efficient Electrocatalyst

The expansion of economical, effective, and durable electrocatalyst for the oxygen reduction reaction (ORR), oxygen evolution reaction (OER), and hydrogen evolution reaction (HER) have significant importance for numerous electrochemical devices, such as fuel cells, water electrolyzer, and rechargeable metal-air batteries. Heteroatom-doped catalyst has arisen as one of the best emergent material to replace the precious metal catalyst for electrochemical reactions, but to control their delicate structure and the lack of atomic-level mechanistic understanding are still a challenging issue. Herein, we advocate the assembly of separate single atoms of Ni and Fe (SSAs-NiFe) dual sites, anchored in defects engineered graphene (DG) substrate prepared by a nitrogen (N) elimination from N-doped precursor by an adsorption-calcination strategy. The concept of defect mechanism showed that the topological defect (*e.g.* multiple pentagon—octagon—pentagon or pentagon—heptagon—pentagon rings) appeared correspondingly in the combination of carbon atoms that avoided random dislocations and disclinations. We believe that these specific combinations of carbon atom rings in DG substrate with the decoration of Ni and Fe SSAs can provide more active sites, better electrical conductivity, and low energy barrier during catalytic reaction. We used the combination of high-resolution transmission electron microscopy/high-angle annular dark field (HAADF) images in scanning transmission electron microscopy (STEM) for studying the desired atomic structure. At a very low resolution (1nm) the individual metal atoms in the field of view are detected through STEM and resolved in graphene lattices. The electron energy loss spectroscopy (EELS) atomic spectra are used to identify the difference between Ni and Fe SSAs on graphene substrate. The chemical structure insights of the material are examined by pre-edge peaks in X-rays absorption near edge structure (XANES) spectra. Furthermore, we demonstrated the electrocatalytic performance of our hybrid material in acidic/alkaline mediums and found it active and stable trifunctional electrocatalyst for ORR, OER and HER. This work offers new prospects and underscores the importance of identifying the active species for multiple reaction catalysts.

556777 Mechanisms for Hydrogen Evolution on Transition Metal Phosphides and Pt

Water electrolysis for renewable energy storage and conversion is limited by the high cost of Pt-based catalysts used for the hydrogen evolution reaction (HER). Earth-abundant transition metal phosphides have emerged as active HER catalysts with lower cost than Pt. However, the catalytically active sites and reaction mechanisms on these catalysts remain unclear. We present models of the HER on four different transition metal phosphide surfaces (Co_2P, CoP, Fe_2P, and FeP) using cluster expansions parameterized by density functional theory. We predict the structures and energetics of adsorbed hydrogen as a function of temperature and applied potential, allowing us to determine the potential-dependent activities of different sites while fully accounting for interactions among adsorbed hydrogen atoms. For comparison, we have used the same approach to model the HER mechanisms on Pt surfaces – including Pt(111), Pt(110), and Pt(100). We demonstrate the effects of potential-dependent surface coverage on HER current density and propose mechanisms that are in good agreements with experiments. The present work provides a general and effective way to study catalytic reaction mechanisms and probe catalytically active sites on surfaces, which can facilitate the design of highly active catalysts.

*Office of Naval Research N00014-16-1-2355

573804 Multi-Task Machine Learning to Predict ORR Catalyst Descriptors and Performance across Surface Composition

Due to the growing global population and industrialization of many countries, demand for energy for social and economic development is increasing. The oxygen reduction reaction (ORR) is an important reaction in fuel cell devices and the lack of sufficiently good ORR catalysts limits the overall efficiency of such devices. Despite the search for novel catalysts and new bimetallic catalysts are regularly discovered or optimized, the ORR activity of materials have not improved significantly. Large-scale screens for optimal ORR composition have focused on single descriptors using the assumption of linear scaling relations between key intermediates. In this work we demonstrate the high-throughput generation of adsorption energy datasets for *O, *OH, and *OOH intermediates on over 3000 possible surfaces including 820 different crystal bulk structure across 44 unique elements. We show that these quantities can be predicted with coordination-based descriptors to approximately 0.2 eV accuracy, sufficient to screen for potential activity across dozens of elements. Further, we show that these predictions can be improved through multi-task non-linear regression to simultaneously include and improve upon the well-known scaling relations. We also identify adsorption configurations that deviate from linear scaling relations by using machine-learning guided high throughput calculations. Finally, we use these methods to predict promising catalysts for experimental ORR validation.

569094 Post-Transition Metal Nanoparticles As Electrocatalysts for Nitrogen Reduction to Ammonia

As an important economical chemical, Ammonia (NH_3) has a wide variety of applications including fertilizers, carbon-free energy carriers and as a potential hydrogen-storage molecule. In nature, NH_3 can be biologically synthesized under ambient conditions by some microorganisms with the help of nitrogenase enzymes. In industry, however, NH_3 production is dominated by the Haber-Bosch process which requires high temperature (150-300 atm) and high pressure (150-300 atm). Electrochemical synthesis of NH_3 is a potential alternative to the Haber-Bosch process because of mild operating conditions and renewable electricity as energy source. As an emerging field, various electrocatalysts have been studied in recent years. However, optimal electrocatalysts with high NH_3 yield rate, Faradaic efficiency and stability have not yet been discovered and understood. As indicated by computational studies, transition metals such as Au and Pt could suffer from severe selectivity problem against the hydrogen reduction reaction (HER) while early transition metals such as Ti, bind N_2 too strongly to achieve high yield rate. Some recent studies on post-transition metals like Bi shed light on how to achieve a high performing electrocatalyst for N_2 reduction Inspired by these progresses, we conduct a series of study on the performance of post-transition metal (Bi, Sn) nanoparticles with various sizes and morphologies to understand of the catalytic behavior of these elements in NH_3 electrosynthesis and inspire design of more advanced electrocatalysts.

571796 Isolated Diatomic Fe-Ni Metal-Nitrogen Sites for Synergistic Electroreduction of CO_2

Pyrolyzed carbon-based M-N_x materials with single-atom structures from MOFs have recently attracted increasing research attention. Compared to the traditional porous materials, M-N_x materials show some specific advantages, such as the structural diversity, easily functionalized, good designability and high specific surface area. However, the fabrication and identification of such active site remain difficult to achieve, and the limited density of isolated M-N_x sites greatly limit the enhance of activity. Here, we report isolated diatomic Fe-Ni sites anchored on nitrogenated carbon as an efficient electrocatalyst for CO_2 reduction. Ni-doped precursor is explored to controlling iron metal dispersity in ZIF-8 precursors. After pyrolysis, Fe-Ni-N_x sites are formed with porous structures. The best catalyst is loaded 10%wt. Fe with 3%wt. Ni, exhibiting high selectivity with CO Faradaic efficiency over a wide potential range from -0.5 to -0.9 V, and robust durability with 90% of its initial selectivity after 10 hours of electrolysis. The high exposed Fe-Ni-N_x sites demonstrate excellent mass activity toward CO_2 reduction, outperforming most reposted Fe-Ni-N_x materials.

571805 Reutilization of $Ba_{0.75}Sr_{0.25}CoNdO_{6-T}$ for Oxygen Reduction Reaction in Alkaline

Perovskite-type oxides (PO) have a wide range of applications in catalytic reactions and photoelectric conversion. Accordingly, a large amount of PO waste is inevitably generated. The effective reutilization of PO waste is of great significance to energy and environment, etc. In this work, we report a novel high activity $Ba_{0.75}Sr_{0.25}CoNdO_{6-\theta}$ (BSCN) catalyst for oxygen reduction (ORR) based on nitrogen-doped carbon (CN). BSCN/CN is prepared by high temperature annealing of low-cost industrial material melamine (as the carbon nitrogen precursor) and BSCN. This facile approach can simultaneously achieve carbonization and heteroatom doping in the synthesis process. The high-resolution N1s spectrum shows that the pyridine-like nitrogen atoms content which promotes the oxygen reduction reaction (ORR) of 0.2g-BSCN/CN (0.2g is the mass of BSCN) reaches 30.49%. Electrochemical characterizations obviously demonstrate that 0.2g-BSCN/CN exhibits excellent catalytic performance and stability in 0.1 $mol \cdot L^{-1}$ KOH. Notably, its onset potential ($\boldsymbol{E_{onset}}$) reaches 0.9V, half-wave potential ($\boldsymbol{\Delta E_{1/2}}$) achieves 0.75V, and limiting current density ($\boldsymbol{J_L}$) accomplishes -5.76 $mA \cdot cm^{-2}$, which significantly exceed the BSCN catalyst. The enhancement of catalytic performance can be attributed to the fact that the incorporation of the pyridine-type nitrogen atoms into the carbon material increases the molecular oxygen absorption capacity of the catalyst. Besides, the electron flows between the BSCN and the nitrogen-doped carbon likely promote the catalytic activity.

575558 Orbitalwise Coordination Number in Search of Metal Nanocatalysts for Oxygen Reduction

Low temperature polymer electrolyte membrane (PEM) fuel cells have great potential as a clean and efficient electric energy generation system. An important obstacle to the commercialization of this technology is the significant voltage loss in converting chemical energy of fuels, e.g., H_2, to electricity, also known as overpotential, mainly due to the sluggish electrochemical kinetics of the oxygen reduction reaction (ORR) at the cathode, $O_2 + 4(H^+ + e^-) \rightarrow 2H_2O$ (E_0: +1.23 V). The state-of-the-art elemental metal electrocatalyst, consisting of 2~5 nm Platinum (Pt) nanoparticles, sacrifices ~300 mV for appreciable current densities[1].

Among many types of materials, multimetallic Pt monolayer electrocatalysts have emerged as a promising alternative. It has been demonstrated that the adsorption energies of oxygen-containing species (e.g., *O, *OH, and *OOH) at an active site are predictive ORR reactivity descriptors. The stability of those intermediates can be tuned by controlling the lattice strain (the bond distance of an active site with neighboring atoms) and the metal ligand (the nature of atoms surrounding a catalytic center)[2-3]. Since a perturbation of a metal site by alloying affects concurrently the ligand and strain, it is not known a priori which metals can be introduced in what geometric arrangements to attain desired catalytic properties. To understand molecule-surface interactions, Bayesian learning model rooted at the Newns-Anderson type model Hamiltonian is used for analysis, which captures trends of chemical bonding of complex species at metal surfaces. Application of the approach can be used for an extensive search of Pt nanocatalysts with enhanced oxygen reduction activity and reduced cost.

[1]Xin, H.; Holewinski, A.; Linic, S. Predictive Structure–Reactivity Models for Rapid Screening of Pt-Based Multimetallic Electrocatalysts for the Oxygen Reduction Reaction. *ACS Catal.* **2012**, *2* (1), 12–16.

[2]Ma, X.; Xin, H. Orbitalwise Coordination Number for Predicting Adsorption Properties of Metal Nanocatalysts. *Phys. Rev. Lett.* 118, 36101 (2017).

[3]Wang, S.; Omidvar, N.; Marx, E.; Xin, H. Overcoming Site Heterogeneity In Search of Metal Nanocatalysts. *ACS Comb. Sci.* 2018 20 (10), 567-572

575515 Conversion of Glucose to Lactic Acid Using and Electrocatalytic Cell System

In 2015, the global market for lactic acid was 1.7 billion pounds, and the worldwide demand of lactic acid is expected to grow to 3.7 billion pounds by 2022.[1] The market for lactic acid will continue to grow due to polylactic acid, the polymerized form of lactic acid, showing promise as an eco-friendly packaging material.[2] Additionally, The United States Department of Energy named lactic acid one of ten potential building blocks of the future.[3] The market for lactic acid is clear, but the manufacturing price of lactic acid needs to reach 0.8 $/kg, roughly half its current cost, to compete with fossil-fuel based plastics.[4]

Currently, fermentation produces 90% of lactic acid.[4] The fermentation process has many advantages, such as low substrate cost and energy consumption.[4] The disadvantage of the fermentation process is low selectivity, causing downstream separation to be expensive.[4] To address this problem, researchers are exploring inorganic catalysts. Application of inorganic catalysts allows selection of active sites to favor the production of lactic acid over the other byproducts.

The majority of work in the field has employed a thermocatalytic system.[5–7] These systems have achieved up to 80% yield of lactic acid but require high temperature or other extreme conditions.[5,6] There are two potential rate-limiting steps in this reaction pathway.[7] The first is the isomerization of glucose to fructose, where this ideally favors fructose.[7] The second is the retro-aldol reaction of fructose to glyceraldehyde (GlycAld).[7] There is agreement that the first problem, favoring fructose in its isomerization with glucose, is solved by a Lewis-acid catalyst.[5–8] Alkaline conditions generally solve the second problem, but the presence of water favors the aldol reaction of GlycAld back into fructose.[7]

No research has been done on this reaction in an electrocatalytic environment. The purpose of this study is to employ an electrocatalytic system to achieve high lactic acid yield, at reasonable reaction conditions, using a copper oxide catalyst. In an electrocatalytic system, the applied voltage should push the equilibrium of the fructose-GlycAld equilibrium to favor GlycAld, increasing the speed of the rate-limiting step. The selection of copper oxide is due to divalent metallic catalysts showing the best yields in thermocatalytic systems.[6]

This study will examine the yields of lactic acid over a voltage range. Initially, copper oxide will be used as the working electrode, followed by trials with deposited copper oxide on a conductive carbon electrode.

References

1. Research and Markets. *Global Lactic Acid Market Size, Market Share, Application Analysis, Regional Outlook, Growth Trends, Key Players, Competitive Strategies and Forecasts, 2015 to 2022*. (2017).
2. Bozell, J. J. & Petersen, G. R. Technology development for the production of biobased products from biorefinery carbohydrates - The US Department of Energy's 'top 10' revisited. *Green Chem.* **12**, 539–554 (2010).
3. United States Deparment of Energy. *Top Value Added Chemicals from Biomass: Volume I--Results of Screening for Potential Candidates from Sugars and Synthesis Gas*. (2004).
4. Komesu, A., de Oliveira, J. A. R., Martins, L. H. da S., Maciel, M. R. W. & Filho, R. M. Lactic acid production to purification: A review. *BioResources* **12**, 4364–4383 (2017).
5. Li, L., Shen, F., Smith, R. L. & Qi, X. Quantitative chemocatalytic production of lactic acid from glucose under anaerobic conditions at room temperature. *Green Chem.* **19**, 76–81 (2017).
6. Choudhary, H., Nishimura, S. & Ebitani, K. Synthesis of high-value organic acids from sugars promoted by hydrothermally loaded Cu oxide species on magnesia. *Appl. Catal. B Environ.* **162**, 1–10 (2015).

7. Marianou, A. A. *et al.* Effect of Lewis and BrØnsted acidity on glucose conversion to 5-HMF and lactic acid in aqueous and organic media. *Appl. Catal. A Gen.* **555**, 75–87 (2018).
8. Zakaria, S. *et al.* Conversion of glucose into lactic acid using silica-supported zinc oxide as solid acid catalyst. *Pure Appl. Chem.* **90**, 1035–1043 (2018).

576576 The Role of Sulfur in $CoSe_{(1-x)}S_x$ for Enhanced Oxygen Electrocatalysis

Advanced electrochemical energy systems like metal air batteries, metal ion batteries, PEM fuel cells, and water electrolyzers have garnered a lot of interest. Limitation in efficiency and cost has been the critical barrier for any commercialization of these systems. Primary reactions occurring in a fuel cell or a battery consists of oxygen electrocatalysis i.e. oxygen evolution reaction (OER) and oxygen reduction reaction (ORR). Tremendous efforts have been made to study different aspects of Transition Metal Chalcogenides (TMDs) like its ORR/OER kinetics, facet interactions with O_{ads} and metal vacancy effects. Employing solvothermal techniques, we synthesized a series of $CoSe_{(1-x)}S_x$. With increasing sulfur incorporation, a phase transformation from hexagonal to cubic crystal structure. The formation and stability of Co_9S_8 comes from the octahedral Co atom linking at the center of the structure as opposed to CoS. Pure sulfide or selenide exhibited excellent current density with a low overpotential (0.767 V). The catalyst also exhibits low tafel slope of 57 and 44 mV/dec respectively. Pure CoSe showed stability upto 2000 cycles of CV. For exploring the fundamental aspects in detail, temperature programmed reduction studies to generate S vacancies leading to altered electron cloud densities at the active sites are in progress. Introduction of S_{vac} can also enhance charge transfer properties in the bulk.

560520 Devisable Bottom-up/ Top-Down Process Directed Facet-Engineered Pt Ni Porous Films As Efficient Oxygen Reduction Electrocatalysts

As the state-of-the-art catalysts for oxygen reduction reaction (ORR), the high cost and scarcity of Pt-based materials remain the major obstacles for their large-scale application in renewable energy industries. To produce efficient Pt-based ORR catalysts with low Pt content, a series of PtNi porous films (PFs) with abundant and facet-selectively exposed Pt active facets were rationally designed. The unique dynamic oxygen-bubble template (DOBT) formed through square-wave potential was utilized in order to realize the control of morphology and exposed Pt crystal planes in PtNi PFs. The generation and adsorption of hydroxyl species (OH_{ads}) on Pt centers can be tuned by the applied anodic voltage, which can affect the oxygen bubbles adsorption on Pt in DOBT. As a result, a (111) facet enriched PtNi PF (donated as $Pt_{3.5\%}Ni$ PF) was obtained and it shows prominent ORR activity with an onset potential of 0.94V and a limiting current of 5.3 mA cm^{-2}

evaluation on PtNi PFs surfaces indicates that less OH^- are adsorbed on $Pt_{3.5\%}Ni$ PF as the existence of rich (111) facets, providing more active sites for initial O_2 adsorption for ORR process, and thus showing superior ORR activity.

561839 Plasmonic Perovskite Semiconductor for Photoelectrochemical Water Splitting

Photoelectrochemical (PEC) water splitting, as an environmental-friendly approach for hydrogen generation, has drawn numerous attentions. For practical hydrogen generation, it is imperative to develop efficient photoanodes with low cost. However, the PEC water splitting performance is still far from practice due to their insufficient light absorption and serious charge recombination during the water oxidation and reduction reactions. Herein, great efforts have been put on the development of photoanodes with expanding light absorptions and enhanced charge separations by virtue of localized surface plasmonic resonance (LSPR). In this work, a nonmetallic photocatalyst, namely, reduced $SrTiO_3$ nanotubes with a crystalline-core and amorphous shell nanostructures, was developed. It exhibits strong LSPR with reversible tunability. Both experimental and theoretical studies illustrate that the abundant oxygen vacancies existing in the amorphous shell induce the high carrier concentration and thus are responsible for LSPR effect. It plays an important role in PEC water splitting. A remarkable photocurrent density of 170.0 $\mu A\ cm^{-2}$ and a maximum incident photon-to-current conversion efficiency of 4.0 % were achieved under visible light excitation by using the novel plasmonic $SrTiO_3$. It is comparable to the state-of-the-art noble metal sensitized catalyst with plasmonic effect. Present work, the nonmetallic plasmonic perovskite semiconductor with LSPR properties, opens a novel paradigm for PEC water lighting in the field of solar energy harvesting.

578367 Exploratory Textual Data Analysis for Understanding the Research Development of Oxygen Reduction Reaction

The oxygen reduction reaction (ORR) has received growing interest in the last decade both in the computational and experimental studies due to the continuous attention on the sustainable power generation technologies, i.e., proton exchange membrane fuel cells (PEMFCs). Electro catalytic materials with novel structures and compositions have shown great promise for catalyzing the ORR electrochemical reaction and the discovery of less expensive and more abundant catalysts is becoming the central theorem of modern research. Even though an overwhelming amount of unstructured textual articles available today provides us a rich source for the ORR catalysis studies. However, it is very challenging and time-consuming to gain a comprehensive understanding of current research status of ORR catalysts development by manually reading through thousands of scholar articles exhaustively from the website. Therefore, developing an semi-automatic web-based search engine that can be used to rapidly process up to thousands of literatures for valuable insights is attractive.

In this work, we demonstrated a web-based search engine to explore a vast amount of literature that are associated with the topic of oxygen reduction reaction. As the core component, the machine learning algorithms based on the task-specific Natural Language Processing (NLP) technique is the key to the success of the web-based search engine. The search engine stars with an HTML web scraper to systematically extract the unstructured textual data from the available website. Then, a text normalization process is applied to filter out the non-informative components of the articles such as HTML syntax, punctuations, stop words. In next step, a series of pre-trained surrogate models (i.e., categorical topic classifiers, Named-entity recognizer, sentiment analyzer) is implemented to process and highlight the most informative keywords for describing the given article such as the types of catalytic material, research scopes and primary techniques. Besides, some document characteristic information, i.e., publication time, impact factor, is also extracted for further analysis. In the end, we perform extensive exploratory analysis for the extracted keywords associating with the documents to draw some insightful knowledge. The exploratory analysis includes the visualization of the histogram distributions of the investigated catalytic materials, the ranking of the research topics versus years and so on. The analytical summary of our textual search engine has huge potential for guiding future research studies and strategic decisions.

564493 Electrochemical Oxidation of Methane at Platinum Electrodes Under Ambient Conditions

Electrochemical oxidation of methane is a potential alternative route towards the production of higher value oxygenates such as methanol, formaldehyde, and formic acid. Compared to traditional catalytic processes such as methane steam reforming and Fischer-Tropsch reactions that operate at high temperatures and pressures, electrochemical oxidation of methane has the benefit of operating at near ambient conditions. However, there are still many challenges associated with electrochemical methane oxidation including catalyst activity and selectivity. Previous work has established platinum as one of the few catalysts able to oxidize methane electrochemically. In the present work the electrochemical conversion of methane to CO_2 on platinum electrodes under ambient conditions was investigated. Through a combination of experimentation, density functional theory (DFT), and micro-kinetic modeling, we have developed an improved understanding of the reaction mechanism and the factors that determine catalyst activity. We hypothesize that the methane activation step is a thermochemical process, not electrochemical, as there is only weak potential dependence, if any. Micro-kinetic models that assume thermochemical methane activation as a rate limiting step match well with experimental results. We expanded our microkinetic model to include other transition metals via a descriptor-based analysis and found platinum to be the most active catalyst for the reaction, in line with previously published experimental observations.

565589 Transitional Metal Coordinated with N-Doped Carbon Nanofiber Integrated with Graphene for the Oxygen Evolution Reaction in Acid

Title: Transitional metal coordinated with N-doped carbon nanofiber integrated with graphene for the oxygen evolution reaction in acid

Author: Chaojun Lei, Lecheng Lei, Yang Hou*

[Chaojun Lei is reportor, Yang Hou is corresponding author]

Abstract: Proton exchange membrane (PEM) water electrolyzer requiring acidic electrolyte has been regarded as the most promising water splitting electrolyzer, however it is still limited due to the oxygen evolution reaction (OER) catalysts with both high activity and stability remain Ru-based and/or Ir-based noble-metal materials in acid medium. Developing high-peromance low cost electrocatalysts for OER in acid is crucial to realize large-scale water splitting to generate hydrogen energy. Inexpensive transitional metal (Fe, Co, Ni) materials are considered as ideal electrocatalysts to replace noble metal catalyst. To overcome the instability and understand the active sites of transitional metal materials under harsh acid electrolyte for OER is of great significance. In this study, we used graphite flake as substrate to load polyamine via electrochemical deposition and then made polyamine to coordinate with transitional metal by impregnation and pyrolysis. The prepared transitional metal-nitrogen coordinated carbon-based electrocatalysts delivered an impressive OER activity. Beyond OER performance, we also consider the role of transitional metal in this carbon-based material. The role is that the content of transitional metal adjusted by the impregnation concentration has significant impact in OER. The transitional metal dispersed in carbon-based material as transitional metal-nitrogen coordinated sites rather than particles showed the best OER performance. Moreover, the pyrolysis temperature also affects performance that the catalysts carbonized under 900°C showed the best OER performance. As a typical example, the active site (FeN_4) was detected by X-ray photoemission spectroscopy (XPS), X-ray absorption near edge structure (XANES), Extended X-ray absorption fine structure (EXAFS) and KSCN poisoning experiment. The transitional metal coordinated with nitrogen was harder than transitional metal oxides to dissolve in acidic electrolyte, exhibiting remarkable stability. Thus, this study opens a new road to rational design and synthesis of highly efficient transition-metal/nitrogen-doped carbon catalysts and identifies the active sites.

565689 Enhancing Organic Electrosynthesis through Artificial Intelligence: The Case of Adiponitrile Electrohydrodimerization

The increase of fossil fuel-derived carbon dioxide (CO_2) emissions has made their impact in the environment self-evident. The chemical sector is a major contributor to this increase, accounting for more than 5.5% of the global CO_2 emissions. The electrification of the chemical industry can represent a major step in the integration of renewable electricity in the industrial sector. Inorganic electrosynthetic processes have been largely implemented in industry (e.g., chloro-alkali and Aluminum production), however electrosynthetic methods for the production of organic chemicals is severely limited. Adiponitrile (ADN) is a precursor to Nylon 6,6 and it can be obtained via the electrohydrodimerization of acrylonitrile (AN). Although the electrochemical production of adiponitrile (ADN) is the largest organic electrochemical process in industry, it faces many challenges in terms of selectivity control. Optimizing the composition of the electrolyte has resulted in significant performance improvements, however mass transport limitations have proven to be a challenge at high current densities, where propionitrile (PN) is the main reaction by-product. Here, we provide insights of the effects of pulsed potential techniques on the mitigation of mass transport limitations and the effective control of product distribution. Moreover, an approach that combines experimental insights and artificial intelligence is proposed for the rapid optimization of operation conditions and the maximization of reaction performance.

The regulation of electron-transfer rates with the use of complex pulse potential waveforms is implemented to control the electrochemical environment surrounding the reaction surface, ultimately favoring ADN production. Electrochemical pulses balance reactant diffusive fluxes to the electrode and the generation/consumption of species in the electrical double layer (EDL), helping mitigate mass transport limitations at high current densities. An improvement of over 250% in the production ratio of ADN:PN (desired to undesired product) was achieved optimizing pulse duration and amplitude. Furthermore, the careful control of overall reaction time and composition of the EDL led to a 20% increase in ADN production with respect to DC operation.

The integrated use of machine learning (ML) algorithms enabled the prediction of ADN formation in a complete landscape of pulse durations. Using ML, a new set of optimal conditions was identified, leading to a 30% increase in ADN production with respect to DC operation for periodic pulses varying from -60 mA cm^{-2} (120 ms) to 0 mA cm^{-2} (5 ms). A new paradigm in organic electrosynthesis research is proposed, where a combination of electrochemical design principles, artificial neural networks, and judiciously designed experimental campaigns are used to uncover optimal electrosynthetic processes.

565853 The Role of Oxygen Incorporation in the Oxygen Reduction Reaction Activity of Molybdenum Nitride Catalysts

Title: The Role of Oxygen Incorporation on the Oxygen Reduction Reaction Activity of Molybdenum Nitride Catalysts

Authors: Melissa E. Kreider[1], Michaela Burke Stevens[1], Yunzhi Liu[2], Alessandro Gallo[1,3], Anton Ievlev[4], Apurva Mehta[5], Robert Sinclair[2], Laurie A. King[1], Thomas F. Jaramillo[1,3]

[1] Department of Chemical Engineering, Stanford University, 443 Via Ortega, Stanford, California 94305, United States

[2] Department of Materials Science and Engineering, Stanford University, 496 Lomita Mall, Stanford, California 94305, United States

[3] SUNCAT Center for Interface Science and Catalysis, SLAC National Accelerator Laboratory, 2575 Sand Hill Road, Menlo Park, California 94025, United States

[4] The Center for Nanophase Materials Sciences and the Institute for Functional Imaging of Materials, Oak Ridge National Laboratory, Oak Ridge, Tennessee 37831, United States

[5] Stanford Synchrotron Radiation Lightsource, SLAC National Accelerator Laboratory, 2575 Sand Hill Road, Menlo Park, California 94025, United States

The development of active, stable, and low-cost catalysts for the oxygen reduction reaction (ORR) is necessary for the viability of polymer electrolyte membrane fuel cells (PEMFCs). Despite decades of study, the required mass loading of platinum-based catalysts remains prohibitively expensive and has limited the growth of fuel cell technology.[1]

Transition metal (TM) nitrides have recently gained attention as active ORR catalysts with promising stability in acidic electrolyte.[2] Typically, TM nitrides are nanostructured and supported on high surface area carbon or nitrided-carbon[3], which can contribute significantly to overall activity and complicate understanding of intrinsic catalyst activity, including the structure and composition of the active material.

In this study, we synthesize well-defined, carbon-free molybdenum nitride (Mo_xN) thin films by reactive sputtering. These catalysts were found to be active in acid electrolyte, with the most active films demonstrating onset potentials between 0.5 and 0.6 V vs RHE. Activity, stability, and selectivity in acid electrolyte were found to depend on the catalyst structure and extent of oxygen incorporation.

To understand the effect of structure of on the catalyst surface and activity, we characterized the catalysts extensively. By varying the synthesis conditions, including temperature, N_2 partial pressure, O_2 partial pressure, and substrate bias, we found that the crystallinity, crystal structure, and nitrogen- and oxygen-content of the films could be modified. Using grazing incidence x-ray diffraction (GIXRD), it was determined that the films were polycrystalline with different bulk structures and different nominal nitrogen contents (e.g. Mo_xN, 1=x=2). Near edge x-ray absorption fine structure spectroscopy and cross-sectional transmission electron microscopy were also used to distinguish between similar crystal structures and to understand better the spatial distribution of crystal structures throughout the film based on the synthesis conditions. Time-of-flight secondary ion mass spectrometry (ToF-SIMS) was used to obtain a depth profile of the Mo, N, and O composition of the film. Differences in bulk oxygen content correlate strongly with the crystal structures determined by GIXRD and with electrochemical activity, as shown in the accompanying figure. Cyclic voltammograms for Mo_xN in 0.1 M $HClO_4$ show the highest ORR activity for the $Mo_{1.5}N$ film, which had the lowest bulk O concentration. We hypothesize that the bulk structure of the catalyst, including the presence of a N gradient and low O content, affects the active surface and thus the ORR activity.

To probe this activity difference, catalyst conductivity was tested using a ferrocyanide/ferricyanide redox couple. No substantial differences were measured, indicating that differences in conductivity cannot explain the differences in activity between these catalysts. Additionally, electrochemical surface area measurements indicate that activity improvement is not correlated with increased surface area. Changes in structure and composition with electrochemical testing were probed through *ex-situ* x-ray photoelectron spectroscopy and x-ray reflectivity characterization and indicate a stable crystal structure with loss of surface oxide character for all catalysts. The similarity of the surface morphologies suggests that differences in catalyst activity for molybdenum nitride structures are related to the bulk properties. This provides a design principle in which catalyst activity can be tuned by changing the bulk structure and total oxygen content.

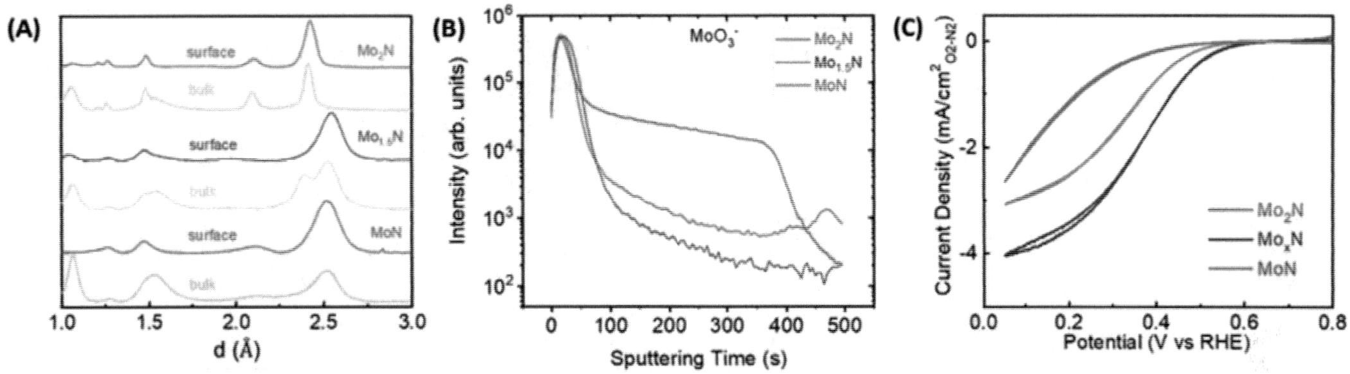

Figure Caption. Structural, compositional, and activity characterization of the three different Mo_xN catalysts (where x = 2, 1.5, 1). (A) GIXRD patterns, showing surface and bulk structures. (B) TOF-SIMS depth profile measuring oxygen content, collected in negative ion mode. (C) Cyclic voltammograms using a rotating disc electrode (1600 rpm) showing the ORR activity in 0.1 M $HClO_4$.

References:

Rabis, A.; Rodriguez, P.; Schmidt, T. J. Electrocatalysis for Polymer Electrolyte Fuel Cells: Recent Achievements and Future Challenges. ACS Catal. 2012, 2, 864–890.

Dong, S.; Chen, X.; Zhang, X.; Cui, G. Nanostructured Transition Metal Nitrides for Energy Storage and Fuel Cells. Coord. Chem. Rev. 2013, 257 (13–14), 1946–1956.

Ota, K.; Ishihara, A. Metal Oxide-Based Compounds as Electrocatalysts for Oxygen Reduction Reaction. In Electrocatalysis in Fuel Cells: A Non- and Low- Platinum Approach; Shao, M., Ed.; Springer London: London, 2013; pp 391–416.

569090 Potential Dependent Kinetic Barriers of Oxygen Reduction Reactions (ORR) on Pt (111)

Oxygen reduction reactions (ORR) have been extensively studied in the past few decades owing to their application in proton exchange membrane fuel cells (PEMFCs). Despite the high efficiency, sustainability and environment friendly operation of PEMFCs, they are known to suffer from the slow rate of ORR at the cathode. Although, an ideal electrochemical reaction should generate 1.23 V_{RHE} per electron, the typical operating potential for electro-catalytic ORR on Pt(111) surface is observed below 0.9 V_{RHE}. First principle thermodynamic atomistic calculations [1-2] using Density Functional Theory (DFT) have shown a presence of high over-potential of ORR on Pt(111) due to the strong binding of the reaction intermediates like O* and HO*. However, the reason behind the sluggish kinetics of ORR remains unclear and ambiguous in the literature.

In this study, we employ a kinetic study using DFT along with the Solvated Jellium (SJ) model [3] – a method developed in our group – to study the potential dependence of the elementary steps of ORR via both dissociative and associative mechanism using density functional theory. The aim of the study is to understand the kinetics of ORR at different potentials and relevant coverages. We further comment on the dominance of either associative or dissociative mechanisms at different operating conditions.

[1] *Norskov, J. K. et Al*, Origin of the Overpotential for Oxygen Reduction at a Fuel-Cell Cathode J. Phys. Chem. B, 2004, 108 (46), pp 17886–17892

[2] Sharma S. et al, Face-centered tetragonal (FCT) Fe and Co alloys of Pt as catalysts for the oxygen reduction reaction (ORR): A DFT study, J. Chem. Phys. 150, 041704 (2019)

[3] Kastlunger, G. et al, Controlled-Potential Simulation of Elementary Electrochemical Reactions: Proton Discharge on Metal Surfaces, J. Phys. Chem. C 122 (24), 12771 (2018)

572143 Mechanistic Insights into Solution Phase Oxygen Reduction Reactions and the Effect of Metal Cation Dopants

Non-aqueous metal-O_2 batteries with high theoretical gravimetric energy storage density stands to be one of the potential future battery technologies to replace current Li-ion batteries. [1-4] This has led to renewed interest into fundamental studies of oxygen reduction reaction (ORR) in non-aqueous electrolytes because they are the central electrochemical processes involved at the metal-air cathodes. Fundamental insights into these reactions are critical to further improvement and eventual commercialization of these technology prototypes.

During discharge, the main fundamental reaction involved is ORR which leads to the formation of discharge products. The discharge products can grow either through a surface mediated mechanism [5] that produces the discharge product directly on the electrode or a solution mediated mechanism [6-7] which leads to the precipitation of nanoparticles. In this talk, we show that the formation and solvation of O_2^- at low discharge overpotentials (shown in Fig. 1a) open up pathways to thermochemical ORR in the solution phase. This leads preferentially to the formation of solid Li_2O_2 vs. solid NaO_2 and KO_2 at low overpotentials. We propose, based on theoretical calculations, the reaction pathways for the formation of alkali superoxide and peroxide in several solvents that explain the observed product selectivity.

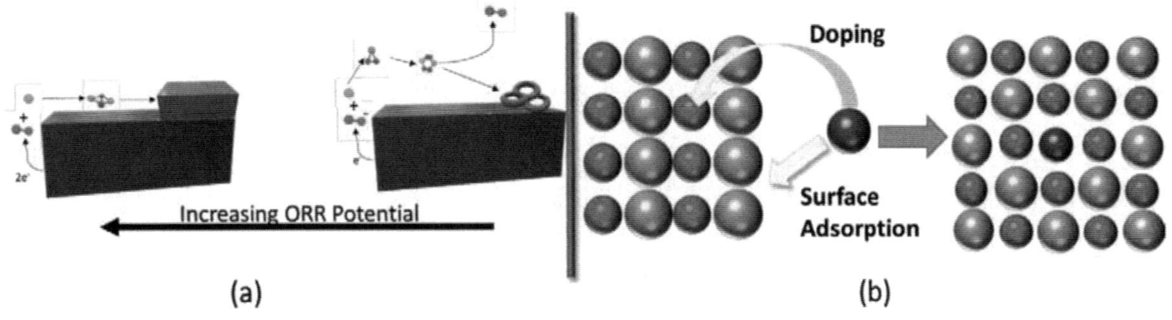

Figure 1: (a) Schematic of surface and solution mediated discharge mechanism governed by discharge overpotential. (b) Schematic of metal cation doping in Li_2O_2 bulk and adsorption on $Li_2O_2(0001)$ surface.

The electrolyte solution for such batteries contain metal cations through added salts. The effect of of these metal cations (dopants) can have morphological effects on the final discharge product through formation of localized domain microstructures as shown in Fig. 1b. This translates to microstructures with different lattice types than the thermodynamic minimum crystalline $P6_3/mmc$ Föppl structure of Li_2O_2 as the discharge product. The formation of these microstructures can aid in the overall OER process. Such kind of incorporation of foreign particles in the form of metal cations into the discharge product has already been shown to be successful experimentally.[8] Using evolutionary algorithms coupled with DFT we explore the possibility of thermodynamic formation of a new phase of bulk solid spurred by the incorporation of high levels of metal cation dopants and its consequential effects on the electronic properties of the resultant bulk solid. We show that these microstructures can aid in improving electron conduction and hence reduce the high overpotential for OER which the bulk discharge product is associated with in Li-O_2 electrochemistry.[9] We consider a only a few commonly used elements with known electrocatalytic properties for presenting a general approach to doping. We also present a theoretical study for the adsorption of metal cations on the $Li_2O_2(0001)$ surface with and without Li/O vacancy defects. We present an extended thermodynamic formulation for the adsorption of these cations to conclude the required conditions that would facilitate the facile adsorption of the concerned adsorbates.

References

1. Bruce, P. G.; Freunberger, S. A.; Hardwick, L. J.; Tarascon, J. M., Li-O-2 and Li-S batteries with high energy storage. Nature Materials 2012, 11 (1), 19-29.

2. Lu, J.; Li, L.; Park, J.-B.; Sun, Y.-K.; Wu, F.; Amine, K., Aprotic and aqueous Li–O2 batteries. Chem. Rev. 2014, 114 (11), 5611-5640.

3. Black, R.; Adams, B.; Nazar, L., Non-Aqueous and Hybrid Li-O2 Batteries. Advanced Energy Materials 2012, 2 (7), 801-815.

4. Luntz, A. C.; McCloskey, B. D., Nonaqueous Li–air batteries: a status report. Chem. Rev. 2014, 114 (23), 11721-11750.

5. Xu, Y.; Shelton, W. A., O2 reduction by lithium on Au(111) and Pt(111). J. Chem. Phys. 2010, 133 (2), 024703.

6. Aetukuri, N. B.; McCloskey, B. D.; García, J. M.; Krupp, L. E.; Viswanathan, V.; Luntz, A. C., Solvating additives drive solution-mediated electrochemistry and enhance toroid growth in non-aqueous Li–O2 batteries. Nat. Chem. 2015, 7 (1), 50-56.

7. Lutz, L.; Alves Dalla Corte, D.; Tang, M.; Salager, E.; Deschamps, M.; Grimaud, A.; Johnson, L.; Bruce, P. G.; Tarascon, J.-M., Role of Electrolyte Anions in the Na–O2 Battery: Implications for NaO2 Solvation and the Stability of the Sodium Solid Electrolyte Interphase in Glyme Ethers. Chem. Mater. 2017, 29 (14), 6066-6075.

8. Matsuda, S.; Uosaki, K.; Nakanishi, S., Improved charging performance of Li–O2 batteries by forming Ba-incorporated Li2O2 as the discharge product. J. Power Sources 2017, 353, 138-143.

9. Hummelshøj, J.; Luntz, A.; Nørskov, J., Theoretical evidence for low kinetic overpotentials in Li-O2 electrochemistry. The Journal of chemical physics 2013, 138 (3), 034703.

573277 Effect of Ionomer Coverage on Pt Electrocatalyst Performance in PEFCs

Polymer electrolyte fuel cells (PEFCs) are an attractive technology for mobile and stationary electric power generation that holds several benefits over conventional internal combustion engines, which are less efficient and emit harmful exhaust. Despite their significant promise, commercially available fuel cell vehicles currently utilize high platinum metal loadings to overcome cell voltage losses. Commonly, perfluorosulfonic acid ionomers such as Nafion are incorporated in the catalyst layer to improve ionic (protonic) conductivity and Pt site utilization thereby improving ORR performance in both RDE and MEA studies. However, high ionomer loadings can be detrimental to performance since the strong interaction between sulfonate ($-SO_3^-$) functional groups and Pt nanoparticles has been linked lower catalytic rates and higher O_2 transport resistances.

It is essential to understand the local distribution of ionomer near catalytic sites in order to accurately assess whether improvements result from increased accessibility or higher intrinsic activity on catalytic sites. Current *ex situ* characterization techniques (USAXS, STEM/EDS, and Nano-CT) cannot provide necessary resolution to determine ionomer coverage on Pt sites so it is difficult to determine how different ink formulations (catalyst, solvent, and ionomer) impact cell performance.

Here, the development and utilization of *in situ* diagnostics designed to probe the local ionomer interactions will be utilized in conjunction with established testing protocols to relate ionomer coverage (on Pt and/or support) with kinetic performance and O_2 transport resistance. Specifically, CO displacement chronoamperometry and electrochemical impedance spectroscopy techniques will be described to determine local ionomer coverages across a series of Pt/C MEAs. These results will promote a deeper understanding of ionomer-catalyst interactions which in turn will inform future materials development and integration necessary to achieve high performance PGM (e.g. Pt/C and Pt alloy/C) electrocatalyst-based electrodes.

574293 Atom-Thick Ir-Rich Skin on Ptir Nanocube Electrocatalysts for Efficient Ethanol Oxidation

The direct alcohol fuel cell (DAFC) as an alternative power source for various applications has attracted great attention due to its high efficiency under ambient operating conditions. Ethanol is more favored as the fuel than methanol due to its higher energy density, lower toxicity, and larger mass-production capacity. However, the state-of-the-art Pt-based electrocatalysts possess limited active sites to break C-C bond and are easily poisoned during ethanol oxidation reaction (EOR), which leads to very low fuel utilization efficiency and hinders the commercialization of DEFCs. Here, we report colloidal preparation of PtIr core-shell nanocubes (NCs) with atom-thick Ir-rich shells to enhance EOR performance. Compared with pure Pt NCs and $Pt_{17}Ir$ NCs (two-atom thick Ir-rich skin), we demonstrate that (100)-exposed $Pt_{38}Ir$ NCs with one-atom-thick Ir-rich skin present excellent EOR activity, improved CO_2 selectivity and long-term stability. The EOR current density of $Pt_{38}Ir$ NCs electrocatalyst is 4.5-times higher than that of Pt/C with a negatively shifted EOR onset potential of 320 mV. Also, it can deliver a CO_2 current density up to 14 times higher than that of commercial Pt/C at 0.85 V. We find that the improved EOR performance can be ascribed to the favorable cleavage of the C-C bond by strongly absorbed $*C_xH_yO/C_xH_y$ species and easily desorption of CO from the PtIr surface. This work highlights the decisive role of surface atom layers on shape-controlled electrocatalysts, and shed light on further development of effective EOR electrocatalysts.

575518 Ternary and Quaternary Microwave Synthesized Electrocatalyst for Ethanol Oxidation

Ternary and Quaternary Microwave Synthesized Electrocatalyst for Ethanol Oxidation

Shelby L. Foster[1], Patricia Means[1], Reid Eakin[1], Francis Osonga[1], and Lauren F. Greenlee[1]

[1]Department of Chemical Engineering, University of Arkansas, Fayetteville, Arkansas

The development of an electrocatalyst for the oxidation of the carbon-carbon (C-C) bond in long chain hydrocarbons would result in a liquid fuel flexible fuel cell. Ethanol is a common hydrocarbon used to investigate the oxidation of the C-C bond. Electrochemical studies show there are two pathways of electrooxidation of ethanol: 1) cleavage of the C-C bond which releases the maximum possible energy from the reaction, and 2) removal of two hydrogen atoms resulting in limited energy generation and overall poisoning of complete electrooxidation to carbon dioxide (CO_2). Several key metallic catalysts surfaces have been investigated to suppress the second pathway of electrooxidation of ethanol. Specifically, PtRhNi ternary nanoparticle catalyst have been shown to convert ethanol to CO_2 effectively with increased reactivity in alkaline environments. The increased reactivity of PtRhNi ternary nanoparticles in alkaline environments for electrooxidation of ethanol is the investigation of the project. Prior to the work by Boettcher and co-authors[1], Ni was believed to be the active site for water oxidation by $Ni(OH)_2$ catalysts in alkaline media. The identification of Fe doping into $Ni(OH)_2$ by electrolyte Fe impurities lead to Fe now attributed as being the active site for water oxidation. Therefore, PtRhNi ternary nanoparticles with Fe doping in alkaline environments will be investigated for electrooxication of ethanol reactivity.

The microwave synthesized PtRhNi ternary nanoparticles reactivity have been investigated in electrolyte free of Fe impurities and electrolyte containing Fe impurities. The electrolyte containing Fe impurities has resulted in higher current densities, indicating the Fe does in fact play a role in the electrooxidation of ethanol. In this presentation, I will present our work on ternary (PtRhNi) and quarternary (PtRhNiFe) nanoparticles for the electrooxidation of ethanol. The composition, morphology, surface characteristics, and electrochemical reactivity of the PtRhNi ternary nanoparticles and the PtRhNiFe quaternary nanoparticles will be discussed.

1. Trotochaud, L.; Young, S. L.; Ranney, J. K.; Boettcher, S. W., Nickel-iron oxyhydroxide oxygen-evolution electrocatalysts: the role of intentional and incidental iron incorporation. *J Am Chem Soc* **2014,** *136* (18), 6744-53.

577302 Multifunctional Vanadium-Doped Cobalt Oxide Layer on Silicon Photoanodes for Efficient Photoelectrochemical Water Oxidation

Due to the limited availability of fossil fuels, there is an urgent demand to develop sustainable energy technologies for efficient utilization of renewable energy sources, particularly solar energy. Photoelectrochemical (PEC) water splitting has emerged as an attractive strategy to convert and store solar energy into hydrogen fuel. However, developing highly efficient photoanodes for PEC cells to meet industrial requirements remains a challenge. Silicon has high carrier mobilities and mature technologies for fabrication and modification, which is regarded as a promising material for PEC electrodes. Here, we present a study of effectively engineering the onset potential of Si photoanodes through doping of V into Co oxide film by a low-cost way of magnetron co-sputtering deposition on $p^{+}n$ junction Si cell (p^{+}nSi/CoVO). Using such a p^{+}nSi/CoVO photoanode, a high photocurrent density of 29.15 mA cm^{-2} (at 1.23 V vs the reversible hydrogen electrode) as well as a high photovoltage of 608 mV are achieved. Moreover, a successful decrease of the onset potential by 40 mV to 1 V (at 1 mA cm^{-2}) and a 3.6-fold increase in incident photons to current conversion efficiency (IPCE) are reached, compared to those of p^{+}nSi/CoO without V-doping. We found that the doped V atoms can decrease the charge transfer resistance of Co_3O_4 film, serve as Lewis acid to increase local pH value, and facilitate the generation of Co(IV) oxo-bridged species to lower the kinetic barrier of oxygen evolution reaction (OER), thus to decrease the onset potential of p^{+}nSi/CoVO in OER. This work suggests a general approach to improve the performance of OER catalysts by engineering active sites and local chemical environment through element doping.

References:

[1] Xing, Z.; Wu, H.; Wu, L.; Wang, X.; Zhong, H.; Li, F.; Shi, J.; Song, D.; Xiao, W.; Jiang, C.; Ren, F. A multifunctional vanadium-doped cobalt oxide layer on silicon photoanodes for efficient and stable photoelectrochemical water oxidation. *J. Mater. Chem. A* **2018**, *6*, 21167-21177.

577762 A Microkinetic Analysis of the CO Electro-Oxidation Reaction on Pt Bimetallics: Understanding the Interplay of Bifunctional and Electronic Effects

The CO electro-oxidation reaction has been one of the most thoroughly studied reaction mechanisms in electrocatalysis, mostly due to its role as a model reaction for the oxidation of larger organics like methanol and ethanol, each of which have important renewable energy applications in direct alcohol fuel cells. Bimetallics of Pt-M (where M is an oxophilic modifier such as Ru or Sn) have been shown extensively to have improved CO tolerance compared to pure Pt. The reason for this enhancement is often attributed to some combination of bifunctional (oxophilic metal provides OH to remove poisoning CO on Pt) and electronic (oxophilic metal weakens the CO binding strength on the Pt sites) effects. However, it has been difficult to determine the extent to which these different effects are operative in a given system.

In this work, we perform a mean-field microkinetic analysis of the CO electro-oxidation reaction in order to elucidate the roles of bifunctional and electronic effects. We look specifically at the classic Pt-Ru bimetallic system, and determine that the rates of bifunctional coupling are generally lower than the rates of coupling on Pt sites electronically modified by Ru. This indicates that, at least for the Pt-Ru system, the electronic effects that weaken CO binding on Pt are more operative than the bifunctional effect. This implies that that the prospects for bifunctional gains in general may be more limited than once thought, and that a more rational design parameter for bimetallic materials may be the electronic effects between the two components.

571293 Analysis of PEFC Cathode Catalyst Layer Based on Rate Dependency on Oxygen Partial Pressure

Introduction

In a polymer electrolyte fuel cell (PEFC), protons generated on the anode are transported through the anode catalyst layer and the proton exchange membrane (PEM) to the cathode. Electrons are generated on the anode and transported through the external circuit to the cathode. On the cathode of the PEFC, the oxygen reduction reaction (ORR) takes place as follows:

$O_2 + 4\ H^+ + 4\ e^-$ ¨ $2\ H_2O$. (1)

Phenomena on the cathode are governed by interaction between the reaction and the transport of oxygen and protons. In our previous study, the effectiveness factor of cathode catalyst layer (CCL) was reported to be a function solely of 4 dimensionless moduli, $M_O{}^{(C)}{}_m$, $M_p{}^{(C)}{}_m$, $P_O{}^{(C)}{}_m$ and y_{Oc},[1,2] which were derived from the isothermal one-dimensional model of the cathode consisting of oxygen balance, oxygen transport, proton transport, and reaction stoichiometry. $M_O{}^{(C)}{}_m$ does not depend on the oxygen partial pressure, while $M_p{}^{(C)}{}_m$ depends on it. In this study, a method to determine the dimensionless moduli which control the effectiveness factor from experimental data measured at different oxygen partial pressures is proposed.

Theory

The ORR rate per unit volume of the CCL, r_{vc}, is expressed as follows:

$r_{vc} = k_{vc}{}^* \exp(-E_c/b_c)\, p_O$ [mol/(m^3·s)], (2)

where $k_{vc}{}^*$ is the reaction rate constant per unit volume of CCL [mol/(Pa·m^3·s)] at $E_c = 0$, E_c is the cathode electromotive force (emf) [V], b_c is the Tafel slope [V] and p_O is the oxygen partial pressure [Pa]. The cathode emf is the difference between the electron potential, f_e, and the proton potential, f_p. The reaction rate constant is not constant since E_c derivates in the CCL from the boundary condition, and r_{vc} depends on the oxygen partial pressure and cathode emf. The current density, i, is a product of the reaction rate without mass transfer resistances, $k_{vcm}\, p_{Oc}$, and the effectiveness factor, $F_e{}^{(C)}$.

$i = 4\ F\ d^{(C)}\ k_{vcm}\ p_{Oc}\ F_e{}^{(C)}$ [A/m^2], (3)

where the subscript m means the PEM–CCL boundary and the subscript c means the CCL–gas diffusion layer (GDL) boundary. $d^{(C)}$ is the CCL thickness and F is the Faraday constant. $F_e{}^{(C)}$ is a function of 4 dimensionless moduli, $M_O{}^{(C)}{}_m$, $M_p{}^{(C)}{}_m$, $P_O{}^{(C)}{}_m$, and y_{Oc}, which have the following meanings respectively:

$M_O{}^{(C)}{}_m$: Ratio of reaction to oxygen diffusion (Thiele modulus),

$M_p{}^{(C)}{}_m$: Ratio of reaction to proton conductivity (Our modulus),

$P_O{}^{(C)}{}_m$: Ratio of convection to oxygen diffusion (Peclet number), and@

y_{Oc}: Oxygen mole fraction.

$M_O{}^{(C)}{}_m$ and $M_p{}^{(C)}{}_m$ are defined as follows:

$$M_O{}^{(C)}{}_m = d^{(C)}(k_{vcm}RT/D_{eO})^{1/2} \text{ and (4)}$$

$$M_p{}^{(C)}{}_m = d^{(C)}[4Fk_{vcm}p_{Oc}/(s_{ep}b_c)]^{1/2}, (5)$$

where D_{eO} is the effective oxygen diffusion coefficient [m^2/s], and s_{ep} is the effective proton conductivity [S/m].

If the temperature, T, is identical, $M_O{}^{(C)}{}_m$ is constant regardless of the oxygen partial pressure since both the reaction rate and the oxygen diffusion rate are proportional to the oxygen partial pressure. On the other hand, $M_p{}^{(C)}{}_m$ changes with the oxygen partial pressure since the proton transport rate is independent of the oxygen partial pressure. Under the reaction control conditions, the current density, i, is proportional to p_{Oc} since $F_e{}^{(C)} = 1$. Under the oxygen transfer control conditions, i is proportional to p_{Oc} since $F_e{}^{(C)}$ is inversely proportional to $M_O{}^{(C)}{}_m$ and $M_O{}^{(C)}{}_m$ is not affected by p_{Oc}. Under the proton transport control conditions, i is proportional to $p_{Oc}{}^{0.5}$ since $F_e{}^{(C)}$ is inversely proportional to $M_p{}^{(C)}{}_m$. Therefore, the slope of log i versus log p_{Oc} is located between 0.5 and 1 when $d^{(C)}$ and E_{cm} are fixed. Eqs. (3) and (5) indicate the relationship between i / p_{Oc} and $p_{Oc}{}^{0.5}$ corresponds to the relationship between $F_e{}^{(C)}$ and $M_p{}^{(C)}{}_m$. The former relationship is obtained from measurements and the latter is obtained from the theory. In case of $P_O{}^{(C)}{}_m = 0$, if $M_O{}^{(C)}{}_m$ and y_{Oc} are given, the theoretical curve of $F_e{}^{(C)}$ versus $M_p{}^{(C)}{}_m$ is determined. Then $M_O{}^{(C)}{}_m$ is obtained by fitting the experimental results with the theoretical curve.

Experimental

The membrane electrode assembly (MEA) was consisted of the catalyst layers (CLs) made of TEC10E50E (Tanaka Kikinzoku Kogyo K.K.) catalyst with Nafion® ionomer and membrane (DuPont NR-211). Pt/C weight ratio was 0.50, ionomer/carbon ratio was 0.90, and the CLfs thickness was 9.0 μm. The cell temperature was 80 °C, the relative humidity (RH) was 100 % and the total pressure was 101.3 kPa. I–V measurement was carried out at the oxygen partial pressure of 3.24, 6.47, 11.3, and 53.9 kPa.

Results and discussion

When i / p_{Oc} was plotted against $p_{Oc}{}^{0.5}$, the slope of log i versus log p_{Oc} obtained from the experiments was between 0.5 and 1. This indicates that the experiments were carried out neither under the proton transport control conditions nor the reaction control conditions.

The slope of the theoretical line of log $F_e{}^{(C)}$ versus log $M_p{}^{(C)}{}_m$ is nearly 0 when $M_p{}^{(C)}{}_m$ is negligibly low. The slope of the line is –1 when $M_p{}^{(C)}{}_m$ is extremely high. When the experimental data are located on a horizontal line, it means that the proton transport resistance is negligible. When the experimental data are located on an $M_p{}^{(C)}{}_m{}^{-1}$ line, it means the data were measured in the proton transport control region. In both the extreme cases, $M_p{}^{(C)}{}_m$ cannot be determined from the log i versus log p_{Oc} plot. For determination of $M_O{}^{(C)}{}_m$ and $M_p{}^{(C)}{}_m$, data should be analyzed where proton transport resistance is not extremely low or high.

By the data fitting it was found that $M_O{}^{(C)}{}_m$ was 1.11 at $E_{cm} = 0.74$ V and $M_O{}^{(C)}{}_m$ was 2.1 at $E_{cm} = 0.7$ V, for instance. By substituting $M_O{}^{(C)}{}_m$, $M_p{}^{(C)}{}_m$, and $F_e{}^{(C)}$ in eqs. (3), (4), and (5), k_{vcm}, s_{ep}, and D_{eO} were determined as follows at 101.3 kPa, 80 °C, RH = 100 %, and I/C = 0.9:

$k_{\mathrm{vc}} = 1.03\sim10 \;\; \exp\left(-E_{\mathrm{c}} / 0.0383\ \mathrm{V} \right)\ \mathrm{mol/(Pa\cdot m\ \cdot s)}$,

$D_{\mathrm{eO}} = 7.33\sim10^{-8}\ \mathrm{m^2/s}$, and

$s_{\mathrm{ep}} = 1.26\ \mathrm{S/m}$.

The obtained Tafel slope, 88.2 mV/decade, is close to the values reported by others for the same catalysts.[3,4] The apparent Tafel slope appears greater than the true value if the transport resistances in the catalyst layer increase. A high value such as 118 mV/dec. at 35 °C[5] was obtained from measurements without model analysis. Although the obtained effective oxygen diffusion coefficient is lower than that reported by other researchers,[6,7] the effect of convective flow was ignored in this and other usual studies, which resulted in the scattered values of D_{eO}. The effective proton conductivity as well as the effective oxygen diffusion coefficient depends remarkably on the pore and agglomerate structure. The value of the effective proton conductivity is in good agreement with the values, 0.9 S/m and 2.0 S/m, measured at similar I/C and similar RH reported in literture.[8,9]

Conclusion

A method to determine the dimensionless moduli governing the PEFC cathode behavior from the dependency of the ORR rate on the oxygen partial pressure measured using a usual cell and MEA is proposed. The electrochemical reaction rate constant and transport properties were successfully determined by the proposed method. Measurements of dependency of the rate on the oxygen partial pressure should be carried out at identical total pressure, temperature, and RH by using a same MEA. In the analysis, only current and voltage data measured under the conditions where proton transport resistance is not extremely low or high can be used for determining the dimensionless moduli, $M_{\mathrm{O}}{}^{(\mathrm{C})}{}_{m}$ and $M_{\mathrm{p}}{}^{(\mathrm{C})}{}_{m}$. From the dimensionless moduli determined by the analysis, the reaction rate constant, k_{vc}, and the transport properties, s_{ep} and D_{eO}, can be estimated.

References

[1] M. Kawase, K. Yamaguchi, M. Kageyama, K. Sato, and G. Inoue, *ECS Trans.* **75**(14), 147–156 (2016).
[2] M. Kawase, K. Sato, R. Mitsui, H. Asonuma, M. Kageyama, K. Yamaguchi, and G. Inoue, *AIChE J.* **63**(1), 249–256 (2017).
[3] S. Takaichi, H. Uchida, and M. Watanabe, *Electrochim. Acta.* **53**(14), 4699–4705 (2008).
[4] W. Zhang, S. Shironita, and M. Umeda, *Int. J. Hydrogen Energy* **41**(15), 6526–6533 (2016).
[5] A. Parthasarathy and C. R. Martin, *J. Electrochem. Soc.* **138**(4), 916–921 (1991).
[6] G. Inoue and M. Kawase, *J. Power Sources* **327**, 1–10(2016).
[7] S. Litster, W. K. Epting, E. A. Wargo, S. R. Kalidindi, and E. C. Kumbur, *Fuel Cells* **13**(5), 935–945 (2013).
[8] R. Makharia, M. F. Mathias, and D. R. Baker, *J. Electrochem. Soc.* **152**(5), A970–977 (2005).
[9] Y. Liu, M. W. Murphy, D. R. Baker, W. Gu, C. Ji, J. Jorne, and H. A. Gasteiger, *J. Electrochem. Soc.* **156**(8), B970–980 (2009).

574373 Electrostatically Enhanced Catalytic Phase Transfer Hydrogenation of Acetophenone Under Low External Electric Field

Phase transfer catalysis (PTC) has been extensively studied as a practical method for organic synthesis, as it not only achieves high reactivity and selectivity but also features operational simplicity, and economic and environmentally benign reaction systems. While having various advantages, PTC processes are not as efficient as homogeneous catalysis due to the mass transfer limitation between the two essentially immiscible liquid phases. Additionally, the novel design of phase transfer catalysts is a significant challenge. Recently, the use of external electric fields to establish electrostatic dispersion has been proved as an energy-efficient method for eliminating mass transfer resistance in liquid/liquid systems theoretically and experimentally in different processes. With the presence of an oriented-external electric field (OEEF), the charged components/species can be selectively transferred from the feed liquid phase to a second liquid phase at an accelerated rate due to the electrostatic forces which enhance the dispersion and droplet mechanics. While phase-transfer catalysts and external electric fields each enhance mass transfer across phase boundaries, the latter is much easier to apply, avoiding lengthy rational design of phase transfer catalysts for targeted reactions.

In the present work, the OEEF promoted phase transfer catalysis reaction was reported for the first time. The reaction of choice for this study was the transfer hydrogenation of acetophenone, with sodium formate as the hydrogen source. The two phase system was established with acetophenone/organic solvent as the less dense organic phase and sodium formate solution as the more dense aqueous phase. An external electric field was applied across the two phases without any agitation. It was demonstrated experimentally that an OEEF can act as promoter or inhibitor for the phase transfer hydrogenation of acetophenone depending on its direction, showing a potential on controlling rate of chemical reactions. With 1-butanol as organic solvent, the conversion can be achieved as high as 86% with an enantioselectivity of 94% under room temperature and pressure. This work presents a methodology for promoting PTC reactions in the absence of a chemical phase transfer agent and shows promise for significantly improving the performance of organic reactions with OEEF in biphasic liquid systems.

THEORETICAL STUDY OF PHOTOCATALYTIC WATER SPLITTING IN A TUBE-IN-TUBE SETUP

Nopphon Weeranoppanant, Patchara Chaichon, Woraphin Saetung*

Department of Chemical Engineering, Faculty of Engineering, Burapha University, 169 Longhard Bangsaen, Saensook, Muang, Chonburi, Thailand, 20131
**Corresponding author: nopphon.we@eng.buu.ac.th*

Abstract

Tube-in-tube setup has been employed for many reactions and processes in flow. The setup uses a Teflon AF-2400 tubing as an inner tubing while the outer tubing is non-permeable. This allows gas-liquid reactions, such as hydrogenation, to occur in flow without the two phases contacting directly. Alternatively, it can be used to remove gas from liquid. In this work, we constructed transport equations and modeled hydrogen production in flow via photocatalytic reaction. The models were formulated from a simple case (without reaction) to a case with reaction kinetics and solved analytically. This work offered insights into a potential use of the tube-in-tube setup for hydrogen production in continuous-flow mode and influence of processing conditions.

Introduction

Water splitting is a potential paradigm for production of hydrogen energy. Water splitting is a non-spontaneous reaction; on the contrary, it requires an energy input to overcome the thermodynamics. It can utilize solar energy, which is a major source of energy. Currently, there are two main light-based approaches to perform water splitting: photochemical and photoelectrochemical. [1]. The energy required for water splitting is about 1.23 eV, which is close to an energy of an electromagnetic wave with 1000 nm in wavelength. Therefore, semiconductors can be used as photocatalysts to absorb lights in order to form a electron-hole pair as well as induce subsequent reduction and oxidations. Significant development of effective photocatalysts have been reported [2-4]. However, the efficiency of the process remains low due to a number of factors. For instance, some photocatalysts are only active in UV-region, so the absorbance in visible light is low [5]. Water splitting in absence of sacrificial agents is also quite challenging [6]. Another problem is that the produced hydrogen can easily recombine with oxygen over photocatalysts [7]. In other words, a reverse reaction of water splitting can occur very fast. As a result, very low net rate of hydrogen production is often observed.

In this work, we are interested in photocatalytic (photochemical) water splitting. To address the recombination problem, we employ a gas-permeable membrane tubing, Teflon-AF, to separate gas from liquid water. This way, hydrogen gas can directly be removed from water, which also contains photocatalysts. Teflon-AF has an excellent gas permeability, with higher selectivity toward hydrogen over oxygen. It also has an excellent optical clarity and low refractive index, an important property for photochemical process [8]. Here, we studied a reactor in tube-in-tube setup [9, 10], in which the Teflon-AF tubing is an inner tubing while the outer tubing is an outer tubing. We developed a theoretical model using mass transport equation to examine effects of process variables.

Methodology

Figure 1 shows methodology to develop the theoretical modelling in this research. We used a conservation equation as a starting point [11]. Then, a number of assumptions were made to reduce the terms in the equation. The partial differential equation (PDE) was simplified by using the concept of "bulk concentration." It was then finally solved analytically. The analytical solutions were used to simulate the process at varying conditions.

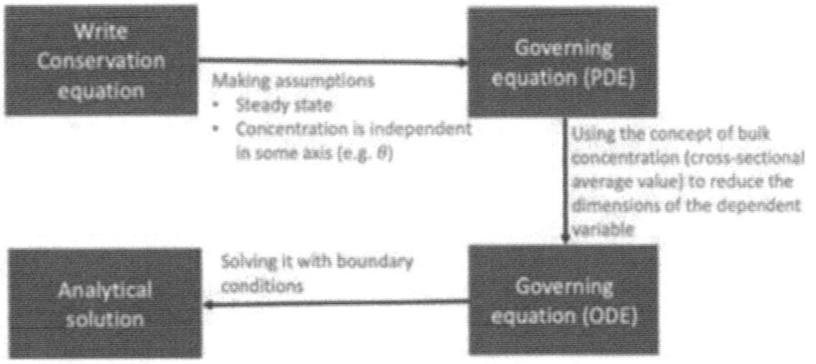

Figure 1. Methodology for formulating and solving theoretical models

In order to understand both physical and chemical phenomena, we divided the studies into two cases, one without a reaction (simply physical mass transfer to the outer tubing) and the other with the water splitting reaction in the inner tubing as depicted in Figure 2.

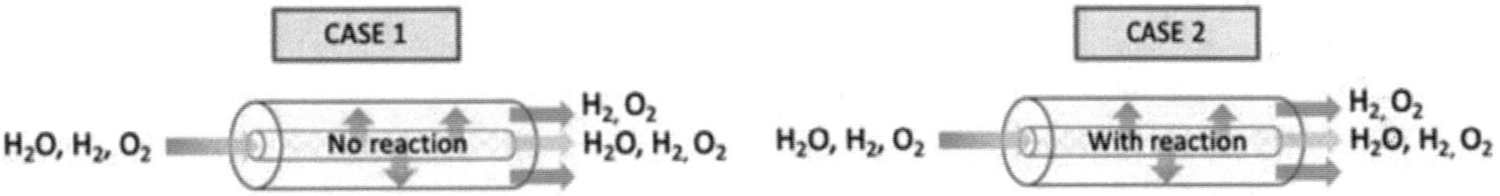

Figure 2. Two cases studied in this research (with and without water splitting reaction)

Theoretical formulation

The tube-in-tube setup consists of two tubing. Inside the inner tubing, water splitting occurs with formation of hydrogen and oxygen. Then, hydrogen and oxygen will selectively permeate out through the inner tubing wall to the outer tubing as shown in Figure 3. The outer tubing can be flushed with fresh air or operated under vacuum.

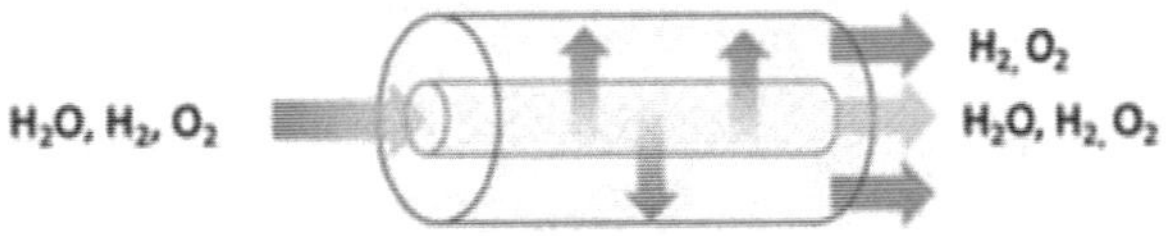

Figure 3. Drawing of the tube-in-tube setup

To formulate hydrogen and oxygen transport in the inner tubing, we used the conservation equation in a cylindrical coordinate (*i* represents a chemical species, H_2 or O_2):

$$\frac{\partial c_i}{\partial t}+v_r\frac{\partial c_i}{\partial r}+\frac{v_\theta}{r}\frac{\partial c_i}{\partial \theta}+v_z\frac{\partial c_i}{\partial z}=D_i\left[\frac{1}{r}\frac{\partial}{\partial r}\left(r\frac{\partial c_i}{\partial r}\right)+\frac{1}{r^2}\frac{\partial^2 c_i}{\partial \theta^2}+\frac{\partial^2 c_i}{\partial z^2}\right]+R_{vi}$$

To simplify the equations, we made a number of assumptions:

- The system is in steady state
- Flow velocity in radial and angular directions are negligible.
- Concentrations of H_2 and O_2 are independent of θ-position.
- Axial diffusion is negligible.

With those assumptions, the conservation equation can be simplified to

Case 1:

$$v_z\frac{\partial c_i}{\partial z}=D_i\left[\frac{1}{r}\frac{\partial}{\partial r}\left(r\frac{\partial c_i}{\partial r}\right)\right]$$

Case 2:

$$v_z\frac{\partial c_i}{\partial z}=D_i\left[\frac{1}{r}\frac{\partial}{\partial r}\left(r\frac{\partial c_i}{\partial r}\right)\right]+R_{vi}$$

It is noteworthy that case 2 has a reaction term since water splitting occurs in bulk. The obtained equations are still in partial differential equation. These can be solved numerically along with boundary conditions. However, in this study, we aimed to solve the equations analytically. Therefore, we simplified the PDEs further by using the concept of the "bulk concentration (C_b)." The bulk concentration is the average value of concentration in cross-sectional area of the tubing.

$$C_b(z)=\frac{\int C_i v_z dA}{\int v_z dA}$$

Also, the diffusion flux can be related to two modes of the mass transfer in this system as shown in Figure 4:

- convective mass transfer in the liquid water (K_c)
- permeation through the tubing wall (K_m)

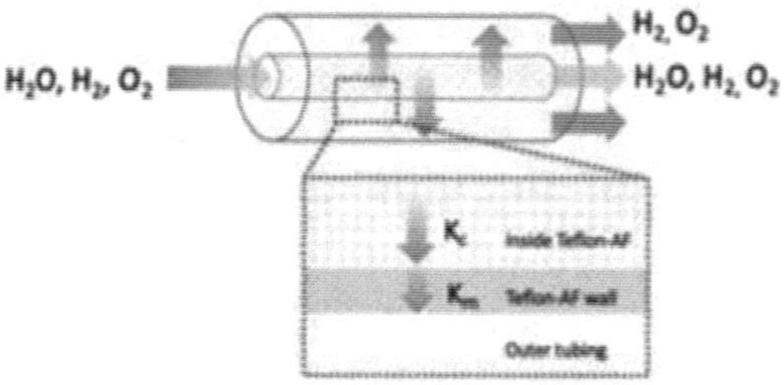

Figure 4. Two modes of mass transfer in the tube-in-tube system

We could converted the diffusion flux into a total flux with K_c and K_m and used this definition of the bulk concentration. The PDEs became the following first-order ODEs:

Case 1:

$$\frac{dC_b}{dz} = \frac{-2}{RU} K_c(z) \cdot (C_b(z) - C_i(z))$$

Case 2:

$$\frac{dC_b}{dz} = -\frac{2}{RU}\left[\frac{K_c K_m}{K_m + K_c}\right] C_b(z) + \left\{\frac{2}{RU}\left[\frac{K_c K_m}{K_m + K_c}\right] C_d + \frac{R_{vi}}{U}\right\}$$

The boundary condition for these ODEs is that initial gas concentration or $C_b(z = 0)$ is defined to be C_0. Solving the ODE with the boundary condition, we obtained the following solutions:

Case 1:

$$\ln\frac{(C_b(z) - C_d)}{(C_0 - C_d)} = \frac{-2z}{RU}\left(\frac{K_c K_m}{K_m + K_c}\right)$$

Case 2:

$$C_b(z) = \left(1 - e^{-\tau K z}\right) C_d + \left(1 - e^{-\tau K z}\right)\frac{R_{vi}}{\tau K U} + C_0 e^{-\tau K z}$$

, Where $\tau = \frac{2}{RU}$ and $K = \frac{K_c K_m}{K_m + K_c}$

Results and discussion

We used the solutions of the conservation equations to simulate concentration of H_2 and O_2 in the tubing, and subsequently evaluated the performance of the tube-in-tube setup at varying conditions. We first studied the effect of liquid flow velocity (U). As shown in Figure 5, we found that in case 1 (without reaction), as U increased, more H_2 and O_2 remained in the inner tubing. However, in case 2 (with reaction), as U increased, less H_2 remained in the tubing. The mass transfer of hydrogen to the outer tubing was faster than the hydrogen evolution rate. In other word, the rate of H_2 production was instead limited by the kinetics of water splitting (reaction rate). On the other hand, in case 2, since O_2 concentration in the inner tubing increased with increasing U, the rate of O_2 production was mass transfer-limited.

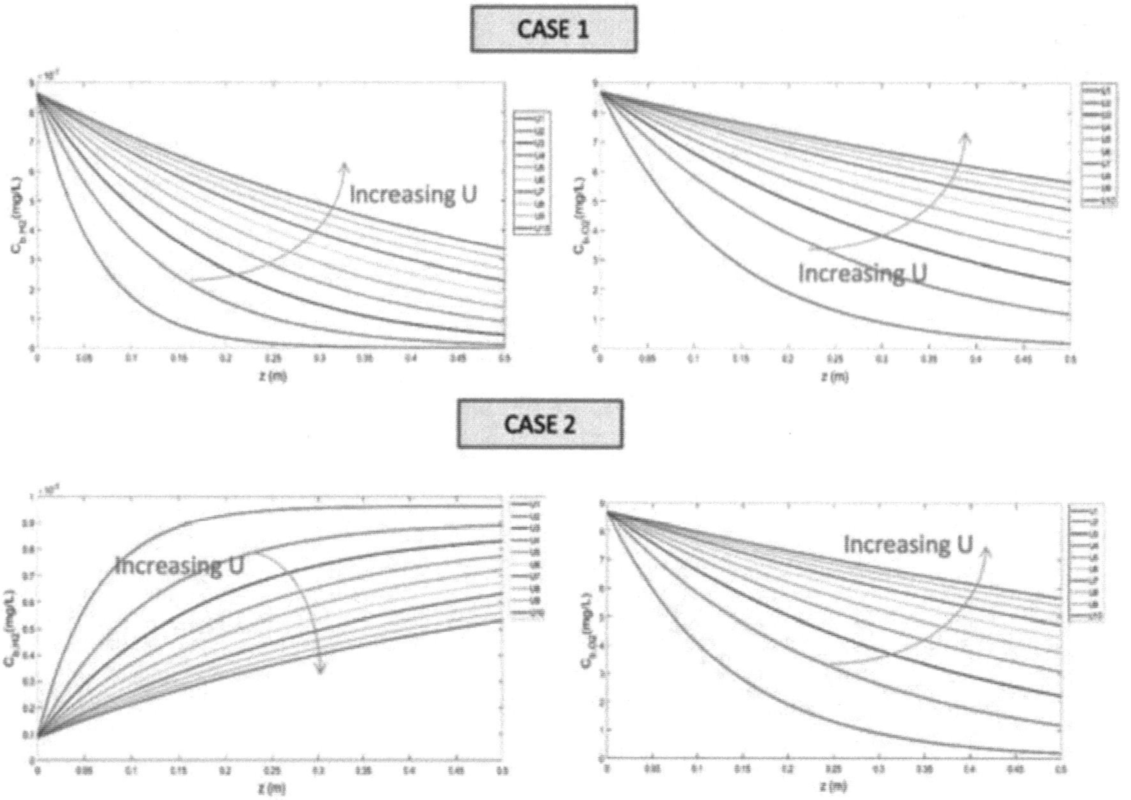

Figure 5. H_2 and O_2 concentration in the inner tubing at different locations (z) and varying liquid flow velocity (U)

We then further examine the performance of the tube-in-tube setup based on their selective separation H_2 and O_2. Ideally, only H_2 will be removed into the outer tubing while O_2 remains in the inner tubing. As a result, we define a performance (η) as follows:

$$\text{Performance } (\eta) = \frac{\%\text{Removed } (H_2)}{\%\text{Removed } (O_2)}$$

% Removed (H_2) and % Removed (O_2) are the percentage of gas amount (for H_2 and O_2, respectively) that is separated into the outer tubing as compared to the initial amount in the liquid water entering the setup (case 1) or the total amount of gas in the inner tubing in the absence gas permeation (case 2). From Figure 6, we found that in both cases, liquid flow velocity (U) had a significant effect on the performance while the effect of the concentration of oxygen in the outer tubing was negligible.

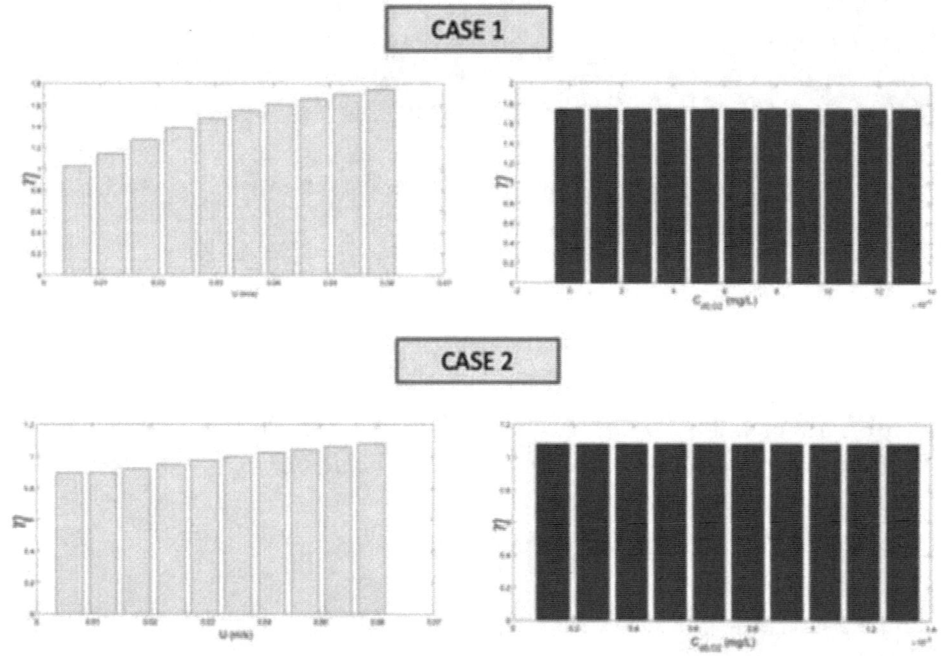

Figure 6. Performance of the tube-in-tube setup based on the selective removal of H_2 and O_2 to the outer tubing at varying conditions of liquid flow velocity (U) and initial oxygen concentration in the outer tubing ($C_{d0,\ O2}$)

Another important consideration is the productivity of H_2 produced. This can be evaluated based on the amount of H_2 in the outer tubing. Since U is only an important factor, the performance (η)and hydrogen concentration in the outer tubing at z = 0.5 m (C_{dH2}) are plotted against U. The opposite trends of the two dependent parameters (η and C_{dH2}) were observed in both cases. In other words, at higher U, the setup had a high selectivity towards H_2 separation, but the productivity was low. Therefore, these two parameters have to be compromised upon the process design of this setup.

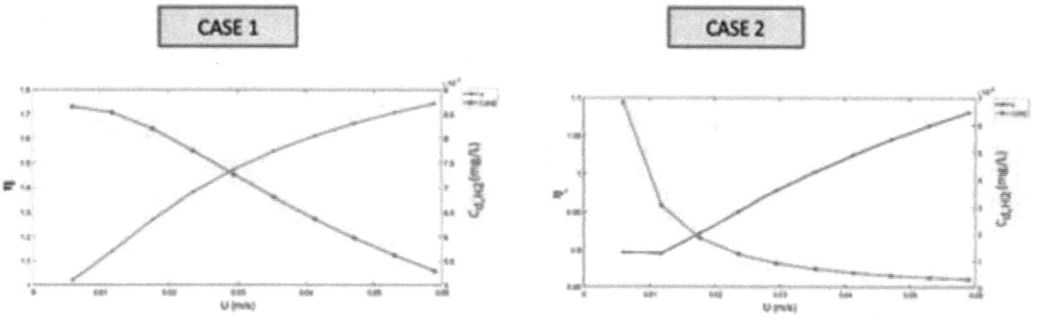

Figure 7. Performance and productivity are plotted at varying liquid flow velocity (U)

Conclusion

- Flow velocity affects significantly hydrogen production in a tube-in-tube setup while the concentration of O_2 in the outer tubing has only a minor effect.
- If reaction is considered, the hydrogen transfer to the outer tubing is mass-transfer limited

while the oxygen transfer is reaction-limited.

- According to the plot of η and $C_{d,H2}$ VS. U, we found that he separation performance has an opposite trend to the productivity. That is, if we choose to use high flow velocity (i.e., high U), then the H_2 gas in the outer tubing has a high purity, but the net amount of H_2 in the outer tubing is low.
- Models with more realistic assumptions (e.g. non-constant density) should further be developed to improve its accuracy in prediction.

Acknowledgment

This research is financially supported by the NSTDA Chair Professor Grant (P-17-52536) and funding from Faculty of Engineering, Burapha University, Thailand.

References

[1] Takata, T., & Domen, K. (2019). Particulate photocatalysts for water splitting: recent advances and future prospects. *ACS Energy Letters*, *4*(2), 542-549.

[2] Ni, M., Leung, M. K., Leung, D. Y., & Sumathy, K. (2007). A review and recent developments in photocatalytic water-splitting using TiO2 for hydrogen production. *Renewable and Sustainable Energy Reviews*, *11*(3), 401-425.

[3] Ye, S., Wang, R., Wu, M. Z., & Yuan, Y. P. (2015). A review on g-C3N4 for photocatalytic water splitting and CO2 reduction. *Applied Surface Science*, *358*, 15-27.

[4] Acar, C., Dincer, I., & Naterer, G. F. (2016). Review of photocatalytic water-splitting methods for sustainable hydrogen production. *International Journal of Energy Research*, *40*(11), 1449-1473.

[5] Moniz, S. J., Shevlin, S. A., Martin, D. J., Guo, Z. X., & Tang, J. (2015). Visible-light driven heterojunction photocatalysts for water splitting–a critical review. *Energy & Environmental Science*, *8*(3), 731-759.

[6] Martin, D. J., Reardon, P. J. T., Moniz, S. J., & Tang, J. (2014). Visible light-driven pure water splitting by a nature-inspired organic semiconductor-based system. *Journal of the American Chemical Society*, *136*(36), 12568-12571.

[7] Gao, W., Zhang, W., & Lu, G. (2017). A two-pronged strategy to enhance visible-light-driven overall water splitting via visible-to-ultraviolet upconversion coupling with hydrogen-oxygen recombination inhibition. *Applied Catalysis B: Environmental*, *212*, 23-31.

[8] Alawajji, R. A., Kannarpady, G. K., Nima, Z. A., Kelly, N., Watanabe, F., & Biris, A. S. (2018). High temperature, transparent, superhydrophobic Teflon AF-2400/Indium tin oxide nanocomposite thin films. *Nanotechnology*.

[9] Brzozowski, M., O'Brien, M., Ley, S. V., & Polyzos, A. (2015). Flow Chemistry: Intelligent processing of gas–liquid transformations using a tube-in-tube reactor. *Accounts of chemical research*, *48*(2), 349-362.

[10] Yang, L., & Jensen, K. F. (2013). Mass transport and reactions in the tube-in-tube reactor. *Organic Process Research & Development*, *17*(6), 927-933.

[11] Deen, W. M. (1998). *Analysis of transport phenomena* (Vol. 2). New York: Oxford university press.

576150 Synthesis and Characterization of Pt-Zn Intermetallic Nanocatalysts for Light Alkanes Oxidative Dehydrogenation

Supported Pt nanoparticle has been commercialized in dehydrogenation of alkane for more than 50 years, for its superior affinity to paraffinic C-H bond over C-C bond. One of the challenges for Pt catalyst in alkane dehydrogenation is large Pt size favor side reactions such as hydrogenolysis and coke formation[1]. It is challenging for traditional synthesis method such as wet impregnation and colloidal chemistry to prepare Pt nanoparticle that is smaller than 2 nm[2]. Recent development in the field pointed out a new direction that effectively reduced Pt particle size by alloying Pt with other metals (e.g., Sn, Cu, In). Pt-based intermetallic alloy can suppress side reactions, due to their advantage in ordered structure with isolated active Pt sites over solid solutions in which atoms are randomly arranged.

In this work silica supported Pt-Zn intermetallic nanoparticles with different Pt-Zn alloy phases were synthesized. Specifically, we use *in situ* XAS to characterize the transition of the pre-catalyst $Pt/ZnO/SiO_2$ to the intermetallic nanocatalyst Pt_1Zn_1/SiO_2 in H_2 with a multi-stage temperature ramping. Pt-Zn intermetallic phases are identified as active sites for light alkanes oxidative dehydrogenation (ODH). It was found that at temperature above 600 ^oC, ZnO was fully reduced, and Pt_1Zn_1 was formed. The oxidation state and structural change of the catalysts under ODH condition was monitored by *in situ* XAS. A relationship between the stability of the Pt-Zn intermetallic phases in the nanocatalysts and the catalytic performance is established, and their deactivation mechanism in ODH reaction is discussed.

References

1. Sattler, J.J.H.B., et al., *Catalytic Dehydrogenation of Light Alkanes on Metals and Metal Oxides.* Chemical Reviews, 2014. 114(20): p. 10613-10653.
2. Lei, Y., et al., *Synthesis of Pt–Pd Core–Shell Nanostructures by Atomic Layer Deposition: Application in Propane Oxidative Dehydrogenation to Propylene.* Chemistry of Materials, 2012. 24(18): p. 3525-3533.

576563 Process Modified UV Assisted TiO_2 Adsorptive Desulfurization Mechanism Study

One-time UV pretreatment on commercial Anatase TiO_2 has been proven effective in enhancing the sulfur heterocycle specificity and saturation capacity of TiO_2 sorbents for over two weeks. Previous efforts in our laboratory have shown the increased densities of surface hydroxyl are the key factors for this enhancement effect. The current study provides further evidence of the previous assertion by tracking the UV-induced conversion of adsorbed water and lattice oxygen into surface hydroxyls as monitored by XPS. Two different commercial TiO_2 adsorbents (@163 and 70 m^2/g) were investigated at various hydration levels as determined by TGA. The deconvolution details from photoelectron peaks demonstrated both the qualitative nature and quantitative conversion of one adsorbed water and one lattice oxygen into two surface hydroxyls. Eight different reaction cases were examined using the 163 and 70 m^2/g sorbents at 4 different hydration levels each. The measured stoichiometry for the above noted conversion varied from 1.83 to 2.15 with an average value of 1.98 and a standard deviation of 0.10. Samples removed from the XPS UHV chamber and stored in laboratory conditions showed that UV-generated surface hydroxyls can persist for over 2 weeks with or without handling.

571952 Optimized Kinetic Parameters of Metallocene Catalyzed Olefin Polymerization through Modelling and Simulation

Single-site catalysts are revolutionizing polyolefin production because they enable more precise control of the polymer molecular architecture, which in turn controls the physical properties of the polymer. The benefit of metallocene catalysts is that they allow the production of tailored macromolecules, leading to new products such as iso- and syndiotactic polypropylene or syndiotactic polystyrene. The enormous structural variety possible in single-site catalysts gives rise to the need for a quantitative method to design a catalyst system to achieve a desired molecular architecture of the polymer. Mathematical modelling of the polymerization process is a crucial step towards developing such understanding and affords opportunities for optimization. It can not only explain the important phenomena during polymerization qualitatively, but also predict the relationships between operating conditions and polymer properties quantitatively. Olefin polymerization kinetics have been investigated for a wide variety of homogeneous metallocene catalyst systems. Many experimental studies are focused on identifying the individual steps of the kinetic mechanism. However, only a few kinetic models have been validated using instantaneous reaction rate and molecular weight data. This work provides a detailed description of developing mathematical models and its validation using experimental data. The kinetic model was based on the key mechanistic steps of the coordination-insertion mechanism of single site metallocene olefin polymerization catalysts. The developed kinetic model is restricted to catalyst systems that undergo a bimolecular chain transfer reaction as the main termination event. Present work also provides a detailed account of different types of catalyst systems that can be used for metallocene catalyzed olefin polymerization process and their functioning.

569721 Catalytic Ethylene Dimerization to 1-Butene Using Automated Construction of Microkinetic Models with Rmg-Cat

The dimerization of ethylene to 1-butene is a reaction of interest as not only is the cost of ethylene low, but also because 1-butene is used in the production of high value plastics and polymers such as polyethylene and polybutene. Unfortunately, this reaction needs to be described in great detail to understand the full chemistry of the system, which makes intimate knowledge of all species both adsorbed and not adsorbed to the catalytic surface necessary, resulting in models that are quite large. Acquiring this level of detail for one industrially relevant reaction on a single catalyst is time consuming and complex, making the need for a quick and accurate high-throughput method for screening novel catalysts for any reaction even greater. Our Reaction Mechanism Generator for Heterogeneous Catalysis (RMG-Cat) [1], built upon the well-established open-source RMG software [2], can provide the complete reaction network from user-defined ignition conditions. It can: propose both elementary reactions and new species by applying reaction templates and recipes; estimate thermodynamic and kinetic parameters using databases of DFT calculations, functional-group decision trees, and heuristics; solve governing rate equations; and decide which reaction pathways should be explored further.

RMG-Cat can estimate adsorbate thermochemistry on a range of hypothetical metal surfaces using linear scaling relations and construct detailed microkinetic models for any surface, with no prior assumption of a rate limiting step. By running simulations with sensitivity analysis, we can determine the rate limiting step on each surface and can plot a “volcano surface” of the degree of rate control of each reaction as a function of elemental binding energies. We demonstrate this new tool on the catalytic dimerization of ethylene to 1-butene and longer olefins.

[1] Goldsmith, C. F., West, R. H. “Automatic Generation of Microkinetic Mechanisms for Heterogeneous Catalysis” J. Phys. Chem. C. *21* (18), 9970–9981 (2017) http://doi.org/10.1021/acs.jpcc.7b02133

[2] Gao, C. W., Allen, J. W., Green, W. H., West, R. H. "Reaction Mechanism Generator: Automatic construction of chemical kinetic mechanisms". Comput. Phys. Commun., 203, 212-225, (2016) http://doi.org/10.1016/j.cpc.2016.02.013

575877 Metal-Promoted Dehydroaromatization of Ethylene over ZSM-5 Catalysts

Industrial production of aromatics heavily relies on catalytic reforming in refineries.[1] Driven by the limited supply of fossil fuels and the recent shale gas revolution, it is highly desirable to develop methane- and/or light olefin-based processes to meet the growing demand for aromatics. To this end, ethylene dehydroaromatization (DHA) over metal-exchanged ZSM-5 is promising given the worldwide availability of ethylene feedstock. ZSM-5 was chosen as a catalyst to study ethylene DHA, because it has been used commercially in the UOP/BP Cyclar process, which converts light alkanes to aromatics.[2] Zeolites with extra-framework metals are conventionally prepared by aqueous ion exchange or incipient-wetness impregnation; however, these methods often result in large fractions of metal oxide deposits on external surfaces due to the slow diffusion of hydrated metal ions into the micropores. Alternative methods have been introduced by Bell and coworkers to incorporate Lewis acids in ZSM-5 pores by vapor-phase exchange.[3] Jones and coworkers demonstrated a direct hydrothermal synthesis using functional silanes to prepare metal-exchanged MFI for propane dehydrogenation (PDH).[4]

In this presentation, we will discuss the preparation and testing of metal-exchanged ZSM-5 catalysts using three different synthesis approaches, including the conventional aqueous ion exchange and direct hydrothermal synthesis (with and without the addition of inorganic structure-directing agents). Our findings reveal dramatic differences in catalyst performance depending on the method of synthesis, with aromatic selectivities enhanced by ~3-fold compared to H-ZSM-5. Collectively, this work provides new insights into the design of metal-containing zeolite catalysts for natural gas upgrading.

[1] aFranck, H.-G.; Stadelhofer, J. W., *Industrial Aromatic Chemistry*. Spring: **1988**; bArpe, H.-J.; Hawkins, S., *Industrial Organic Chemistry*. 5 ed.; **2010**.

[2] Gosling, C. D.; Hamm, D. A. Process for the production of benzene from light hydrocarbons **1992**.

[3] Phadke, N. M.; Van der Mynsbrugge, J.; Mansoor, E.; Getsoian, A. B.; Head-Gordon, M.; Bell, A. T., *ACS Catalysis* **2018,** *8*, 6106-6126.

[4] Choi, S. W.; Kim, W. G.; So, J. S.; Moore, J. S.; Liu, Y. J.; Dixit, R. S.; Pendergast, J. G.; Sievers, C.; Sholl, D. S.; Nair, S.; Jones, C. W., *Journal of Catalysis* **2017,** *345*, 113-123.

576452 Noble Bi-Metallic (Pt-Ru) Catalyst for Diesel Autothermal Reforming – Activity Tests and Characterizations

A diesel reforming is a promising method for supplying hydrogen to mobile fuel cell systems. However, due to the diesel is a complex mixture of heavy hydrocarbons which contains long-chain of paraffins and aromatics, it is easy for the reforming catalyst to be degraded. As a result, fuel is not fully converted to hydrogen, carbon monoxide and carbon dioxide, and unwanted by-products such as ethylene are produced. Noble metals are well-known active metals due to its high activity and coke resistance. Nevertheless, it is still not verified for long-term performance for diesel reforming. In our previous study, Pt-Ru bimetallic catalyst had introduced as an autothermal reforming (ATR) catalyst and it showed great stability and coke resistance. In this study, we investigated the reasons that enhance the durability of the bi-metallic catalyst. For monometallic(Pt,Ru) and bimetallic(Pt-Ru) supported on CGO, activity tests catalyst and various characterization methods were performed. Temperature programmed reduction(TPR), CO-chemisorption, and transmission electron microscopy(TEM) were used to analyze redox property, metal dispersion and morphology of the active metal, respectively. Finally, we conducted 1,000 hour long-term test at 800oC, SCR 3.0 and OCR 0.8. All catalyst were synthesized by glycine nitrate process. As results, we found strong synergetic effects between Pt and Ru which compensated both of the low activity of Pt at low temperature and poor durability of Ru. Moreover, agglomeration effect of the active metal was reduced. This performance enhancement came from the formation of Pt-Ru clusters demonstrated by TEM images and TPR profiles. The Pt-Ru catalyst showed excellent reforming performance for 1000 hours without any degradation signals.

564819 An Interesting Four-Stage Model of Au-Pd/TS-1 Bimetallic Catalysts for Direct Propylene Epoxidation with H_2 and O_2

Au-Pd bimetallic nanoparticles (NPs) has been proven to be active for the direct synthesis of hydrogen peroxide (H_2O_2), which has attracted significant research interest. In this study, supported Au-Pd NPs on titanium silicalite-1 (denoted as Au-Pd/TS-1), prepared by the alcohol reduction method, shows activity for direct propylene epoxidation with H_2 and O_2. Generally speaking, the propylene epoxidation reaction is performed at 200 °C over Au/TS-1. But for Au-Pd/TS-1, when the reaction was carried out at 100 °C, the catalytic performance exhibited an interesting phenomenon. It can be divided into four stages, denoted as stages I, II, III and IV. In the stage I, propylene hydrogenation dominants and then the epoxidation dominants in the stage II. Afterwards, the two reactions achieve a balance in the stage III and the activity decreases gradually in the stage IV. Based on these, we propose an interesting four-stage model for the first time. The insights reported here reveal the competition between hydrogenation and epoxidation reactions of propylene, as well as may guide us to understand the reaction mechanism of propylene epoxidation over Au-Pd/TS-1 bimetallic catalysts.

571205 Kinetic Modeling of Catalytic Cracking of Paraffinic Naphtha with Molecular Mechanism

Title: kinetic modeling of catalytic cracking of paraffinic naphtha with molecular mechanism.

Author: Ha-Nui Jo[1], KiWoong Kim[2], Jinsu Kim[1], Yoojin Han[3], Jae-Wook Shin[2], Young-Seek Yoon[3]* and In-Beum Lee[1]**

1) Department of Chemical Engineering, Pohang University of Science and Technology, Pohang, South Korea

2) Korea Research Institute of Chemical Technology, Daejeon, South Korea

3) Graduate Institute of Ferrous Technology, Pohang University of Science and Technology, Pohang, South Korea

*E-mail: ysyoon@postech.ac.kr
**E-mail: iblee@postech.ac.kr

Light olefins such as ethylene and propylene are the basic component in the petrochemical industry. These light olefins can be obtained by catalytic cracking of naphtha. However, the pathway of Naphtha-to-Olefin (NTO) reaction has not yet been defined because naphtha consists of various compositions of hydrocarbons and the reaction rate is very fast. For the interpretation of the NTO reaction, mechanistic mechanism and molecular mechanism can be used. The mechanistic mechanism is a reaction path that covers all of the detailed reaction paths including carbenium and carbonium ions, which are intermediates formed on the catalyst surface. Based on this mechanism, intrinsic reaction rates can be derived, which does not change depending on the experimental conditions. However, this mechanism has more than 2000 reaction pathways, which not only includes unobservable intermediate such as carbenium ion but also increases the computation load for estimating reaction rate parameters. In addition, this makes it difficult to derive accurate reaction parameters because vast amounts of experimental data are required to estimate all of the reaction parameters accurately for more than 2000 reactions. As an alternative to this, the molecular mechanism can be considered, which configures the reaction pathway only with observable molecules. In the existing studies, simple reaction pathways are presented, which focuses on the high predictability of light olefins. However, these reaction paths do not include all the actual reaction paths and they can yield results that are inconsistent with the experimental data at different reaction conditions.

In this paper, we proposed a reaction path that is constructed with the molecular mechanism by converting the mechanistic mechanism, which is the actual reaction mechanism. Therefore, this reaction mechanism included all the reaction pathways while it was composed with observable molecules. It made the estimated reaction parameters to have robustness depending on the experimental conditions such as feedstock composition and temperature. In addition, we initially used experimental data with single hydrocarbon feed to estimate reaction parameters with high accuracy, which is less complex than a normal naphtha cracking. By using these parameters as a criterion to determine other parameters at a complex cracking reaction of normal naphtha, the difference of reaction rate by carbon number and reaction type was reflected as constraints. This approach can derive the reaction rate parameters with physical meaning, not derive the numbers simply fitting the values of the experimental data and model.

Acknowledgement

This work was supported by the National Research Council of Science & Technology (NST) grant by the Korea government (MSIT) (No. CRC-14-01-KRICT).

571322 Investigation of Ceria, Vanadia/Ceria and Gold/Ceria Catalysts By *Operando* Raman-Online FTIR during Toluene Oxidation

Toluene oxidation on three ceria catalysts, CeO_2, VO_x/CeO_2 and Au/CeO_2, was investigated by an *operando* Raman-online FTIR reactor cell. The reactive surface oxygen sites were preferred sites for vanadium and gold deposition. The deposited vanadium existed as V^{5+}, while most of the gold was in Au^+ state and roughly a third of ceria was in the reduced Ce^{3+} state. The Au/CeO_2 and CeO_2 catalysts (T_{50}^*=260 and 290°C) were more active and selective toward the complete oxidation of toluene than VO_x/CeO_2 catalyst (T_{50}^*=370°C), where T_{50}^* refers to the temperature for 50% of CO_2 yield. The η^2-peroxide $O_2^{2-}{}_{ads}$ was detected on Au/CeO_2 and CeO_2 catalysts, where toluene molecules preferentially adsorbed parallel to the surface *via* π-bonding. Au/CeO_2 gave complete combustion producing mainly CO_2. On the other hand, η^1-superoxide $O_2^-{}_{ads}$ was found on VO_x/CeO_2 catalyst and the toluene molecule adsorbed *via* σ-bonding forming carbenium ions on the vanadia Brønsted acid sites. This catalyst produced significant amount of benzaldehyde as partial oxidation byproduct and CO (< 25%). The nature of the active sites, configurational adsorption of toluene and the reactive oxygen species play an important role in the catalyst activity and selectivity leading to a large contrast in the catalytic behavior of the CeO_2, VO_x/CeO_2 and Au/CeO_2 catalysts.

557068 Research on Bifunctional Catalysts for Hydrogen Production from Dimethyl Ether Steam Reforming

Faced with the crisis of energy and environment, polymer electrolyte membrane fuel cells (PEMFCs) have a bright prospect on account of its characteristics. Currently, hydrogen supply is still one of the main obstructions of PEMFCs' commercialization. Steam Reforming of Dimethyl Ether (SRDME) is a popular method in hydrogen production, and it can efficiently solve the hydrogen supply problems of PEMFCs. Current research revealed that two kinds of active sites are required in dimethyl ether steam reforming: one for the hydrolysis of dimethyl ether and one for steam reforming of methanol. Therefore, the study on bifunctional catalysts and the reaction mechanism are the most essential for the study of SRDME. Tranisition metal carbides (TMCs) have shown great promise for many important reactions such as methane dry reforming (DRM), water gas shift reaction (WGS), and steam reforming of dimethyl ether due to its noble-metal-like characteristics.

The new bifunctional catalyst Pt-MoC/Al_2O_3 was prepared for dimethyl ether steam reforming. Pt-MoC/Al_2O_3 exhibited a stable behavior and good performance on SRDME reaction. HRTEM results showed that Pt particles loaded on MoC surface uniformly. Pt particles had a strong interaction with MoC due to MoC's capacity of dispersing and stabilizing metal particles. The performance test showed that conversion rate of dimethyl ether reached nearly 100% at 350°C, H_2 production rate and H_2 yield reached 1600 $mmol \cdot min^{-1} \cdot g^{-1}$ and 85% respectively. MoC had a strong ability to activate water at 350°C, and the O species produced by water could deactive MoC. SRDME reaction exhibited a stable behavior because of Pt-MoC/Al_2O_3 catalysts have the excellent resistance to oxygen posioning by coordinating the reactions in the SRDME reacting system.

571998 Assessing the Role of Zr-Al Interactions in Oxide-Grafted Zr Precursors for Ethylene Oligomerization Reactions

Assessing the role of Zr-Al interactions in oxide-grafted Zr precursors for ethylene oligomerization reactions

Galiya Magazova, Joshua D. Wright, Neha Mehra, William F. Schneider, and Jason C. Hicks*

Department of Chemical and Biomolecular Engineering, University of Notre Dame, 182 Fitzpatrick Hall, Notre Dame, IN 46556

Many Group IV transition metal catalysts have been reported to selectively dimerize or trimerize ethylene. However, these homogeneous catalysts 1) require an expensive co-catalyst, 2) are difficult to separate from reaction media and require significant reactor/reaction engineering for continuous operation in a flow reactor, and 3) are not thermally stable to temperatures > 300 ^{o}C in many cases. In this presentation, we will discuss our ongoing hypothesis that Group IV catalysts grafted on a solid support can grow ethylene chains effectively while avoiding liquid-based activating agents. We also hypothesize that the interaction between the grafted Zr species and the support is essential in providing a suitable active center for oligomerization.

It was recently reported that zirconium hydride catalysts supported on dehydroxylated silica-alumina are selective for 2–butene dimerization.[1] We will report our findings on the appropriate synthesis conditions required for the effective oligomerization of ethylene by grafted zirconium species in the absence of an added co-catalyst. Background oligomerization activity is observed with the silica-alumina support. However, the grafted and H_2 activated zirconium species enhance the reactivity by a factor of two. The product distribution follows a Flory-Schulz distribution, with dimers as the major products followed by trimers and heavier hydrocarbons. The ethylene consumption rate increases with an increase in the zirconium loading on the support, which indicates that zirconium centers play a role in activating ethylene. However, without the presence of Al in the oxide support, no activity is observed, which implies a cooperative or synergistic relationship between Zr and Al. Therefore, we will also show the correlation between the Zr/Al ratio and the catalytic performance.

References

(1) Szeto, K.C.; Merle, N.; Rios, C.; Rouge, P.; Castelbou, J.L.; Taoufik, M. *Catal. Sci. Technol.* **2015,** *5*, 4765.

570754 Mechanistic Understanding of the Role of Ga in the Dehydrogenation of Ethane on Ga/Al_2O_3 Catalyst

Owing to abundant shale gas resources, catalytic production of light alkenes from shale gas is an alternative to the conventional steam cracking process. Ga-modified Al_2O_3 has been found to be a potential catalyst for dehydrogenation of light alkanes to alkenes with selectivity of 85 % or higher.[1-2] However, the role of Ga is not mechanistically known especially since the Al_2O_3 support is also known to be active for alkane dehydrogenation.[3]

In this work, we use density-functional theory calculations and first-principles microkinetic modelling to investigate the mechanism of ethane dehydrogenation on Ga-modified γ-Al_2O_3 (110) surfaces and elucidate the role of Ga sites. We consider γ-Al_2O_3 (110) surfaces with Ga atoms either as surface adatoms (Ga/Al_2O_3) or as same-valence dopants ($GaAl_2O_3$). The Ga/Al_2O_3 surface is constructed by condensation of $Ga(OH)_3$ with surface hydroxyls of γ-Al_2O_3 (110). The doped $GaAl_2O_3$ surface is constructed in standard manner, *viz.* by replacing the undercoordinated and tetrahedral Al sites on the surface of Al_2O_3 with Ga.

The dehydrogenation of ethane entails sequential activation of two C-H bonds and H_2 recombination. We investigate these elementary steps on numerous sites of the two Ga-modified alumina (110) surfaces and compare with the energetics of the reaction on pure alumina (110). The first C-H bond activation invariably entails proton transfer to a basic O atom of the surface while engaging a vicinal Lewis acidic metal atom to stabilize the excess negative charge on the resulting carbanion ($C_2H_5^-$). Of the possible channels for the second C-H bond activation, proton transfer to a second O atom is energetically unfavorable as it would entail recombination of two protons. Instead, energetically more favorable pathways are those that involve either hydride formation on a Lewis metal center followed by proton-hydride recombination, or direct hydride-proton recombination without mediation by a Lewis acid site.

We find that on both Ga and Al sites, the transition states in the stepwise hydride pathway are lower in energy than in the direct hydride-proton recombination pathway. Reaction path analysis obtained from microkinetic model shows a higher reaction flux for the stepwise hydride-proton recombination pathway over the direct hydride-proton recombination pathway. Furthermore, although the Al sites are more Lewis acidic than the Ga sites, the Al-O-Ga bridges are more basic that the Al-O-Al ones. As a result, the Al-O-Ga sites stabilize surface intermediates more strongly. O atoms bridging the Ga and Al exhibit higher Lewis basicity than those bridging two Al atoms, resulting in significantly lower energy intermediates on Ga-O site pairs of Ga/Al_2O_3 when compared to the Al-O site pairs of Al_2O_3. Interestingly, Ga-O site pair stabilizes the intermediates more than the transition states. This observed stabilization of the intermediates due to the addition of Ga can potentially influence the reaction rates by changing the surface coverages of the reaction intermediates. The stable intermediates on Ga can also potentially generate new catalytic cycles thereby unravelling new active sites.

References:

1. Xu, B., Zheng, B., Hua, W., Yue, Y., and Gao, Z. "**Support Effect in Dehydrogenation of Propane in the Presence of CO2 over Supported Gallium Oxide Catalysts**" *Journal of Catalysis* 239, no. 2 (2006):470–477.

2. Chen, M., Xu, J., Su, F. Z., Liu, Y. M., Cao, Y., He, H. Y., and Fan, K. N. "**Dehydrogenation of Propane over Spinel-Type Gallia-Alumina Solid Solution Catalysts**" *Journal of Catalysis* 256, no. 2 (2008):293–300.

3. Rodemerck, U., Kondratenko, E. V, Otroshchenko, T., and Linke, D. "**Unexpectedly High Activity of**

Bare Alumina for Non-Oxidative Isobutane Dehydrogenation" (2016): 12222–12225.

559087 Effects of P and Metal Composition on the Performance of Mono- and Bimetallic Phosphides for Light Alkane Dehydrogenation

Catalytic dehydrogenation of light alkanes to alkenes is a practical route to create platform molecules for the production of polymers, chemicals, and fuels. Current catalysts used for this chemistry deactivate from coking and sintering and are based on expensive and/or environmentally unfriendly materials such as Pt and Cr, respectively. In this presentation we will discuss our recent results comparing Ni_2P and Ni for ethane dehydrogenation. Specifically, we have applied experimental and computational methods to understand the effect of P on the dehydrogenation ability of these materials. We are interested in transition metal phosphide materials because their geometrical and electronic properties can be tuned to provide a level of control over their catalytic properties.[1-4] Flow reactor studies, including time on stream measurements, have been performed with both ethane and propane as feeds to compare Ni_2P/SBA-15 and Ni/SBA-15. An alkene selectivity enhancement for both ethane and propane is observed for Ni_2P/SBA-15. Our results support the hypothesis that the incorporation of P provides a different preferred pathway for adsorbed ethylene on Ni_2P than Ni due to the presence of 3-fold Ni ensembles on Ni_2P, resulting in the improved selectivity and stability of Ni_2P over Ni. Diffuse reflectance infrared Fourier transform spectroscopy (DRIFTS) was used to interrogate the interaction of ethylene with Ni and Ni_2P surfaces. Ethylene adsorption on Ni shows features associated with the formation of carbon precursors on the surface, whereas Ni_2P shows reversible adsorption of ethylene. Lastly, we will discuss the effect of metal composition using supported bimetallic phosphides for propane dehydrogenation.

560706 Mechanism of Oxidative Dehydrogenation of Ethane at High Temperatures

The large availability of low-cost shale gas reserves has generated interest for the production of high value-added chemicals. Although methane is the main component in shale gas (often more than 85%), ethane is also available. Olefin production has traditionally happened via steam cracking, a high energy and capital intensive process. It is crucial to develop on-purpose catalytic approaches that will not only help to reduce the carbon footprint of the chemical industry, but also enable the development of energy efficient processes. The short contact time reactor is a promising concept for autothermal conversion of light paraffins into their corresponding olefins in compact reactors at high temperatures. It has been experimentally shown that ethylene yields, approximately 60%, comparable to steam cracking can be obtained by partial oxidation of ethane at high temperatures with a contact time of ca. 1 ms using a platinum-tin catalyst.[1] Different mechanisms from purely heterogeneous to combination of both heterogeneous and homogeneous and purely homogeneous reaction mechanisms have been proposed in the literature. Fundamental understanding of the reaction mechanisms is required for the optimization of the reactor and hence commercialization of the process. In this work, we use detailed modeling which combines both catalytic and gas-phase reactions to understand the process. The latter includes more than 100 species and over 2,500 reactions. For the former, comprehensive reaction network of 311 reversible elementary reactions steps and 147 surface and 31 gas species was generated via a Reaction Network Generator (RING).[2] Simulations show that ethane oxidation occurs that produces water, CO, and CO_2. In later stages, ethylene is produced initially via dehydrogenation of ethane and consumed via steam reforming and hydrogenolysis reactions. Detailed analysis of the results reveals the substantial role of gas-phase chemistry. Overall, the interplay between catalytic and gas phase reactions is demonstrated and recommendations for improved yield are made.

1. Bodke, A. S., Olschki, D. A., Schmidt, L. D., & Ranzi, E. L. I. S. E. O. (1999). High selectivities to ethylene by partial oxidation of ethane. Science, 285(5428), 712-715.
2. Rangarajan, S., Bhan, A., & Daoutidis, P. (2012). Language-oriented rule-based reaction network generation and analysis: Description of RING. Computers & Chemical Engineering, 45, 114-123.

575194 Self-Assembled Polymer Nanoreactors for Cascade Reactions

Confining catalyzed reactions to the hydrophobic core of micellar nanoreactors can facilitate organic phase reactions in a bulk aqueous environment with improved reaction kinetics due to enhanced localized reagent concentrations. The focus of this work was evaluating use of self-assembled core-shell polymer nanoreactors for the oxidation of benzyl alcohol to benzaldehyde followed by the reductive amination of benzaldehyde with 4-nitrophenol. Using polystyrene as a model hydrophobic microenvironment, the effect of base, oxidant, and reducing agent selection on benzyl alcohol oxidation were investigated. Potassium carbonate was found to provide higher benzyl alcohol conversion and benzaldehyde selectivity than organic, hydrophobic bases. Use of hydrogen peroxide as an oxidant resulted in a 10-fold increase in benzyl alcohol conversion compared to air. The effect of reducing agent, base selection, and relative reagent concentrations on the reductive amination of benzaldehyde will be discussed.

576236 Towards Safe Process Intensification of Partial Oxidation Reactors: Theoretical Insights into Selective Ethylene Oxide Formation on Ag Catalysts

Process intensification (PI) is a design approach that leads to significantly cleaner, safer, and more energy-efficient process technology. Catalytic partial oxidation processes are excellent candidates for PI efforts. These chemical conversions are responsible for a wide range of industrial chemicals, plastics, and intermediates used at extremely large scales. Moreover, these processes are relatively expensive to perform, and typically operate at high thermodynamic inefficiencies because these reactions are very temperature sensitive and also non-selective when even minor temperature excursions occur within the reactor. Specifically, the selective oxidation of ethylene to ethylene oxide (EO) will be the focus of our study. Approximately 26 million tons of ethylene oxide have been produced globally by selective oxidation of ethylene in 2013, and the scale of production increased to 34.5 million tons in 2016. [1] EO is mainly used to produce ethylene glycol, and it is produced commercially via an exothermic, vapor-phase reaction of ethylene and oxygen over Ag catalysts. There exists a strong driving force toward unselective reaction to complete combustion, and conversion must be kept low to ensure high selectivity to EO.

In the recent years, the experimental studies on the surface catalytic mechanism of EO process has attracted much attention. However, these experimental methods are still seeking the support from theoretical analysis. [2] In our study, periodic plane-wave Density Functional Theory (DFT) methods are used to analyze the reaction mechanism of the EO process on the Ag(111) surface facet with low coverage. Binding energies, reaction energies, and activation barriers have been calculated for both desired and undesired reaction pathways to allow for rational catalyst design efforts to improve selectivity and yield of commercial catalysts. Our DFT analysis will be discussed in the context of microkinetic modeling techniques which provide detailed information on surface coverage and gas phase species concentrations as a function of reactor process conditions and feed ratios. Key reactive surface species have been identified which are responsible for the branching ratio of ethylene oxide and acetaldehyde, which ultimately decides the selectivity of ethylene partial oxidation. Our results are consistent with previous kinetic modeling efforts in the literature that did not employ DFT analysis. [2] Lastly, our study demonstrates how fundamental theoretical investigations and multi-scale modeling techniques are currently impacting rational catalyst design and process intensification efforts in the chemicals industry.

[1] Transparency Market research (2011). Global Ethylene Oxide and Ethylene Glycol Market by Applications, by Geography, Raw Materials, Price Trends and Forecasts. Transparency Market research, TMRGL431.

[2] C. Stegelmann, N.C. Schiødt, C.T. Campbell, P. Stoltze, Microkinetic modeling of ethylene oxidation over silver, Journal of Catalysis 221 (2004) 630-649.

562590 Microwave Catalytic Reactor for Converting Light Alkane to Aromatics

Microwave (MW) catalytic processing has been applied to science discovery in several research fields. The potential of direct ethane dehydroaromatization under MW irradiation was explored. Meanwhile, the interaction between catalyst and microwave was studied through model simulation. It was observed that under MW irradiation, significant enhancement in ethane conversion to aromatics was achieved at 400 °C which otherwise would take up to 600 °C in a thermally heated fixed-bed reactor. The role of MW irradiation in activating ethane over the catalyst and the reaction mechanism were elucidated. The modeling results reveal the existence of hotspots and dipoles on the catalyst surface due to MW irradiation, which facilitate the activation of C-H bonds in ethane molecules and subsequently form reaction intermediates. This activation process has significantly lowered activation energy, allowing the reaction to proceed under lower bulk temperature with enhanced aromatic yield.

562099 Synthesis of High Purity Norbornene in a Microreactor Under Supercritical Conditions

Synthesis of High Purity Norbornene in a Microreactor under Supercritical Conditions

Xin Xu, Yin Wu, Bei Yuan, Zhen Yao, Kun Cao*

(State Key Laboratory of Chemical Engineering; College of Chemical and Biological Engineering, Zhejiang University, Hangzhou, P.R. China)

* Corresponding author: kcao@che.zju.edu.cn. This work was supported by the National Natural Science Foundation of China (No. 21676237).

Abstract:

Norbornene (NB) is obtained by Diels-Alder reaction of cyclopentadiene (CPD) and dienophile ethylene, while producing by-products tetracyclododecene (TCD) and tricyclopendiene(TCPD). CPD is unstable at room temperature and exists in the form of dimer dicyclopentadiene (DCPD), which is one of the main components in the by-product C5 fraction in the petroleum cracking process. In recent years, China's petrochemical industry has developed rapidly. Millions of tons of C5 fractions are produced each year without efficient way to be utilized. Studying the Diels-Alder reaction of CPD is conducive to promoting the fine and comprehensive utilization of C5, which is of great significance to the upgrading of China's petrochemical industry.

Since ethylene does not contain electron-withdrawing substituents and the energy level difference between the matching orbitals is large, the Diels-Alder reaction between ethylene and CPD is carried out at high temperature and pressure with serious safety concerns. For this reason, a microreactor has been employed in this work. Supercritical states were introduced by changing the reaction temperature and pressure. The two substances (E, CPD) participating in the main reaction are concentrated in the supercritical phase, and the side reaction reactants (CPD and DCPD or NB) are separated in different phases, which enhances the main reaction and inhibits the side reaction. The experimental results have shown that DCPD conversion rate and NB selectivity reached 99.39% and 99.34%, respectively, at 25MPa and 300°C.

Keywords: Norbornene; Microreactor; Supercritical state

567471 Preparation, Characterization, and Performance Evaluation of High Performance Mesoporous Materials Based Hydrodesulfurization Catalysts

A series of nickel and cobalt incorporated and molybdenum impregnated mesoporous materials based MCM-41 (metal-MCM-41) type of catalysts for hydrodesulfurization (HDS) was prepared by hydrothermal synthesis method. Catalysts with Si/Al = 10, Si/Mo = 10 and Si/Co and Si/Ni as 50 were prepared, characterized, and evaluated for HDS activity. Then molybdenum was impregnated on Ni-MCM-41 and Co-MCM-41 by incipient wetness impregnation method. MCM-41 based catalysts were characterized by X-ray diffraction (XRD), Brunauer-Emmett-Teller (BET) for surface area, pore volume, pore-size distribution, and temperature programmed reduction (TPR) techniques. Hydrodesulfurization activity of the prepared catalysts at atmospheric pressure and at different temperatures was evaluated for model compounds such as thiophene and benzothiophene in pulse reactor. HDS of dibenzothiophene as a model compound, dissolved in dodecane, was also performed in a batch autoclave reactor at high pressure and temperature. The results of catalysts evaluation for the hydrodesulfurization activity indicate that molybdenum impregnated metal-MCM-41 showed better performance as compared to commercial catalyst. Highest conversion, per mole of metal in hydrodesulfurization of thiophene and benzothiophene, was achieved in the case of Mo-CoMCM-41 catalyst. The results of catalyst performance were also compared with a sample of NiMo-Y zeolite (similar to the commercial catalyst) catalyst prepared in the laboratory.

571468 Isobutane Alkylation Kinetics with Mixed C4 Olefins Catalyzed By Sulfuric Acid

Isobutane Alkylation Kinetics with Mixed C4 Olefins Catalyzed by Sulfuric Acid

Ling Zhao, Weizhong Zheng, Piao Cao, and Weizhen Sun

State Key Laboratory of Chemical Engineering, East China University of Science and Technology, Shanghai 200237, China

The alkylate, has the advantages of high-octane number, low vapor pressure, and zero content of olefins and aromatics that allow it to become a desirable blending component for high-quality gasoline. It is well accepted that the isobutane alkylation reaction proceeds following the classic carbonium ion mechanism. Considerable efforts have been made to fully understanding the mechanism, providing the fundamental insight of the isobutane alkylation. However, due to simultaneously considerable reactions occurring in the alkylation system and the exist of more than three dozen isoparaffins which are hard to detect owing to the lack of the effective analysis method, few literature was devoted to the isobutane alkylation kinetics with sulfuric acid as catalyst to explain the formation pathway of several key components in alkylate, such as trimethylpentanes (TMPs), dimethylhexanes (DMHs), and heavy ends (HEs).

In this work the alkylation kinetics of isobutane with mixed C4 olefins using sulfuric acid as catalyst were investigated under conditions of industrial interest. The kinetic model was established on the basis of the carbonium ion mechanism with three key components in alkylate being measured, including TMPs, DMHs, and HEs. Meanwhile, the reaction pathways for polymerization of C4 olefins to produce HEs were modified and the isomerization reaction between butenes was fully considered in this model. Furthermore, the transition state and activation energy of the isomerization reaction between three C4 olefins were calculated using DFT method and further compared with the predicted results by the kinetic model. Using MD simulation, the diffusion coefficients of three C4 olefins were calculated to demonstrate their fast diffusion and self-assembly at interface before alkylation reaction.

According to the result of alkylation experiments carried out under the temperature range of 276.2-285.2 K, the concentration changes of three key components in alkylate, i.e. TMPs, DMHs, and HEs, were well predicted by the kinetic model with satisfactory agreement between experiments and model calculations. The rate constants of transformation rate of 1-butene into 2-butene and 2-butene into isobutene are two and one order of magnitudes larger than that of the corresponding reversible reaction, respectively. The reliability of the kinetic model was verified by the isobutane alkylation experiments with mixed C4 olefins with different compositions, confirming the good transferability of the model regardless of the species of butenes in the feed. Furthermore, MD simulations show that butenes can diffuse more easily than isobutane in the H_2SO_4, facilitating the fast polymerization reaction and the growth of heavy ends (HEs).

574400 Influences of Metal-Modification and Lamellar Zeolite Structure on Ethylene to Liquid Aromatics Conversion Reaction Using MFI Catalysts

Efficient conversion of short chain olefins such as ethylene to liquid aromatics using zeolites as catalysts is industrially important. The challenges of this reaction include the formation of undesired polyaromatic compounds and solid carbonaceous species (coke) resulting in deactivation of the catalyst. Addition of mesopores in zeolite structures provides higher mass-transfer for reactants and products and might increase the selectivity of the reaction for desired aromatics. Metal modification of the zeolites has been shown to decrease the coke formation rate in these catalysts. Therefore, the effects of meso-/microporous structure and metal modification with gallium or zinc on catalytic performances of lamellar MFI zeolites in ethylene conversion reaction to liquid aromatics were investigated. The characterization results showed that the zeolite structure is a pivotal factor for controlling the type, size, and distribution of metal dopant species forming on zeolite. Adding metal dopants to lamellar zeolite structures changed the acidity properties of the catalysts and increased the liquid aromatics yield and selectivity for mono-benzene alkylated aromatics while decreased the coke formation rate compared to their microporous commercial MFI analogies. The metal modified lamellar MFI zeolites as bifunctional catalysts open an avenue for converting biomass derived molecules to desired products.

575765 Assessment of the Coke Deposited on Lamellar Metal-Modified MFI Zeolite Catalysts in the Transformation of Ethylene to Liquid Aromatics

The effects of zeolite structure and its metal modification on the coke deposition during ethylene conversion reaction to liquid aromatics have been studied by combining the results of multiple analytical and spectroscopic techniques. Gallium or zinc-loaded lamellar MFI zeolites were used as catalysts. The nature, location, and amount of coke in the spent lamellar catalysts were determined by FTIR, UV-Vis, argon adsorption-desorption, and MS-temperature programmed oxidation (MS-TPO) and compared to their commercial microporous zeolite counterparts. FTIR-TPO has also been used to investigate the changes in the nature of the coke during the catalyst regeneration through combustion. The results show that the total amount of coke and the heavy coke fraction are reduced in the lamellar zeolite compared to its solely microporous analogue due to its meso-/microporous structure that provides lower diffusion limitations for bulky coke precursors and facilitates the formation of higher amount of the external coke. Metal-modification of the zeolite changes the acidity properties of the catalyst and as a result decreases the coke deposition rate during the reaction. Therefore, the coke formation on the zeolite catalyst and as a result its regenerability can be tuned by modulating both the textural and acidity properties of the zeolite structure.

573539 Synthesis, Characterization and Mechanistic Study of Structured Ga and Gapt Promoted ZSM-5 in Ethane Dehydroaromatization

The existence of ethane in shale gas pipeline causes condensation, and the separated ethane is often referred to as stranded gas. Therefore, the utilization of stranded ethane has drawn significant interest by researchers due to its economic value. Previously, we have investigated GaPt/ HZSM-5, Pt/ HZSM-5, and Ga/ HZSM-5 for ethane dehydroaromatization (DHA) in a conventional fixed bed reactor. The GaPt/HZSM-5 showed the best performance by having good aromatic and hydrogen selectivity. It also showed a higher activity with a slower deactivation than the other catalysts. In comparison an additional set of Ga catalysts were made using a different preparation method. The galloaluminosilicate ZSM-5 (GaMZ-5) was prepared by hydrothermal in-situ synthesis method. When compared with the GaPt/ HZSM-5 prepared by dry impregnation method, the GaMZ-5 shows higher ethane conversion and aromatics selectivity in ethane DHA using a fixed bed reactor. Similar trend was observed on platinum promoted GaMZ-5. Furthermore, these two catalysts system are tested in microwave reactor to compare their performance under electromagnetic excitations. Pyridine DRIFTS, surface area analysis, NH_3 temperature programmed desorption and other technologies were applied to characterize both dry-impregnated and hydrothermal in-situ synthesized Ga-Pt catalysts.

576266 Core-Shell Structured Mixed Metal Oxides for Oxidative Dehydrogenation of Ethylbenzene Under a Cyclic Redox Scheme

Catalytic dehydrogenation (DH) of ethylbenzene is a mature technology, accounting for 90% of the styrene produced worldwide. This technology relies on co-feeding superheated steam with ethylbenzene in the presence of a promoted, iron oxide based catalyst. The high reaction endothermicity and steam requirements lead to significant energy consumption and CO_2 emissions. In addition, styrene yield is limited by thermodynamic equilibrium. To address these limitations, we propose an autothermal redox oxidative dehydrogenation (RODH) scheme, which has the potential to both reduce the energy consumption and eliminate equilibrium limitation for ethylbenzene DH reactions. In RODH, the lattice oxygen from a mixed metal oxide redox catalyst is used to convert ethylbenzene and air into separate streams of styrene and oxygen depleted air under a cyclic redox process. In the present study, we report a core-shell structured mixed metal oxide, where shell material is both catalytically active and selective in DH reaction and core material can effectively provide active lattice oxygen to combust hydrogen into steam. As such, an ethylbenzene yield significantly higher than thermal equilibrium value for regular DH reaction was obtained. The RODH reaction was determined to be highly dynamic in terms of product distributions. In-situ XRD, XPS, LEIS and TEM were performed to determine the changes in redox catalyst surface and bulk properties and to link the catalyst surface/structural changes with its catalytic properties. DFT simulation was also conducted to reveal the reaction pathway. The simulation results are supported by experimental data obtained from isotope exchange experiments using FTIR-TPSR and mass spectroscopy. Overall, the promising performance using the core-shell redox catalyst supports the potential feasibility and attractiveness of the RODH scheme.

565230 Selective Hydrogenation of Acetylene to Ethylene over Bi-Metallic Catalysts

Selective Hydrogenation of Acetylene to Ethylene over Bi-Metallic Catalysts

Qingyuan Li[1], Yuxin Wang[1], George Skoptsov[2], Jianli Hu[1]*

[1]Department of Chemical and Biomedical Engineering, West Virginia University, Morgantown, West Virginia, 26506, USA
[2]H Quest Vanguard, Inc.

*Corresponding Author Email: john.hu@mail.wvu.edu

Selective hydrogenation of acetylene is one of the most effective ways to remove this undesired by-product generated during conventional petrochemical processes such as naphtha cracking. Concentration of acetylene reported in such processes is typically lower than 1%. Plasma processes have been shown to be effective for direct conversion of methane into hydrogen and acetylene, and hence offer new pathways to use methane as a chemical feedstock. In our study, a mixture of gas (CH_4, C_2H_2, C_2H_4, H_2 and argon) with concentration of C_2H_2 up to 15% was used as the feedstock. Initial experiment was carried out using conventional supported Pd-based catalysts, such as Pd/Al_2O_3 and $Pd\text{-}Ag/Al_2O_3$. About 100% conversion of C_2H_2 and 80% selectivity to C_2H_4 were obtained. To reduce the cost of catalyst, bimetallic Pd-Ni catalyst was synthesized and tested. Though the conversion of C_2H_2 was nearly 100%, the selectivity of C_2H_4 was very low. Efforts were focused on developing non-precious supported Ni-based catalysts including Ni/Al_2O_3, $Ni\text{-}P/Al_2O_3$, $Ni\text{-}Zn/Al_2O_3$ and $Ni\text{-}Ga/Al_2O_3$ catalysts. Under the same conversion, $Ni\text{-}Ga/Al_2O_3$ catalyst showed highest selectivity to ethylene, which was comparable to Pd-based catalyst. Besides, we found the selectivity to ethylene is significantly affected by the ratio of Ni to Ga and the catalysts preparation process. Specifically, with the increasing of Ga loading, the selectivity to ethylene was increased significantly. Furthermore, the calcination temperature and duration had strong impact on the performance of the as-synthesized catalyst.

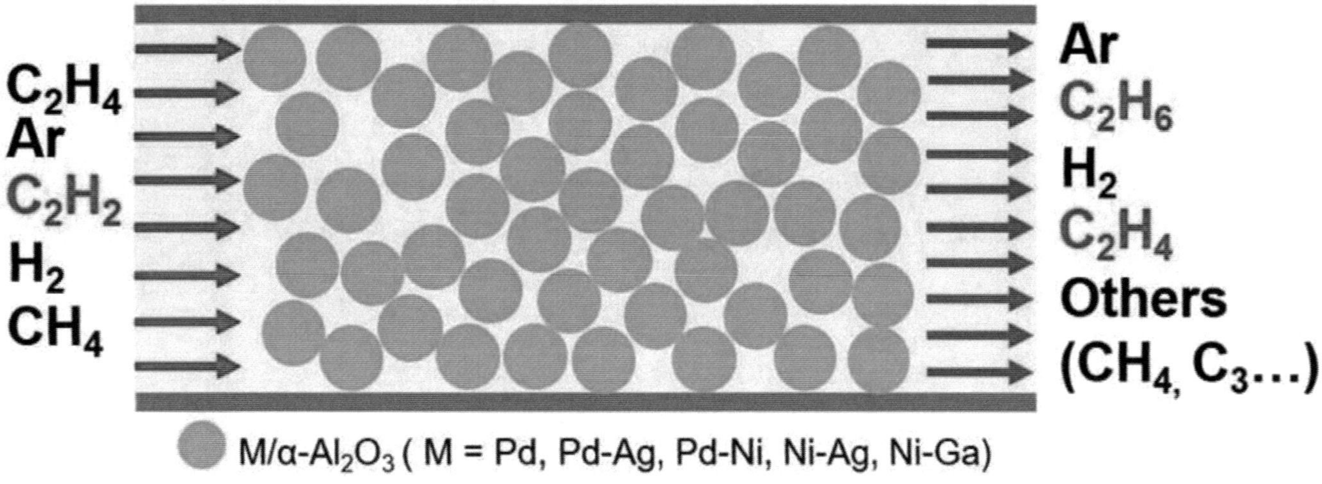

Acknowledgments:

H Quest Vanguard, Inc. is a privately held technology company, based in Pittsburgh, Pennsylvania, focused on the development and commercialization of novel hydrocarbon conversion technologies utilizing microwave plasma. This material is based on work supported by the Department of Energy, Office of Science through sub-award agreement no. 58495899with H Quest Vanguard, Inc. under the Prime Award DE-SC0017227 Phase II SBIR.

572577 Catalyst Development for Oxidative Coupling of Methane in a Gas-Solid Vortex Reactor

Catalyst development for oxidative coupling of methane in a gas-solid vortex reactor

Saashwath Swaminathan Tharakaraman[1], Guy B. Marin[1], Mark Saeys[1*]

1. *Laboratory for Chemical Technology, Ghent University, Technologiepark 125, 9052 Gent, Belgium;*

**Corresponding author: Mark.Saeys@ugent.be*

Oxidative coupling of methane (OCM) is a promising one-step reaction for the direct conversion of natural gas to C_2 hydrocarbons according to the following reactions.[1]

$$2CH_4 + O_2 \rightarrow C_2H_4 + 2H_2O \qquad \Delta H_r^0 = -282 \text{ kJ mol}^{-1} \qquad (1)$$

$$2CH_4 + 0.5O_2 \rightarrow C_2H_6 + H_2O \qquad \Delta H_r^0 = -177 \text{ kJ mol}^{-1} \qquad (2)$$

The presence of oxygen and the high reaction temperature can facilitate the overoxidation of reactants and products to CO and CO_2, typically limiting the C_2 hydrocarbon yields to below 30%.[2]

$$CH_4 + O_2 \rightarrow CO_2 + 2H_2O \qquad \Delta H_r^0 = -810 \text{ kJ mol}^{-1} \qquad (3)$$

The high reaction exothermicity makes isothermal operation challenging on an industrial scale, and quasi-adiabatic reactors are therefore evaluated. Experimental studies have shown that adiabatic operation can be sustained using feed at ambient temperature by employing catalysts that combine high activity and good selectivity to C_2 products.[3] Bifurcation studies indicate that reactors with good thermal back mixing combined with limited species back mixing can potentially improve C_2 yields.[4] The gas-solid vortex (GSVR) reactor satisfies both of the aforementioned conditions and hence emerges as a reactor technology to enhance C_2 yields.[5]

In the GSVR, a dense rotating bed of solids is obtained by tangential injection of gas through 1 mm slots at velocities exceeding 100 m s^{-1}. Centrifugal forces on the solid particles counteract the drag forces, resulting in the formation of an uniform bed of solids with high gas-solid slip velocities. The very high gas-solid slip velocities intensify interfacial transfer of mass, energy and momentum, allowing a reduction in gas phase residence time.

The high reaction temperature, the high solid velocities (exceeding 6 m s^{-1} [5]) and the low space times in the GSVR also require the development of catalysts with a combination of high mechancial and thermal stability and high catalytic activity. Among OCM catalysts, strontium promoted lanthanum oxide has been shown to be highly active for OCM reaction.[6] In this work, we report the development of rare earth oxide based catalysts with a high mechanical stability under elevated temperatures and examine the role of strontium as a catalyst promoter.

To elucidate the promoting effect of Sr on lanthanum oxide, a series of Sr/La_2O_3 catalysts with varying Sr loadings (1 to 11 wt%) were prepared[7] and the catalytic performance of the catalyst was evaluated in a fixed bed reactor operating under intrinsic kinetic conditions.[8] Figure 1 shows the kinetic performance of the tested catalysts. It can be seen that the promotion of lanthanum oxide with strontium enhances the CH_4 activity of the catalyst. Additionally, the selectivity towards C_2 hydrocarbons is improved when the Sr loading is more than 1 wt%. This effect is more pronounced when the reaction temperature exceeds 800 °C. For all the tested catalysts, the C_2 selectivity increases with temperature and stabilizes above 800 °C. Interestingly, addition of 1 wt% Sr slightly reduces the selectivity of the catalysts. To understand the promoting effect of Sr on lanthanum oxide, the catalysts were characterized using XRD, CO_2-DRIFT, CO_2-TPD and XPS showing a connection between the high temperature CO_2 desportion temperature (Figure 2) and the catalyst selectivity.

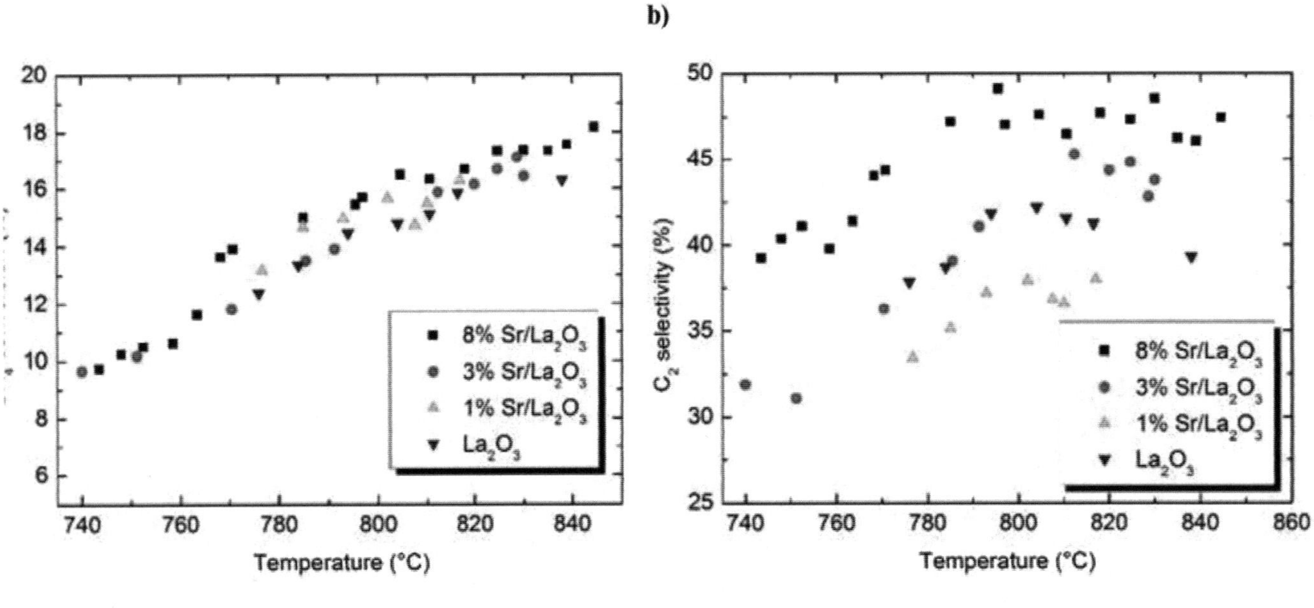

Figure 1: a) Activity and b) C_2 selectivity of unsupported Sr/La_2O_3 catalysts vs. temperature. Other experimental conditions: P_{total} =1.8 bar (50% He), inlet CH_4:O_2 ratio =4:1, space time = 0.2 kg_{cat} s mol^{-1} $_{CH4,0}$.

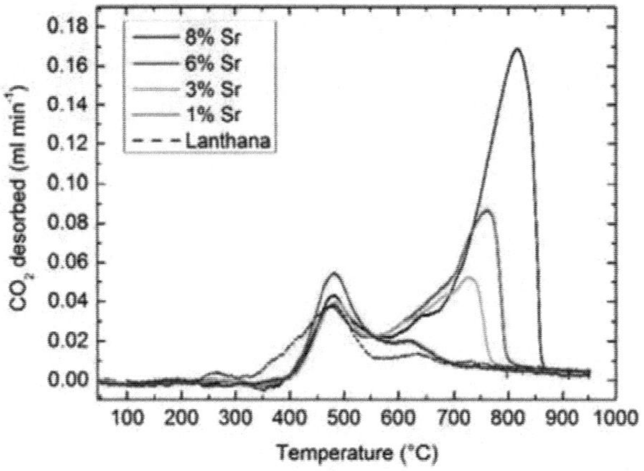

Figure 2: CO_2 TPD profiles of various lanthana and Sr-promoted La_2O_3 catalysts. Experimental conditions: CO_2 desorption was carried out in a He flow rate 60 ml min^{-1} and temperature ramp of 10 °C min^{-1}. CO_2-adsorption was carried out 50 °C.

In the next step rare earth oxide based catalysts were synthesized and their mechanical and thermal stability was tested in the GSVR at a temperature of 700 °C with an inlet air flow of 26 kg h^{-1}. A stable rotating bed of solids with a thickeness of 8 mm was sustained for 1 h at these high temperatures. The introduction of standard Sr/La_2O_3 catalysts in the GSVR resulted in immediate attrition and entrainment.

Parallely, the new catalyst was tested in a fixed bed reactor under quasi-isothermal conditions at very short space times. It shows a reasonable CH_4 conversion ranging at low space times and a good C_2 selectivity. Experiments to test this promising catalyst under reactive OCM conditions in the GSVR are currently underway and will be reported.

Acknowledgements:

This project has received funding from the European Union's Horizon 2020 research and innovation programme under grant agreement No 680777.

The assistance of ir. Manuel Nunez, ir. Shekhar Kulkarni and ir. Sepher Madanikashani while running GSVR experiments is gratefully acknowledged.

References:

1. Keller, G. and M. Bhasin, *Synthesis of ethylene via oxidative coupling of methane: I. Determination of active catalysts.* Journal of Catalysis, 1982. **73**(1): p. 9-19.

2. Zavyalova, U., M. Holena, R. Schlögl, and M. Baerns, *Statistical analysis of past catalytic data on oxidative methane coupling for new insights into the composition of high-performance catalysts.* ChemCatChem, 2011. **3**(12): p. 1935-1947.

3. Sarsani, S., D. West, W. Liang, and V. Balakotaiah, *Autothermal oxidative coupling of methane with ambient feed temperature.* Chemical Engineering Journal, 2017. **328**: p. 484-496.

4. Vandewalle, L.A., I. Lengyel, D.H. West, K.M. Van Geem, and G.B. Marin, *Catalyst ignition and extinction: A microkinetics-based bifurcation study of adiabatic reactors for oxidative coupling of methane.* Chemical Engineering Science, 2018.

5. Gonzalez-Quiroga, A., P.A. Reyniers, S.R. Kulkarni, M.M. Torregrosa, P. Perreault, G.J. Heynderickx, K.M. Van Geem, and G.B. Marin, *Design and cold flow testing of a Gas-Solid Vortex Reactor demonstration unit*

for biomass fast pyrolysis. Chemical Engineering Journal, 2017. **329**: p. 198-210.

6. Alexiadis, V., J. Thybaut, P. Kechagiopoulos, M. Chaar, A. Van Veen, M. Muhler, and G. Marin, *Oxidative coupling of methane: catalytic behaviour assessment via comprehensive microkinetic modelling.* Applied Catalysis B: Environmental, 2014. **150**: p. 496-505.

7. Deboy, J.M. and R.F. Hicks, *Kinetics of the oxidative coupling of methane over 1 wt% SrLa2O3.* Journal of Catalysis, 1988. **113**(2): p. 517-524.

8. Berger, R.J., E.H. Stitt, G.B. Marin, F. Kapteijn, and J.A. Moulijn, *Eurokin. Chemical Reaction Kinetics in Practice.* CATTECH, 2001. **5**(1): p. 36-60.

575454 Impact of Ni and Mg Loadings on Dry Reforming Performance of Pt/Ceria-Zirconia Catalysts

This work investigates the impact of metal loadings on a series of low temperature Pt/$Ce_{0.6}Zr_{0.4}O_2$ dry reforming catalysts. These catalysts containing either Ni and/or Mg were synthesized using the coprecipitation and incipient wetness method. Characterization techniques such as powdered X-ray diffraction, N_2-pysisorption analysis, H_2-temperature programmed reduction (H_2-TPR) and CO_2-temperature programmed desorption (CO_2-TPD) were used to identify the catalyst properties. The H_2-TPR profiles showed that increasing Ni content enhanced the reducibility of the metal oxide support while the opposite effect was observed with increasing Mg. The CO_2-TPD indicated that the presence of Mg improved the catalyst basicity. The catalyst activity for dry reforming was tested using temperature programmed experiment and the results obtained indicated that Ni enhances CH_4 conversion at lower temperatures while Mg decreases CH_4 conversion and improves the product selectivity (H_2: CO). Among the catalysts investigated, the optimal loading of Ni and Mg were found to be 2.7 wt% and 0.50 wt%, respectively, when tested without the other. The coking rate of the 2.7Ni-CZO after a 10 h time on stream (TOS) study was 5.5E-4 g/gcat.hr whereas no coke was observed for the Mg-based catalyst. These results led to the synthesis of a combined catalyst (2.7Ni-0.50Mg-CZO). Coke formation was not detected on the combined catalyst due to the presence of Mg which improved the catalyst basicity, favored CO_2 dissociation and provided reactive oxygen species for carbon oxidation. The results obtained with the 2.7Ni-0.50Mg-CZO catalyst sample showed that the presence of both Ni and Mg in the catalyst composition is essential to obtain a balance between reactant conversions, H_2:CO ratio, and catalyst stability. The temperature required for 10% CH_4 conversion of the combined and isolated metal catalysts followed the trend; 2.7Ni-CZO < 2.7Ni-0.50Mg-CZO < 0.50Mg-CZO. The activation energy of the samples followed the trend; 2.7Ni-0.50Mg-CZO < 2.7Ni-CZO < 0.50Mg-CZO and were lower or comparable to those in literature. These samples gave low temperature activity with 10 % CH_4 conversions at temperature < 500 °C and this indicates their potential application in an intensified process involving both reforming and Fischer Tropsch synthesis.

561163 Kinetics of O_2 Activation over Cu-Exchanged Zeolites: Implications for Partial Methane Oxidation

The direct low temperature conversion of methane to methanol is considered a "holy grail" within the catalysis community, as current technologies are indirect routes that involve steam reforming of CH_4 to form syngas which is subsequently converted to CH_3OH over Cu-based catalysts [1]. Cu-exchanged zeolites (e.g., CHA, MFI, MOR) have received renewed attention as candidate materials for partial methane oxidation, often requiring stoichiometric cycles involving activation in O_2 at high temperatures (~723 K) followed by reaction with CH_4 at low temperatures (~473 K) to form chemisorbed methoxy-derived species that can be extracted by H_2O [1]. Currently, Cu-exchanged zeolites do not give acceptable combinations of rates and selectivities for continuous catalytic operation, because the requirements for high selectivities (i.e., low temperature CH_4 activation, stoichiometric step-wise reactions) are orthogonal to those that enable high rates (i.e., high temperature, isothermal continuous catalytic reaction). This is due, in part, to the lack of understanding of each step along the step-wise process (i.e., O_2 activation, CH_4 reaction, methanol extraction with H_2O). Several studies have investigated the reaction between O_2-activated Cu-zeolites with CH_4 using *in situ* UV-vis spectroscopy, which have shown relatively fast kinetics (complete reaction in <0.5 h, 473 K) in comparison to the time required for the entire step-wise process (>6 h) [2]. The lack of precise details regarding the Cu-based active sites and mechanisms for O_2 activation over Cu-zeolites have prevented extending empirical findings regarding optimizing CH_3OH production on one material to broader classes of materials. Here, we utilize *in situ* resonance Raman spectroscopy and UV-Visible spectroscopy to monitor the kinetics of O_2 activation over Cu-exchanged zeolite CHA (Cu-CHA) to develop a mechanistic understanding for this process.

The activation of O_2 over Cu-CHA produces bis(μ-oxo) dicopper(II) and μ-(η^2:η^2)peroxo dicopper(II) intermediates that are Raman active and are often invoked as the intermediates responsible for CH_4 activation [2-5]. Multivariate curve resolution-alternative least squares was used to extract high-quality Raman spectra and time-resolved contributions from transient measurements to investigate the kinetics of O_2 activation. The kinetics for the reaction between O_2 and dimeric Cu active sites (Cu_2) closely resemble a first-order reaction, because in all cases O_2 is in large excess. The rates of O_2 activation are independent of temperature above 698 K; yet, decrease exponentially with temperature below 698 K, which suggests that there may be two distinct reference states of Cu_2 that depend strongly on temperature. On-going work seeks to extend this methodology for studying O_2 activation to other zeolitic frameworks and explore the dependence on additional reaction conditions (e.g., O_2 pressure, H_2O pressure).

568208 Effect of Promoter Loading on Iron Supported Fischer-Tropsch Synthesis Catalyst at High-Temperature

Effect of Promoter Loading on Iron Supported Fischer-Tropsch Synthesis at High-Temperatures

Yajing Chang*, Sreya Mariya Seby, David Weber, Babu Joseph, John N. Kuhn

Department of Chemical & Biomedical Engineering, University of South Florida

*Presenting author. Tel: +1 727-488-4670; E-mail address: yajingchang@mail.usf.edu

ABSTRACT

To create an intensified process which combines reforming and FTS reactions, a high-temperature FTS catalyst is needed [1]. Based on prior research, iron (Fe) based catalysts have wide temperature adaptability (230-350°C). Selection of the support can also impact the performance of iron-based FTS catalysts. Also, various promoters have been used to improve the catalyst performance.

In this work, the effects of promoters on Fe based catalysts were investigated at high temperatures and low pressures. The catalysts were prepared by incipient wetness impregnation with 10-15 weight % of Iron. The catalysts were characterized using XRD, TPR, N_2 Physisorption, and SEM. The Fe/Al_2O_3, Fe/SiC, and Fe/SiO_2 were activated at 350 °C with hydrogen and FTS reaction was conducted at 290°Cwith syngas having 1:2 CO:H_2 ratio. Fe/SBA-15 was activated at 430°C with syngas and then tested for FTS reaction at the same temperature with syngas. To improve the selectivity and conversion, copper, potassium, and manganese have been investigated as promoters for the Fe-SBA15. Catalysts with manganese promoter showed a decrease in carbon dioxide selectivity from 76% to 50%, an increase in C2-C4 selectivity from 9.6% to 30% and an increase in C5+ selectivity from 0.2% to 2.2%. The catalysts with potassium promoter showed a negative effect on CO conversion. Copper promoter increase the selectivity of the carbon dioxide. Currently, we are evaluating the effect of the different amount of Mn loading on Fe with various supports.

[1] M. Johns, P. Collier, M.S. Spencer, T. Alderson, G.J. Hutchings, Combined steam reforming of methane and Fischer-Tropsch synthesis for the formation of hydrocarbons: A proof of concept study, Catalysis Letters 90(3-4) (2003) 187-194.

[2] Zhang J, Abbas M, Chen J. The evolution of Fe phases of a fused iron catalyst during reduction and Fischer–Tropsch synthesis. Catalysis Science & Technology. 2017;7(16):3626.

564210 Mechanistic Origins of the High-Pressure Inhibition of Methanol Dehydration Rates in Small-Pore Acidic Zeolites

Mechanistic Origins of the High-Pressure Inhibition of Methanol Dehydration Rates in Small-Pore Acidic Zeolites

John R. Di Iorio,[1] Alexander J. Hoffman,[2] Claire T. Nimlos,[1] Steven Nystrom,[2] David Hibbitts,*[2] Rajamani Gounder*[1]

[1] *Charles D. Davidson School of Chemical Engineering, Purdue University, 480 Stadium Mall Drive, West Lafayette, IN 47907, USA*

[2] *Department of Chemical Engineering, University of Florida, 1030 Center Drive, Gainesville, FL 32611, USA*

Turnover rates of methanol dehydration to dimethyl ether (415 K, per proton), a well-studied probe of acid strength and confinement in zeolite catalysts [1], increase as the fraction of protons in paired configuration (Al-O(-Si-O)x-Al; isolated: x = 3, paired: x = 1, 2) increases in small-pore, 8-membered ring (8-MR) zeolites (e.g., CHA, AEI) [2]. Precise kinetic interpretation of the origin of turnover rate differences between isolated and paired protons in small-pore zeolites, however, is convoluted by the onset of kinetic inhibition at high methanol partial pressures (>10 kPa, 415 K), a phenomenon not observed in medium (e.g., MFI) or large-pore (e.g., BEA) zeolites. High pressure inhibition of methanol dehydration rates (per proton) diminishes at elevated reaction temperatures (>450 K) and coincides with the disappearance of IR vibrational bands characteristic of methanol trimers and larger clusters (~3370 cm^{-1}) measured from *in situ* IR spectra (343-473 K, 10 kPa CH_3OH). These results suggest that the formation of large methanol clusters (>2 CH_3OH per cluster) inhibits methanol dehydration rates and that inhibition is not caused by internal mass transport limitations, which would become exacerbated at elevated reaction temperatures. Methanol adsorption isotherms on CHA zeolites (293 K) indicate that micropore filling (~2 CH_3OH per proton) occurs at similar gas-phase CH_3OH Gibbs free energy, which accounts for differences in the stability of equilibrated surface species at different temperatures, to those calculated at the onset of dehydration rate inhibition (415 K, per H^+). These observations are supported by IR spectra measured during methanol dosing onto CHA zeolites (0.01-3 CH_3OH/H^+; 293 K) that shows the appearance of methanol clusters at coverages >0.7 CH_3OH/H^+, which increase in intensity with increasing methanol coverage. Collectively, these observations implicate methanol clusters as inhibiting species in methanol dehydration catalysis in small-pore CHA zeolites, but kinetic interpretation of measured dehydration rates is still convoluted by multiple candidate rate laws that can be derived from different rate limiting steps, which functionally conform to the measured rate data. Density functional theory is used to calculate the Gibbs free energies of various methanol complexes (1-16 CH_3OH per cluster) confined at protons within CHA as a function of temperature and methanol pressure, which indicate that methanol trimers are preferentially stabilized in CHA at methanol partial pressures that are consistent with experimentally observed inhibition (>10 kPa, 415 K). DFT-predicted methanol dehydration rates (415 K) are computed and compared with prior experimental and computational observations to further distinguish between plausible reaction mechanisms. These results highlight the critical role that the zeolite framework plays in stabilizing reactive intermediates and transition states, and how different frameworks preferentially stabilize inhibitory species.

References:

[1] Carr, R. T., Neurock, M., Iglesia, E., *J. Catal.*, **2011**, *278*, 78-93

[2] Di Iorio, J. R., Nimlos, C. T., Gounder, R., *ACS Catal.*, **2017**, *7*, 6663-6674

561027 Fischer-Tropsch Synthesis on Natural Zeolite-Supported Fe Catalysts

Natural zeolites have fascinating properties that makes them attractive as supports in Fischer-Tropsch synthesis (FTS) because of their ubiquitous in nature and relatively low costs. Detailed mineralogical knowledge and profound characterization of natural zeolites are essential for fitting chemical composition to use. Si/Al ratios are very import as well as the other contaminates. In this study natural zeolite called clinoptilolite (a type of zeolite found in South Africa) was used as a catalyst support. This support proved to be effective for low-temperature Fischer–Tropsch synthesis (LTFTS) aiming liquid fuel production, as well as chemical feedstock. Synthesis of this highly active catalyst was by loading of iron on clinoptilolite through the wet impregnation method. The prepared catalyst was then characterized by XRF, BET surface area analyzer, XRD and SEM. The catalyst was then loaded into the reactor and reduced with hydrogen prior to FTS. The effects of its use as support in FTS were investigated in a fixed bed reactor. From the XRF results the molecular ratio SiO_2/ Al_2O_3 of the Clino-support was 5.86. The average crystal size of the particles from both HRTEM and XRD ranged 9.83 -11.649 nm and around 10.10nm for used and fresh catalyst. It was found that the CO consumption rate of 1.02 x 10^{-4} mol/min.gcat of which 7.24 x 10^{-5} mol/min.gcat was the actual Fischer Tropsch rate with the remaining 2.93 x 10^{-5} mol/min.gcat consumed by the WGS reaction. The product distribution of the gaseous phase analysed were more olefinic than paraffinic. The product distribution for our experimental condition agreed with a one alpha ASF distribution with an alpha value of 0.86. These findings may permit the development of new effective support materials, which are cost effective for clean fuel production via FTS process.

567824 A One-Step Method of Durene Synthesis Directly from Syngas Using Integrated Catalyst of $Cu/ZnO/Al_2O_3$ and Co-Nb/HZSM-5

In recent years, the demand for durene has been increased as the raw material of polyimide. However, traditional synthesis methods of durene require a large number of steps with lower efficiency. In this study, a novel synthesis method of a bifunctional integrated catalyst had been utilized, in which the methanol synthesis catalyst $Cu/ZnO/Al_2O_3$ (CZA) and the aromatization catalyst Co-Nb/HZSM-5 were combined, to synthesize durene from syngas in one step. Syngas was converted to methanol and dimethyl ether (DME) followed by the production of different aromatics with a high concentration of durene through the shape selectivity of zeolite. The optimal reaction conditions of 330°C, 4MPa, $200h^{-1}$, 3% Co loading, CZA/HZSM-5=3:2 and a unique pseudo-physical coating (PC) integrated method were applied to increase durene selectivity up to 39.71% in aromatics. Different techniques such as XRD, BET, XPS and NH_3-TPD were employed to characterize different catalysts, and a variety of experiments were proceeded to compare different routes of methanol and FT synthesis. These results explain that the high selectivity of tetra-methylbenzene (tetra-MB) was due to the alkylation of trimethylbenzene (tri-MB) induced by methanol on HZSM-5 surface. This work demonstrated that the integrated catalyst could be advantageous for direct conversion of syngas to tri-MB and finally to the desired durene and it would open up the new avenues in future perspectives.

569514 Metal-Containing ZSM-5 Catalysts for the Dry Reforming of Methane: Exploring Synergistic Effects

Metal-containing ZSM-5 Catalysts for the Dry Reforming of Methane: Exploring Synergistic Effects

Megan Hoffman, Raj Thakur, Justin Smith, Carlos A. Carrero*

Department of Chemical Engineering, Auburn University, Auburn, Alabama 36840

Meh0088@auburn.edu

The abundance of natural gas in the U.S. brought by the development of new fracking technologies for its extraction has triggered a notable renaissance of the U.S. chemical manufacturing industry. Natural gas abundance and availability has made the United States one of the lowest-cost chemical producer in the world, and currently it is an appealing feedstock for producing value-added chemicals and fuels. The selective activation of C-H bonds in the most abundant component in natural gas, methane, is one of the greatest challenges in catalysis. Methane can be converted into raw materials or commodity chemicals as follow: i) indirectly via syngas or ii) directly via, for instance, selective oxidation or dehydroaromatization. Along with methane, emphasis has been placed on the utilization of CO_2 to produce chemicals and fuels do to their abundance and negative impact to the environment. Both CH_4 and CO_2 are the main contributors of the greenhouse effect. As a result, Dry Reforming of Methane (DRM) is an attractive route to convert CH_4 and CO_2 into syngas with an H_2/CO ratio of 1:1, which is the optimal ratio to produce liquid hydrocarbons through the Fischer-Tropsch process. However, among other factors, the multiple undesired side reactions such as methane thermal degradation, carbon monoxide disproportionation, and water gas shift has, have limited the potential application of DRM at an industrial scale. These reactions adversely affect the H_2/CO ratio and the overall catalyst's performance. The major challenge to address, as a result of the mentioned side-reactions, is the rapid catalyst deactivation due to the formation of coke.

In this study, we aim to improve DRM using a series of Fe/Ni/ZSM-5 catalysts. First, various methodologies have been used to prepare these materials trying to determine the most important variables influencing the final bulk and surface structure of the catalyst. We use various characterization techniques such as N_2-physisorption, X-Ray diffraction (XRD), X-Ray photoelectron spectroscopy (XPS), Scanning Electron Microscopy (SEM), and Raman spectroscopy. Reactivity tests are performed in a home-made reaction setup equipped with various mass flow controllers, a stainless-steel reactor, a tubular furnace with a temperature controller, and a gas chromatograph (GC) for the analysis of the products. Steady-state kinetics has shown a stable catalytic activity and selectivity towards syngas for Fe/Ni/ZSM-5 compared to the Ni/ZSM-5 catalyst, indicating that Fe is suppressing the formation of coke during the reaction. We aim to determine the variables and the optimal catalyst composition to enhance the catalytic performance of conventional Ni-based catalysts. We will also perform in-situ/opernaro Raman studies to compliment the kinetic data and stablish a more conclusive structure-reactivity/selectivity/stability relationship.

574043 Long-Term Stability of Modified Ferrite Catalysts for High-Temperature Water-Gas Shift Reaction at Elevated Pressures

Long-term stability of modified ferrite catalysts for high-temperature water-gas shift reaction at elevated pressures

Devaiah Damma and Panagiotis G. Smirniotis*

Chemical Engineering, College of Engineering and Applied Science, University of Cincinnati, Cincinnati, OH 45221-0012, USA

*panagiotis.smirniotis@uc.edu (P.G. Smirniotis)

The water gas shift (WGS) reaction is an important operation for H_2 production from the CO-rich gas stream (synthesis gas) that generated from fossil fuels, agricultural and forestry biomass, and municipal wastes via the gasification route. Especially, the high temperature and high-pressure WGS reaction in a single stage catalytic membrane reactor is a promising technology because of the advantages of faster reaction rate and enhanced H_2 transport through the membrane that is necessary to obtain complete CO-conversion and total H_2 recovery. Industrially, Fe–Cr-based spinels had been proven to be a promising catalyst for HT-WGS. However, excess steam (H_2O/CO ˜ 5) is used in the industry to maintain the catalyst stability by preventing over-reduction of the catalyst and suppressing side reactions (methane formation). Therefore, the commercial catalyst could not be suitable for the WGS reaction based on gasification process because of the presence of CO-rich gas in the reactant stream that leads to catalyst deactivation. In this context, some catalysts have been developed to satisfy the requirements for the HT-WGS reaction using gasification-derived synthesis gas. However, most of the reported catalysts tested at atmospheric conditions, which is not representative for the industrial application (20–30 bar) regarding deactivation of the catalyst and side product formation (especially methane). Development of catalysts at high pressures and temperatures can meet the stringent requirements of membrane reactors operating conditions. The present study is aimed at developing the catalysts for HT-WGS at elevated pressures which represent the membrane reactor conditions. For this purpose, we have reported various modified ferrite catalysts prepared by co-precipitation method for WGS at elevated pressures and high temperatures. Our catalysts showed a stable performance in terms of CO conversion without significant loss in the activity for a long period of time. Generally, methanation reaction can cause the catalyst deactivation during the WGS reaction at elevated pressures. Interestingly, no significant formation of methane was observed in the present investigation. The catalysts in fresh and spent states were also characterized by using various structural and surface techniques in order to establish structure-activity relationships. These interesting results will be discussed in the presentation.

561400 Size Effect of Co_3O_4-Supported Pd Nanoparticle on the Catalytic Activities of CO Oxidation and Water-Gas Shift

It has been reported that nanoparticles with high surface-to-volume ratio dispersed on the support could efficiently enhance the catalytic activity. However, the fundamental understanding of this phenomenon is still not clear. In this study, we report an investigation of the size effect on CO oxidation and low temperature water-gas shift (LT-WGS) reaction over a series of Pd/Co_3O_4 catalysts. We controlled the size of Pd nanoparticle loaded on Co_3O_4 by changing the calcination temperature, and thus could tune the particle sizes from 2.5 nm to 10.6 nm. For both reactions, the catalysts with the smallest Pd particles showed the best activities.

In order to reveal the underlying reason for these results, we employed both experimental and theoretical approaches. Firstly, for CO oxidation, we experimentally found that smaller Pd particles have higher turnover frequency (TOF) compared to the larger particles. To elaborate this observation, DFT calculations were performed with taking Pd(211) and (111) facets as the representatives of small and large particles, respectively. The results suggested that CO adsorption is more favorable on Pd(111) than (211), indicating that CO poisoning effect is stronger for larger particles, thereby causing the lower activity. In contrast, Pd(211) has the lower reaction barrier of CO oxidation. Based on these results, it is thought that the active sites for CO oxidation are supposed to be on noble metal; since CO dominantly adsorbs the flat surfaces such as Pd(111) in the large particles, there are not much sites for O_2 to adsorb dissociatively to form CO_2. In contrast, the smaller particles provide both flat plates for CO adsorption and enough edges and/or vertexes for O_2 dissociative adsorption to result in the higher activity in CO oxidation.

On the other hand, for WGS reaction, we found that the TOF normalized by total area of interface was almost constant regardless of the particle size. This implies that the better WGS activity with smaller particles can be attributed to more metal-support interface area that is the active site in WGS. Therefore, we can conclude that for both reactions, Pd particle size effect showed the same activity trend, but their active sites are actually different. We believe that our results can provide a potential approach for developing highly efficient environmental catalysts.

569225 Hydrogenation of Dimethyl Oxalate to Ethylene Glycol over Mesoporous Silica Supported Copper Catalysts

Hydrogenation of dimethyl oxalate is the final reaction step in syngas to ethylene glycol (StEG) process. Mesoporous silica supported copper catalysts are prepared via ammonia evaporation method for the conversion. The effect of preparation temperature on the structure of the catalysts and the catalytic performance are investigated. Various characterizations are conducted to elucidate the origin of the effect of temperature.

570402 High Yield to Lower Hydrocarbon in FT Synthesis: Based on Melting Effects on Znfe Spinel

Direct conversion of syngas represented one of the most important routes bridging the coal- or natural gas-based industry between the oil- based industry. In our research, the sodium promoted zinc iron spinel was developed, which is effective in converting syngas to lower hydrocarbon (mainly olefins). 95% conversion was achieved and the products were mainly in lower hydrocarbon, in which the selectivity below C5 reached 70%. The selectivity of methane was less than 17% and ratio of olefin to alkane reached 8. Combined with the characterization of DSC and in-situ XRD, it is concluded that the catalyst melts during the reaction and therefor achieves a higher mass transfer rate. According to the theoretical calculation, the sodium promotion changes the surface electronic environment, which facilitates the desorption of olefins during the chain growth process. Achieving higher reaction conversion and low carbon olefin yields provides favorable conditions for olefin enrichment and utilization.

578347 Fischer-Tropsch Synthesis in Microscale-Based Reactor, Experimental and Mathematical Modeling

Fischer-Tropsch Synthesis (FTS) is a surface-catalyzed polymerization process in the presence of cobalt, iron or ruthenium that converts synthesis gas into a wide range of long chain hydrocarbons (olefins, paraffins, isomers), oxygenates, CO_2 and water. The adsorbed synthesis gas (H_2 and CO) molecules react on the surface of a catalyst to form monomers (CH_x). These monomers act as building blocks that further react to give a wide range of products. The products of the reaction include jet fuel, naphtha and diesel.

The goals of this study is to design and construct a microchannel reactor that demonstrates a low selectivity to methane and high yield of naphtha and diesel range hydrocarbons. In addition, a modeling developed to simulate the experimental system. The model incorporates chemical reaction kinetics to represent all elementary reaction kinetics and process model includes the fluidics, mass transport and reaction kinetics for the microscale reactor.

Our investigation of FTS in a microreactor examines the effects of operating conditions including pressure, temperature, residence time and flow ratio on the performance of the reactor. In this contribution, we will present the effort to improve low methane and high range hydrocarbons selectivity, and will also discuss some important features of the microreactor FTS system. Included is a computational model of the system and its outputs.

571406 Kinetic Monte Carlo Study on Multi-Facet Copper Catalytic Surface for Efficient Methanol Synthesis

In the research field of C1 gas chemistry, there have been many studies on the production of more valuable materials such as methanol, ethanol, and acetic acid which are less dependent on crude oil price fluctuations. Methanol is an important compound among the products of C1 gas conversion. Methanol is generally produced over Cu/ZnO catalyst. However, it is controversial that which surface is mainly activated among the Cu (111), Cu (110), Cu (100) or Cu (211) during the reaction process. In this study, we've designed the catalytic surface where methanol synthesis reaction takes place to capture the phenomena more naturally. The catalytic surface consists of a combination of several configurations of surface rather than a single configuration. The kinetic Monte Carlo (kMC) simulation using rate constants on each surface of the catalyst was performed to validate the tendency of adsorption, desorption, and surface reaction with experimental data and to propose which surface the main active surface is. As a result of this study, a well-designed morphology of surface structure will be suggested and it will shed light on manufacturing more effective Cu/ZnO catalysts structure.

Reference

Grabow, L. C., and M. Mavrikakis. "Mechanism of methanol synthesis on Cu through CO2 and CO hydrogenation." *Acs Catalysis* 1.4 (2011): 365-384.

Ovesen, C. V., et al. "Kinetic implications of dynamical changes in catalyst morphology during methanol synthesis over Cu/ZnO catalysts." *Journal of Catalysis* 168.2 (1997): 133-142.

562981 Methane Pyrolysis in Molten Metal/Salt Two-Phase Reactors

Methane is the lowest cost source of hydrogen today. Although steam methane reforming is the most cost effective technology for hydrogen production today, it generates more than 5 kg of CO2 byproduct for every kg of hydrogen. Catalytic methane pyrolysis, catalyzed by conventional heterogeneous catalysts, produces hydrogen without CO2 but the carbon coproduct poisons the catalyst. This can be avoided by using a molten metal alloy catalyst. Performing methane pyrolysis on molten metal catalysts can eliminate catalyst coking. In molten metal bubble columns, the separation of the produced carbon in a continuous process is possible as the low-density solid carbon floats on the surface of the liquid. However, using molten metals, it is difficult to produce high purity carbon without metal contamination. In this study, the production and separation of solid carbon from pyrolysis of methane with reduced metal contamination is achieved using a two-phase bubble column with a catalytic molten NiBi alloy covered with molten salt. The immiscibility of the molten metals and molten salts provided the conditions to produce a multiphase metal/salt/carbon system to separate carbon from the molten liquid efficiently. In the rising gas bubbles containing reactant and product gases with metal vapors, the solid carbon gas can entrain NiBi vapor from the liquid metal reactor region into the molten salt phase where it can condense as a metal film. It is proposed that when the gas-filled bubbles pass through the metal-salt interface, the surface metal film ruptures and metal particles are produced which return to the dense liquid metal phase and the low-density carbon is lifted to the top of the column. HRTEM analysis of the carbons revealed structural differences between carbon generated in the NiBi single-phase and NiBi/salt two-phase reactors. The carbons produced in the metal/salt two-phase reactors contained carbon black aggregates and carbon nanotubes structures while in the NiBi reactor graphitic carbon is observed. The metal contamination in the carbon produced from NiBi/salt two-phase reactor is less than 5 wt. % compared to 83 wt. % obtained from reactors with only NiBi.

573784 DFT Modeling of Liquid Solvent Effects on the Catalytic Surface Reactions in Fischer-Tropsch Synthesis

For decades, the process of Fischer-Tropsch synthesis (FTS) has been studied and applied to produce valuable hydrocarbon products, such as gasoline, from synthesis gas (Syngas). The early commercial designs of this heterogeneous catalytic process were developed under gaseous-phase conditions. Later the process was industrialized under liquid-phase conditions to overcome the drawbacks of gaseous process, such as catalyst deactivation and high yield of methane, as one of the main undesired products. Majority of the literature discussed that the observed enhancements in the liquid-phase FTS are principally due to a better heat distribution in the reactor and the selective solvation properties of the liquid solvent. We have performed first-principle density functional theory (DFT) calculations to study the elementary reactions of FTS on a flat cobalt catalyst surface. A rigorous methodology has been applied from both energetic and entropic aspects to evaluate the reaction rate parameters of the elementary surface reactions to address the challenges in determining the dominant mechanism of reactants' activation and hydrocarbon chain growth*, which have been a matter of debate between the computational and the experimental communities in this field. Ultimately, using the continuum-based conductor-like screening model (COSMO), liquid hexane solvent was included in our DFT study. We found that presence of the liquid solvent also affects the surface reactions; particularly in the intermediate reactions with oxygenate species**. The solvent effects on the kinetics and thermodynamics of these elementary reactions enhanced additional chain growth pathways besides the dominant mechanism, resulting in a product distribution that is in agreement with experimental observations.

References:

*Asiaee, A. and Benjamin, K. M. *Molecular Catalysis* 436 (2017) 218-227

** Asiaee, A. and Benjamin, K. M. *Molecular Catalysis* 436 (2017) 210-217

563266 An Effective Microkinetic Modelling Strategy for Direct DME Synthesis from Syngas over Hybrid Cza/FER Catalyst

In this study, an effective microkinetic model approach was applied to direct synthesis of dimethyl ether (DME) from syngas over Cu-ZnO-Al_2O_3/FER (CZA/FER) which is a hybrid bifunctional catalyst consisting of CZA and ferrierite (FER). The possible reaction network for syngas to methanol reaction suggested by Grabow and Mavrikakis[1], composed of CO and CO_2 hydrogenations, and water-gas shift (WGS) reactions was considered, and both associative and dissociative pathways were included for methanol to DME reaction. In particular, the reaction energetics for methanol to DME reaction were calculated based on Møller-Plesset (MP) perturbation theory which is one of quantum chemistry post-Hatree-Fock ab initio methods in order to consider dispersion forces occurring in the reaction system. Meanwhile, pre-exponential factors to be fitted to experimental data were selected on the basis of sensitivity analysis, reducing the computational load without calculating the partition functions of translational, vibrational and rotational modes. Based on relative reaction rates, the most likely reaction pathways were searched out with the parameter-fitted model, and the rate-limiting step was suggested by calculating the degree of rate control. Also, appropriate operating conditions were proposed considering the effects of pressure, temperature and feed composition on DME synthesis rate. Conclusively, the effectively developed microkinetic model could be utilized to obtain valuable information on the reaction mechanisms, kinetics and optimal operating conditions for direct synthesis of DME from syngas over the hybrid catalyst.

REFERENCE

1. Mavrikakis M, Grabow LC. Mechanism of methanol synthesis on Cu through CO_2 and CO hydrogenation. *ACS Catal*. 2011;1:365-384.

563316 Combined DFT and Microkinetic Study of Dry Reforming of Methane on Ni and B Promoted Ni Surfaces

Dry reforming of methane (DRM) has incited significant academic and industrial attention in the past couple of decades. Although Ni-based catalysts have shown good activity for DRM, deactivation due to carbon deposition is a serious concern. Several strategies are proposed, including doping by other metals and metalloids, to improve the stability of Ni, and hence Boron doped Ni is investigated in this work as a potential catalyst for DRM. The effect of Boron doping on the catalytic activity and selectivity can be evaluated by identifying the elementary reaction steps and by computing reaction energetics in the conversion of CO_2 and CH_4 on Ni and NiB surfaces. In this work, combined DFT and microkinetic modelling are performed to identify the dominant reaction pathways and kinetically relevant steps of the DRM reaction system on Ni and NiB surfaces. The accuracy of DFT methods is crucial and hence benchmarking study was first performed to assess the accuracy of DFT functionals and rPBE-vdW functional, with a correction of 28 kJ/mol for gas phase CO_2 energy, was found to predict the adsorption and reactions accurately. We considered a detailed reaction network involving multiple CO_2 dissociation routes, CH_4 dissociation routes, side reactions (H_2O formation, Boudouard reaction) and desorption of products on Ni and NiB surfaces. Our calculations suggest that both Ni and NiB share similar CO_2 (direct dissociation to CO*) and CH_4 (CH_4 dehydrogenates to C* and then C* is oxidized by OH* to COH* which dissociates to CO*) activation routes. Compared to Ni, the CO_2 activation barrier on NiB surface is higher but the barriers in CH_4 activation routes are significantly lower on NiB. Besides this, CO formation by Boudouard reaction (CO_2*+C*→2CO*) also has a lower barrier on NiB compared to Ni. The dominating reaction pathway on Ni involves direct dissociation of CO_2* to CO* and atomic O*, and CH_4 dissociation to C*, followed by OH* oxidation and finally COH* decomposition to form CO* and the rate-limiting step is C* oxidation by OH*. On NiB, the dominant reaction pathway is CH_4 dissociation to C* and then C* reacts with CO_2 to produce CO*. Here, the rate-limiting step is C* reacting with CO_2*. Subsequently, the DFT calculated energies were used to derive kinetic rate constants of the elementary reaction steps. These rate constants were employed to develop a microkinetic model and the DFT results were further extended to calculate conversions and coverages under reaction conditions. From the current study, we establish that the presence of B alters the DRM reaction pathway with a considerable reduction in the rate limiting barrier. Consequently, by making the alternate pathway which consumes the formed carbon fast enough, this NiB surface can make a better catalyst. Thus, we believe that NiB can be a potential catalyst that can prevent deactivation with an improvement in the catalytic activity for the DRM process.

582744 Determination of Kinetics for Methane Dry Reforming Using Plasma

Determination of Kinetics for Methane Dry Reforming Using Plasma

Presenting author: Cassiane Oliveira Martins (cassiane.oliveira-martins@univ-pau.fr)

Co-authors: Frédéric Marias (frederic.marias@univ-pau.fr), Jean-Paul Robert-Arnouil (jprarnouil@europlasma.com), Stéphanie Moyal (smoyal@europlasma.com)

Keywords: Reactor Design, Kinetics, CFD, Modeling, Plasma Reactor, Methane Dry Reforming

The steel industry is crucial for global development. The worldwide steel demand is forecasted to reach 1,735 Mt in 2019 and all scenarios foresee strong additional growth over the coming decades, considering the rapid development of the BRICS countries. Although in Europe the forecasted growth is lower than the developing countries, since most of its infrastructure is already established, the European steel demand still increases proportionally with population growth.

Since steel production is mainly made from iron ore using carbon-based reduction processes, the steel industry is reasonably confronted with the major challenge of climate change. It is estimated that nowadays the steel industry is responsible for 4-6% of global anthropogenic greenhouse gas emissions and for 25-30% of industrial emissions.

In this scenario, the IGAR (Reducing Gas Injection) project has been devised by European steel industry, in particular by ArcelorMittal, to reduce the carbon dioxide emission. This project is based on the proposition of directly injecting reduction gases, consisting primarily of carbon monoxide and hydrogen, in the blast furnace to substantially reduce its solid carbon requirement and consequently the steel industry CO_2 emissions.

The IGAR project considers that steel gases, produced in large quantities in the integrated plants, will be treated by a plasma reactor where the dry reforming of methane ($CH_4 + CO_2 \rightarrow 2H_2 + 2CO$) will take place. The reducing gas produced by the plasma reactor will be then injected into the blast furnace, as shown by the figure below.

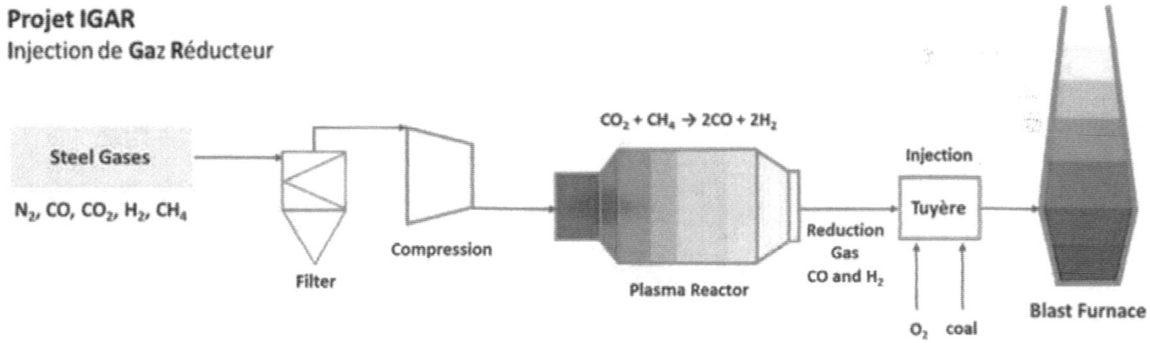

Hence, the IGAR Project has two main objectives:

To test on a pre-industrial scale a plasma conversion system for iron and steel gases to ensure their heating and conversion by the methane dry reforming.

To validate, under industrial conditions, a tuyère specifically designed for the injection of gas produced with pure oxygen and coal, under the operating conditions of the blast furnace process.

The plasma reactor will be installed in ArcelorMittal plant in Dunkerque-France and might be used during a long test test period for pre-industrial validation. In order to design the reactor several studies are being carried out including the computational fluid dynamic (CFD) model of the reactor and a CFD model of a small-scale pilot that will be constructed by Europlasma Company. This pilot will be installed in Europlasma research and development plant in

Morcenx-France.

The CFD model of the pilot was developed by using the ANSYS 19.2 software package and its geometry was constructed in the ANSYS Space Claim Direct Modeler 19.2, which is presented by the figure below. Due to confidential issues, the pilot dimensions are not shown in the figure.

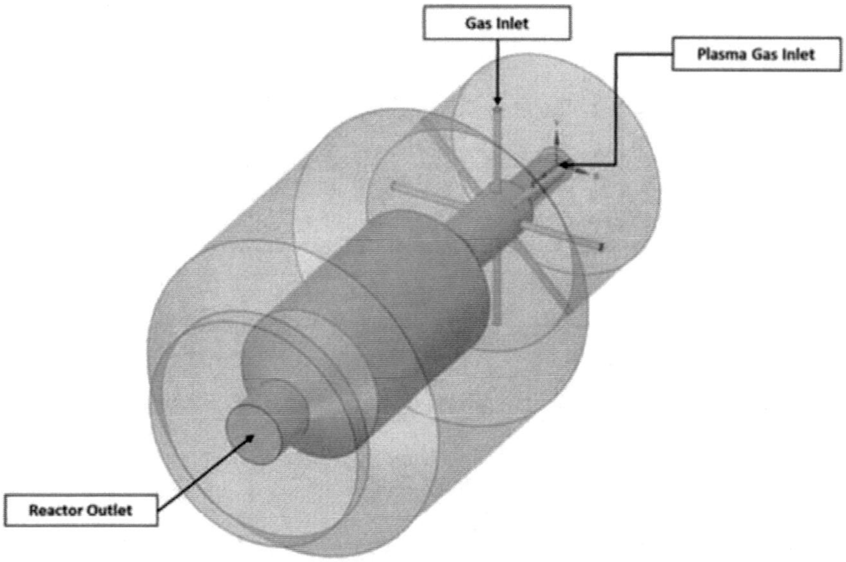

The mesh of the CFD model was composed by 7.362.606 tetrahedral elements and a system of 12 homogeneous reactions was considered to describe the reactional dynamic in the CFD simulation, as presented below:

1	$C_{(s,soot)} + H_2O \rightarrow CO + H_2$	7	$CH_4 + 2O_2 \rightarrow CO_2 + 2H_2O$
2	$CH_4 + H_2O \rightarrow CO + 3H_2$	8	$CH_4 + 0.5O_2 \rightarrow CO + 2H_2$
3	$H_2 + 0.5O_2 \rightarrow H_2O$	9	
4	$CO + 0.5O_2 \rightarrow CO_2$	10	$2CH_4 \rightarrow C_2H_2 + 3H_2$
5	$CO + H_2O \rightarrow CO_2 + H_2$	11	$CH_4 \rightarrow C_{(s,soot)} + 2H_2$
6	$CO_2 + H_2 \rightarrow CO + H_2O$	12	$C_2H_2 \rightarrow 2C_{(s,soot)} + H_2$

One of the challenges of the IGAR project involves the determination of the reactions kinetics considering the plasma environment. The CFD model was developed by using the Arrhenius kinetics, announced in the scientific literature for a non-plasma environment, for each one of these 12 reactions. However, considering the presence of ions and free radicals in the plasma, the intensity of these reactions might be stronger; therefore, this model should be adjusted.

To adjust the kinetic model, experiments will be carried out by using the small-scale pilot that will be constructed in the

Europlasma R&D plant. As a first step, in order to have a better comprehension of how to modify the kinetic parameters to adjust de model, a study was performed to analyze the effect of their variation on the molar fraction of the main products (CO_2, H_2 and $C_{(s)}$) at the outlet of the pilot.

The kinetic parameter chosen for this study was the activation energy (E) of Arrhenius equation. The approach used for the study was to impose a variation on the activation energy of the most important reactions of the system and then run a series of CFD simulations of the pilot's 3D model.

Thus, to follow this approach it was firstly necessary to determinate the most important reactions, the method used for this purpose is described as follow:

The weight of each reaction on the overall conversion of every species was calculated by using the following expression:

$$SR_{i,j} = \frac{\int_V |v_{i,j}| r_j dV}{\sum_{j=1}^{12} \int_V |v_{i,j}| r_j dV}$$

Where:

- V : plasma torch volume,
- $v_{i,j}$: stoichiometric coefficient of the species i on the reaction j,
- r_j : reaction rate.

This calculation considered the reaction rate results obtained with the 3D CFD simulation of the pilot by using the kinetic parameters issued from the literature for a non-plasma environment for each one of the 12 reactions.

For each reaction the species average weight was calculated by the following expression:

$$\overline{SR_j} = \frac{\sum_{i=1}^{n} SR_{i,j}}{n}$$

In which n is the number of species of the system.

The number of species with weight $SR_{i,j}$ greater than zero was counted.

The reactions with largest species average weight $\overline{SR_j}$ and highest number of species with weight $SR_{i,j}$ greater than zero are ranked as the most important in the system. Thus, by using this method it was possible to determine the three reactions that most affect the results in the pilot outlet.

The methodology chosen to study the effect of a variation in the activation energy of each one of these three reactions on the simulation results at the pilot outlet was the 2^k factorial design, which means that for each reaction the effect of a variation of the activation energy will be studied in two levels: high and low.

Taking into account the presence of ions and free radicals in the plasma, the intensity of the reactions might be greater, which can be considered as a lower activation energy in the Arrhenius equation. Following this reasoning, the activation energy announced in the literature in the non-plasma environment was considered as the high level. The low level, in turn, was chosen considering a result obtained for a preliminary study that have shown that the lower the activation energy value used in the pilot's CFD simulation, the higher the necessary number of iterations to achieve convergence. Hence, the first 2^3 factorial design was made with a variation of 10% between the low and high levels ($E_{low}^{j} = 0{,}9E_{high}^{j}$).

Therefore, the complete factorial design was made with 8 CFD simulations contemplating all the possible combinations of the activation energy of the three most important reactions in the low and high levels. The response variable was considered as the main products' molar fraction at the reactor outlet and the main effects were used to construct a regression model such as:

$$\hat{X}_i^{FD,RO} = \beta_0 + \beta_1 x_1 + \beta_2 x_2 + \beta_3 x_3 + \beta_{12} x_1 x_2 + \beta_{13} x_1 x_3 + \beta_{23} x_2 x_3$$

Where:

$\hat{X}_i^{FD,RO}$: molar concentration of the specie i on the reactor outlet,

β_k : regression coefficients,

x_k : coded variable calculated by $x_k = \frac{E^j - (E_{low}^j + E_{high}^j)/2}{(E_{high}^j + E_{low}^j)/2}$,

FD : Factorial Design,

RO : Reactor Outlet.

The interest of this regression model is to create an expression, as simple as possible, to associate the activation energy of the main reactions directly to the molar fraction of the main products at the reactor outlet. Thus, with the results that will be obtained during the experimental campaigns at the pilot, it will be possible to have a first estimation of the activation energy values that must be used in the CFD kinetic model to obtain simulation results that match the experimental ones. The adjusted kinetic model will be then used to design the industrial reactor that will be installed in the ArcelorMittal plant in Dunkerque-France.

This research was supported by the Investissement d'Avenir program funded by the Agence de l'environnement et de la maîtrise de l'énergie (ADEME).

564071 Core-Shell Pt/Al_2O_3@Cu/ZSM-5 Catalyst for Ammonia Slip Catalyst in Diesel Aftertreatment

Core-Shell Pt/Al_2O_3@Cu/ZSM-5 Catalyst for Ammonia slip catalyst in Diesel Aftertreatment

Rajat Subhra Ghosh*, M. P. Harold*, Thuy Le*, J.D. Rimer*, and D. Wang**

**Dept. of Chemical and Biomolecular Engineering, University of Houston, Houston, TX 77204-4004, USA*
***Cummins Inc., 1900 McKinely Av., MC50197, Columbus, IN 47201, USA*

Introduction

The ammonia slip catalyst (ASC) is an essential aftertreatment unit in modern diesel vehicles. The purpose of ASC is to prevent the release of unreacted NH_3 leaving a selective catalytic reduction (SCR) unit. The state-of-the-art ASC for selective oxidation of NH_3 to N_2 has a dual-layer architecture comprising a Pt/Al_2O_3 bottom layer (PGM) and a Cu/Fe-exchanged zeolitic (M-Z) top layer [1]. The dual-layer structure enables a high N_2 selectivity at high NH_3 conversion. It does so by converting NO, one of the main oxidation products of the PGM, through reaction with counter-diffusing NH_3. While recent studies demonstrate the effectiveness of the dual layer design [2], further advances are needed to reduce the PGM loading and ASC volume while enhancing low temperature activity.

In this work, the dual-layer concept is scaled down to the level of a single catalyst particle with the intent to meet the above-mentioned challenges. A core-shell (CS) particle comprising a PGM core and a M-Z shell, specifically Pt/Al_2O_3@Cu/ZSM-5 (Figure 1a), was synthesized. Reaction testing of the CS catalyst showed enhanced low temperature activity without compromising N_2 selectivity on comparing with a physical mixture of catalyst having same Pt and Cu loading. The CS catalyst had comparable light-off to a Pt/Al_2O_3 catalyst having 3x Pt loading. The enhanced activity is attributed to the de-stabilization of platinum oxide (PtO_x) exposing the more active metallic platinum in the CS catalyst, supported by independent temperature programmed reduction (H_2-TPR) measurements.

Materials and methods

Core-shell particle synthesis: A Pt/Al_2O_3[Pt(0.05 w/w)/Al_2O_3] catalyst serving as the core particle was synthesized by incipient wetness impregnation using tetraamine platinum nitrate. Synthesis of the shell involves seeding of core particles, followed by secondary growth [3]. The CS particle was sequentially ion exchanged with NH_4NO_3 and $Cu(NO_3)_2$ to obtain the Cu-ZSM-5 shell layer.
Reaction testing: A bench flow fixed-bed reactor, described elsewhere [2], was used to evaluate the CS powder catalyst. The feed containing 500 ppm NH_3 + 5% O_2 (balance Ar) had a space velocity of 280k h-1. A FT-IR was used to measure effluent NO, NO_2, N_2O, and NH_3.

Results and Discussion

Catalyst characterization: The Si/Al ratio (SAR) of the zeolite obtained from XPS and EDX data was found to be ~30. Figure 1b shows a SEM image of prepared CS particles which are 40-50 microns in diameter with a shell thickness of 1-1.5 microns. From EDX, the Cu weight percent in the shell is ~ 2.9. Powder XRD confirmed the MFI lattice structure of the shell.

NH_3 oxidation: NH_3 oxidation was carried out on 4 samples: Pt(0.15)/Al_2O_3 [P(0.15)-A], Pt(0.05)/Al_2O_3 [P(0.05)-A], physical mixture of P(0.05)-A + 'x' wt% Cu(2.9%)-ZSM-5 [CuZ, x = 10, 20], and the core-shell Pt/Al_2O_3@Cu(2.9%)-ZSM-5 [CS]. The NH_3 oxidation light-off temperatures (T50) were as follows: P(0.15)-A(260°C); P(0.05)-A(310°C); P(0.05)-A+CuZ(~300°C) and CS (250°C), as shown in Figure 1d. The T50 of 250°C for the core-shell sample was 50°C lower than the samples having the same amount of Pt. Remarkably, the CS light-off temperature was ~5°C less than the sample containing 3x higher Pt(P(0.15)-A). Moreover, the CS achieved 100% conversion at much lower temperature (325°C) than P(0.05)-A (450°C). The lower light off for CS catalyst was due to the increase in NH_3 oxidation activity in the core of the catalyst. Reaction testing for steady state ammonia oxidation over different stages of synthesis (core→seeded core→CS) as seen in Figure 1f. The seeded core (P(0.05)-AS) had a light off at 215°C, which is 95°C lower than P(0.05)-A. The CS catalyst with fully developed shell of ZSM-5 lowered the light off to 250°C due to the diffusion resistance which is still lower than P(0.05)-A. The P(0.05)-A and P(0.05)-AS have platinum dispersions of 39 and 25 %, respectively. The dispersion measurement does not explain the lowering of light off. H_2-TPR showed the formation of H_2O peaks at lower temperature for P(0.05)-AS compared to P(0.15)-A (Figure 1g). The oxides of Pt are destabilized in seeded core, making it more susceptible to reduction compared to P(0.05)-A. This exposes more metallic Pt in the case of seeded core making it more reactive [4]. The destabilization effect of platinum oxides is due to the modified metal-support interactions [5,6]. The combined factors of destabilizing effect of platinum oxides, suppression of platinum oxide formation due to the modification of γ -Al_2O_3 surface leads to the higher activity of the seeded core.
N_2 selectivity: The CS catalyst shows a very high N_2 selectivity, achieving a maximum N_2 yield of 95% at 300°C, with negligible N_2O formation (yield < 3%) (Figure 1c). At higher temperature (500°C), N_2 yield drops to ~ 90 % which is much higher than P(0.05)-A + CuZ physical mixture (Figure 1e). The high N_2 selectivity of CS catalyst is due to the SCR chemistry taking place in the Cu-ZSM5 shell [2].

These findings not only confirm that the dual-layer ASC concept can be achieved at the single particle level, but there is an apparent added advantage of lower light-off at reduced PGM loading not obtained previously.

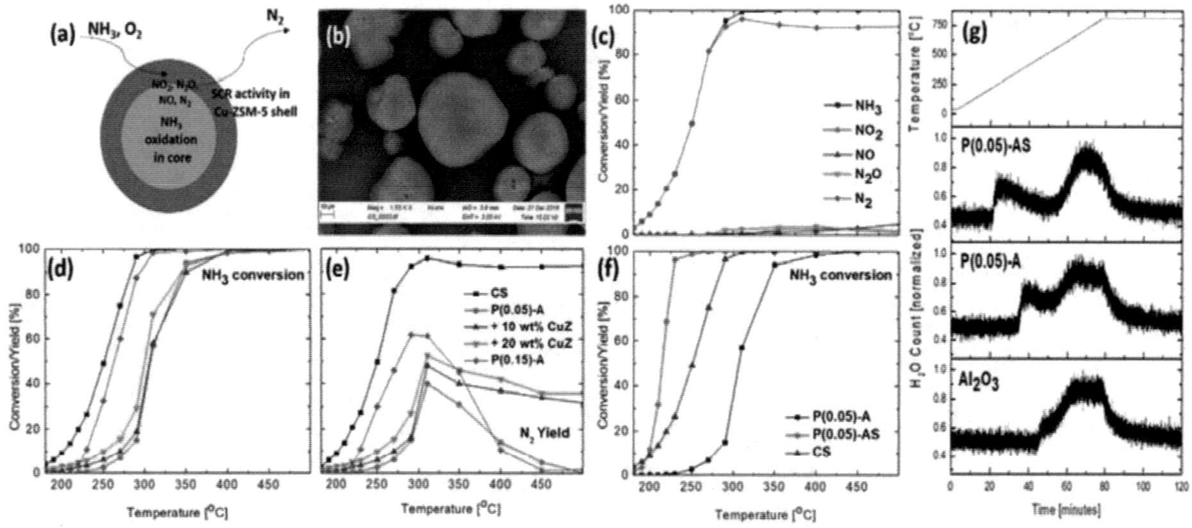

566878 The Role of Noble Metals (M=Rh,Pd,Ag,Au,Pt) Promoting Active Oxygen Enhancing Soot Oxidation in CeO2 Vacancy : Experiment with Tight Contact and a DFT Study

A series of CeO2-based catalysts were investigated for soot oxidation with O2. The initial activity of the catalyst is determined by the number of catalyst–soot contact points rather than the catalytic surface, while the catalyst stability is closely related to the concentration of surface oxygen vacancies. Especially, many of literatures are dealing oxygen vacancy with active oxygen such as superoxide (O2-) and peroxide (O2-) species. When the CeO2 surface is partially reduced, a surface oxygen vacancy forms, and two Ce4+ ions of substrate are reduced to Ce3+ones. However, the original reasons for 'how the active oxygens are generated' is not clearly revealed yet. In this study, we suggest the main cause determining superoxide and peroxide in surface oxygen by comparing the ground state energy with DFT calculation. Moreover, we found outstanding catalytic performance in Rh/CeO2 catalyst with tight contact thus, we suggest the linear tendency between calculated binding energy of active oxygen with soot oxidation rate experimentally measured. Also, we performed surface bader charge transfer to observe the role of metal as electron supplier at catalytic surface.

568004 An Experimental Verification of Pressure Drop for Removal of VOC

New trends in the design of individual pieces of equipment in chemical plants include a gradual decrease discharges of pollutants and investment and operating costs together with an increase in the overall effectiveness of the plant. For the proper and reliable design of such integrated multifunctional apparatuses, it is necessary to have a suitable and experimentally verified calculation tools for their design and analysis. Our work is focused on experimental acquisition of pressure drop data a obtaining mathematical dependencies of catalyst pressure losses for flue gas emission cleaning equipment. This paper presents the results of the first stage of our experimental verification focused on the analysis of the selected type catalysts for removal VOC (Volatile organic compounds) from a hydraulic characteristic point of view.

Organic Compounds (VOCs) are considered harmful on their own due to their negative respiratory, carcinogenic, neurological or other effects. Industrial waste gases contain many different organic substances. For this reason, industrial waste gases are purified that contains VOCs before are released to the atmosphere. Legislation sets and monitors VOC emission limits in most countries around the world. The photochemical reaction of VOCs with NOx creates harmful compounds that can lead to adverse effects of climate and public health.

The removal of VOC pollutants is accomplished by several possible technological processes. High VOC concentrations and large flow rates are commonly disposed of by thermal oxidation, eg in the chemical industry. This is burning the organic vapours with natural gas in the combustion chambers. In many cases of the industrial off gas treatment containing VOC is possible to change the technology from thermal oxidation into catalytic oxidation and reduce operating costs in terms of difference in the savings because the reduced natural gas consumption and the price of used catalyst.

There are many types of ceramic catalyst carriers, from the most well-known HoneyComb type honeycomb structures to various bulk beds such as Pall, Lessing or Raschig rings, various saddles such as Berl or Intalox, Interpack bodies, or spheres. All these catalyst supports produce a significant pressure loss in the process. Flue gas or propulsion fan creates a pressure drop to overcome resistance in pipes and equipment. These resistances are generated by friction or local and filling (catalyzers). Pressure losses need to be minimized and optimized from an energy and thus economic point of view. For this purpose, it is also necessary to perfectly describe pressure losses by mathematical equations. There are several mathematical equations and methods for calculating pressure losses for different types of catalyst supports that need to be verified by data obtained by experimental measurement.

Measurements of some types of materials forming the regenerative or catalitic bed were performed for different fluid flow rates (air) in range of 7 - 50 m^3/h. The pressure drops are introduced for packed bed formed by ball particles of various diameters and honeycomp of various grid. To measure the pressure drop we used the test apparatus located in our institute described. This equipment allows the results of measured pressure drops to be automatically stored in the PC. The material which formed the bed was placed on a sieve inside a square 150x150 mm tube. The height of the bed was different for various diameters and was designed to cover as much as possible the measuring range of the differential pressure sensor. The pressure drop of the sieve was also determined and subsequently subtracted from the total measured pressure drop of the bed.

For the purpose of verification of feasibility in industrial case studies with catalyst and catalyst testing, the modular design pilot plant was proposed and implemented. This unit is scalable and can be used to test the sprinkled catalysts and catalysts in form of the monoliths both in lab or real industrial conditions. This unit is primarily intended to remove nitrogen oxides from the flue gas, and accurate measurement in the reactor allows measurement of the pressure loss of various catalysts.

577892 Impact of Copper Exchange Protocol on Cu-CHA SCR Activity

One of the major milestones in environmental catalysis is the abatement of environmentally harmful compounds such as hydrocarbons, nitrogen oxides (NO_x), sulfur oxides (SO_x) and carbon monoxide, produced by automobiles and power sources. The discovery of Cu-CHA as a catalyst for selective catalytic reduction (SCR) of NO_x was a breakthrough and lead to a rapid commercialization of the technology for the lean-NO_x emissions control. It is a relatively new technology and requires more research on identification of active sites of the catalyst in order to develop next generation catalysts with improved performance, enhanced stability and low cost.

There have been considerable research and progress in molecular-level understanding of the nature of active Cu species and the SCR reaction mechanisms. Unfortunately, many studies were conducted using Cu-CHA prepared via different synthesis methods. Current work investigated four methods of preparation of copper exchanged zeolites reported in the literature and the NOx SCR activities were examined using a packed bed flow system[1-4]. The SCR activities of each sample were considerably different at lower temperature that is relevant to the lean-burn engine operational temperature (Figure 1.). The differences between catalysts prepared via different protocols were further studied by CO binding experiments where it was revealed that each protocol results in different copper species within the zeolite framework. DFT calculations were coupled with CO binding and catalytic results to identify active sites. The poster will show synthesis, characterization and catalytic testing details and results.

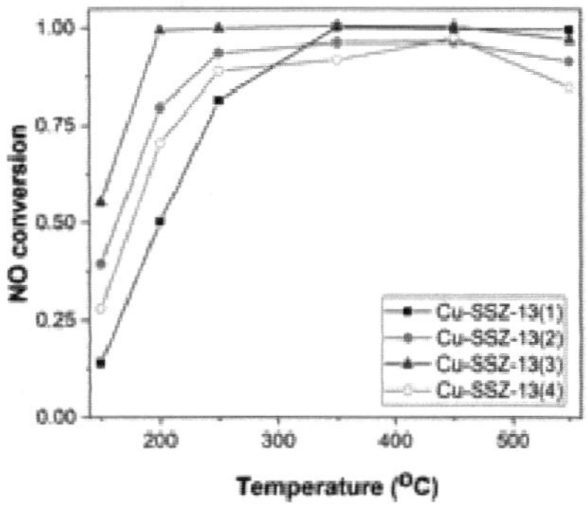

Figure 1. NO conversion profiles for Cu-SSZ-13 samples. Cu-SSZ-13(1-4) were prepared by synthesis protocols reported by Albarracin-Caballero *et al.*, Fickel *et al.*, Grundner *et al.*, and Kwak *et al.*, respectively.

1. Albarracin-Caballero, J. D.; Khurana, I.; Di Iorio, J. R.; Shih, A. J.; Schmidt, J. E.; Dusselier, M.; Davis, M. E.; Yezerets, A.; Miller, J. T.; Ribeiro, F. H.; Gounder, R., Structural and kinetic changes to small-pore Cu-zeolites after hydrothermal aging treatments and selective catalytic reduction of NOx with ammonia. *React Chem Eng* **2017,** *2* (2), 168-179.

2. Fickel, D. W.; Fedeyko, J. M.; Lobo, R. F., Copper Coordination in Cu-SSZ-13 and Cu-SSZ-16 Investigated by Variable-Temperature XRD. *J Phys Chem C* **2010,** *114* (3), 1633-1640.

3. Grundner, S.; Markovits, M. A. C.; Li, G.; Tromp, M.; Pidko, E. A.; Hensen, E. J. M.; Jentys, A.; Sanchez-Sanchez, M.; Lercher, J. A., Single-site trinuclear copper oxygen clusters in mordenite for selective conversion of methane to methanol. *Nat Commun* **2015,** *6*.

4. Kwak, J. H.; Tonkyn, R. G.; Kim, D. H.; Szanyi, J.; Peden, C. H. F., Excellent activity and selectivity of Cu-SSZ-13 in the selective catalytic reduction of NOx with NH3. *J Catal* **2010,** *275* (2), 187-190.

570650 First-Principles Studies of CO Oxidation on $MgAl_2O_4$ Supported Iridium Single Atoms

Single-atom catalysts (SACs) can maximize the efficiency of the metal atom. Previously, a number of supported noble metals (e.g., Pt, Ru) have been identified as active SACs for CO oxidation[1,2]. In this poster, we will discuss our recent studies of CO oxidation on supported Ir single atoms using density-functional theory (DFT) calculations and in-situ/operando experimental measurements.

[1] Zhang, Z. *et al.* Thermally stable single atom Pt/m-Al2O3 for selective hydrogenation and CO oxidation. *Nat. Commun.* **8**, 16100 (2017).

[2] Li, F., Li, L., Liu, X., Zeng, X. C. & Chen, Z. High-Performance Ru1 /CeO2 Single-Atom Catalyst for CO Oxidation: A Computational Exploration. *Chemphyschem* **17**, 3170–3175 (2016).

578286 Photocatalytic Degradation of NO_x Using CNT-TiO_2 Nanocomposites

Photocatalytic Degradation of NOx Using CNT-TiO2 Nanocomposites

Bailey McAuley, Brian Everhart, Ahmed Abdulrazzaq Qasim Al Mayyahi, Placidus Amama

Tim Taylor Department of Chemical Engineering, Kansas State University, Manhattan, KS 66506

Photocatalytic oxidation (PCO) has shown promise in eliminating several gaseous pollutants including volatile organic compounds (VOCs), NO_x, and subsequently tropospheric ozone. PCO is typically conducted with a semi-conductor catalyst, titanium dioxide (TiO_2) where under ultra violet light irradiation, with sufficient energy, the catalyst will activate and oxidize pollutants into less harmful forms. The wavelength of light needed for this reaction is typically in the ultra violet (UV) region for light. Unfortunately, sunlight has only ~5% UV rays, limiting the practical application of TiO_2. Coupling carbon nanotubes (CNTs) to TiO_2 improves photocatalytic efficiency, possibly due to enhanced charge separation. Using a newly constructed laminar flow reactor, NO_x degradation has been tested on catalyst surfaces with and without CNTs under UV light. This study investigates the effect of variables such as humidity, substrate type, temperature, and exposure time on the degradation of NO_x over CNT-TiO2 to better understand their effects.

576128 A Temporal Analysis of Products (TAP) Study of Passive NO_x Adsorption (PNA) on 1% Pd/SSZ-13

Most engine exhaust after-treatment catalytic processes such as selective catalytic reduction (SCR) are effective at temperatures above 200°C. Temperatures below 200°C are considered to be the cold start period when the majority of NO_x and hydrocarbons slip through the after-treatment system. To meet strict NO_x emission standards, trapping NO_x upstream of the SCR catalyst during the cold start period and releasing the NO_x at higher temperatures to enable its reduction is a potential way to lower NO_x emissions further. A material with demonstrated capability for this purpose is Pd-exchanged zeolites, such as ZSM-5, BEA, and SSZ-13. While the PNA concept has been demonstrated in a group of recent studies, the nature and role of the various active sites determining their performance and NO_x absorption chemistry is inadequately understood despite intensive research efforts.[1] In this investigation, we characterize the active sites of a model passive NO_x adsorber (PNA), Pd/SSZ-13 using Temporal Analysis of Products (TAP). The data are used to further develop a working mechanism comprising reaction steps that describe the NO_x trapping and release over a range of temperatures. The TAP data are interpreted with the help of help of flow reactor and computational methods.

The NO_x trapping and release properties on the 1% Pd/SSZ-13 were tested using feeds with different compositions. The experimental results showed the potential of this material to be an effective PNA at low temperatures. Literature shows multiple Pd species co-existing in the catalyst: (i) atomically dispersed or 'naked' Pd in the cationic sites in the pores of the zeolites, and (ii) PdO_x particles on the external surfaces. Forms of active Pd sites present in the catalyst are referred to as Pd^{II}, Pd^{I}, Pd^{II}-OH and $Pd^{II}O$.

The 1% Pd/SSZ-13 catalyst was prepared using an incipient wetness impregnation method. Powder catalysts of 250-400 μm were used for experiments. A generation-1 TAP reactor[2] was used for conducting transient experiments. Controlled pulse experiments were carried out over 5 mg of the catalyst at a defined state to unravel the reaction mechanism. The products from the reaction were measured by a high resolution 100C UTI mass spectrometer. The catalyst was pretreated with 1500 pulses (10^{15} molecules/pulse) of O_2/Ar (1:1) mixture at 500°C before each experiment. Experiments were conducted at temperatures between 100°C and 450°C with mixtures of isotopic and non-isotopic NO, with and without O_2 and CO under dry and wet conditions to study the participation of different Pd species and the effects of H_2O, CO on NO_x uptake and release. The interpretation of experimental data was supported through periodic, spin-polarized density functional theory (DFT) calculations using the Vienna *ab-initio* simulation package (VASP) with the BEEF-vdW exchange correlation functional. Different active site motifs of Pd were modeled in bulk unit cells of SSZ-13 with one or two framework Al atoms.

A DFT investigation on the stability of the above-mentioned sites and the NO adsorption energies on them suggest that NO binds strongly on Pd^{I} sites, the formation of which can be explained through the two step site modification mechanism as in Eq (1).

$$2\ Z\ [Pd^{II}OH] \rightarrow Z\ [Pd^{I} - O - Pd^{I}]\ Z + H_2O \rightarrow 2\ ZPd^{I} \quad (1)$$

H_2-temperature programmed reduction was performed in the TAP reactor and the H_2 consumption at temperatures below 300°C is indicative of the amount of cationic Pd present in the catalyst. It was estimated to be ~85% of the total Pd present in the catalyst. A $^{16}O_2$ pretreated catalyst pulsed with a $^{15}N^{18}O$ and $^{16}O_2$ resulting in the production of $^{15}N^{16}O^{18}O$ and $^{15}N^{16}O_2$ is consistent with the formation of the Pd^{+} sites from two adjacent $Pd^{2+}OH$ sites:

$^{15}N^{18}O + 2\ Pd^{II}\text{-}^{16}OH \leftrightarrow {}^{15}N^{16}O^{18}O + 2\ Pd^{I} + H_2{}^{16}O$ (2)

$^{15}N^{18}O + Pd^{II}\ {}^{16}O \leftrightarrow {}^{15}N^{16}O^{18}O + Pd^{II}$ (3)

The two step modification mechanism for the formation of Pd^{+} sites was confirmed by additional isotopic experiments with $C^{18}O$ on $^{16}O_2$ pretreated catalyst by monitoring the production of $C^{16}O^{18}O$.

The presence of $^{18}O_2$ and N_2 in the product outlet provide evidence for possible NO dissociation. A more detailed insight into the reaction mechanism is provided by the observation of oxygen scrambling between the feed ^{18}O and the catalytic ^{16}O.

$^{15}N^{18}O + ZPd^{II} \leftrightarrow {}^{15}N^{18}O\text{-}\ ZPd^{II}$ (4)

$2\,{}^{15}N^{18}O\text{-}\ ZPd^{II} \leftrightarrow {}^{15}N_2 + 2\ ZPd^{II}\text{-}{}^{18}O$ (5)

$2\ ZPd^{II}\text{-}{}^{18}O \leftrightarrow 2\ ZPd^{II} + {}^{18}O_2$ (6)

$Pd^{II}\text{-}{}^{18}O + Pd^{II}\text{-}{}^{16}O \leftrightarrow 2Pd^{II} + {}^{18}O^{16}O$ (7)

The effects of O_2, H_2O and CO in the feed were also studied for PNA performance of Pd/SSZ-13. The substantial reduction of NO_x storage in the presence of H_2O was primarily attributed to the loss of NO uptake on Brønsted acid sites, which are easily poisoned by H_2O. NO uptake on Pd^{I} is only mildly impeded by H_2O, as suggested by DFT results. Considering that NO_x uptake over Pd zeolites is enhanced in the presence of CO,[3] our results suggest that the oxidation or reduction potential of the gas phase alters the relative availability of active storage sites. In particular, the strongly exothermic nature of CO oxidation renders the reduction from $Pd^{II}OH$ to Pd^{I} thermodynamically favorable.

This study aims to identify the various active sites for passive NO_x adsorption using Pd/SSZ-13 catalyst, which is the most promising material to date. An atomic-level understanding of the ad-/desorption and reaction steps on these active sites will greatly accelerate the design and development of improved PNA materials.

References

[1] Y. Zheng, L. Kovarik. M. H. Engelhard, Y. Wang, F. Gao and J. Szanyi (2017). *The Journal of Physical Chemistry*

[2] J. Gleaves, G. Yanblonsky, X. Zheng, R. Fushimi, P. L. Mills (2010). *Journal of Molecular Catalysis A: Chemical*

[3] A. Vu., J. Y. Luo, J. H. Li; W. S. Epling (2017); *Catalysis Letters*

571877 The Effect of Polyethylene Glycol Modification on CrO_x/TiO_2 Catalysts for NO Oxidation

CrO_x was loaded on different amounts of polyethylene glycol modified TiO_2 (CrO_x/yPEG-TiO_2) by impregnation method and successfully utilized for NO oxidation. The structure-property relationships of the catalysts were explored through different characterization techniques including XRD, BET, Raman, XPS, O_2-TPD and NO-TPD. The results pointed out that the polyethylene glycol modification could enhance the interaction between CrO_x with TiO_2. Among these CrO_x/yPEG-TiO_2 (y = 0, 30, 50) catalysts, the CrO_x/50PEG-TiO_2 had the strongest interaction between CrO_x with TiO_2, the strong interaction resulted high content of Cr^{3+}. Cr^{3+} was the activate oxygen site and associated with the amount of chemical adsorption oxygen. Meanwhile, CrO_x/50PEG-TiO_2 contained a certain amount of Cr^{6+}, which was the NO adsorption site. Chemical adsorption oxygen and Cr^{6+} participated in the reaction of NO oxidation, and CrO_x/50PEG-TiO_2 exhibited the highest catalytic activity due to the synergy effect between Cr^{6+} and chemical adsorption oxygen, which reached 55% at the temperature of 350 °C. Moreover, the CrO_x/50PEG-TiO_2 still exhibited wide activity temperature windows of 225–425 °C and good stability.

573702 Catalytic Oxidation of Volatile Organic Components (VOCs) over Silver Nanoparticles Dispersed in Monodisperse Titania Nanospheres

Volatile Organic Compounds (VOCs) are a class of organic chemicals with high vapor pressure at room temperature. They are emitted from a variety of sources and it is crucially important to treat them properly in order to avoid potential environmental and health issues. Complete oxidation of these VOCs to CO_2 and water at low temperatures with the aid of a suitable catalyst is a promising way to reduce the risks associated with VOCs. Recent developments in the field of catalysis have facilitated some interesting advancements in catalysis architecture. One such development is encapsulating the active metal by a metal oxide nanostructure. This morphology, while helping maintain a constant size distribution, also promotes strong metal-support interactions by maximizing the number of interstitial sites. However, the synthesis of encapsulated catalysts is often complicated and results in an inconsistent yield, which has hindered extensive research using this architecture. In the present study, a simple, one-pot and scalable synthesis of Ag nanoparticles dispersed in mesoporous TiO_2 nanosphere has been explored by controlling the hydrolysis and condensation rates of TiO_2 precursor in a mixture of water and alcohol at room temperature. The method offers independent tunability of the size of silver nanoparticle and the TiO_2 mesopores. These catalysts have been characterized by several physicochemical techniques: XRD, N_2 adsorption, TPR, XPS, HR-TEM, SAED and EDX. The activity of the synthesized catalyst was tested by conducting oxidation reactions of n-butanol. Under n-butanol: oxygen molar ratio of 1:500 and space velocity of 240000 ml $g^{-1}h^{-1}$, n-butanol was completely oxidized into CO_2 and H_2O at 210°C over the 50mg catalyst. The intimate contact between the silver and Titania at the interface makes these materials an interesting candidate for low temperature VOC oxidation. Due to high efficiency, good stability, low cost, and ease of preparation, Silver dispersed in TiO_2 nanosphere is a promising catalyst for the practical removal of volatile organic compounds.

576134 Mechanistic Study of Water Oxidation to Ozone on Rutile SnO_2 (110) with Computational Chemistry

Producing clean water to sustain the population remains challenged with increasing demand. Ozone disinfects water sources from certain bacteria with little environmental trace due to its low residence time in water. While reliable, ozone is difficult to produce. Electrochemical ozone production (EOP) by means of direct water-to-ozone conversion is a potential competitor to cold corona production, a common but inefficient approach, but EOP also faces limited efficiency due to high overpotentials. Understanding the mechanism of water-to-ozone conversion on heterogeneous catalysts gives insight into the overpotentials in EOP and directs searches for proper catalysts to optimize this process. Here, we investigate thermodynamic descriptors of viable reaction steps on rutile SnO_2 (110) surface models with Kohn-Sham density functional theory (KS-DFT) calculations according to the computational hydrogen electrode model. These descriptors are used to build Pourbaix diagrams that show the stability of each step, as a function of pH and applied potential, and suggest routes to improve the catalytic active sites within SnO_2. Finally, we apply and assess our mechanistic findings on SnO
for improved EOP.

577274 Enhanced CO and C3H6 Conversions on Spinel Catalysts: Impact of Catalyst Architecture

Enhanced CO and C_3H_6 Conversions on Spinel Catalysts:

Impact of Catalyst Architecture

Zhiyu Zhou, Michael P. Harold*, Dan Luss**
Department of Chemical and Biomolecular Engineering, University of Houston, Houston, TX 77204

* *mharold@uh.edu*, ** *dluss@central.uh.edu*

Three-way catalysts (TWC) are used in gasoline vehicles to simultaneously eliminate CO, NOx and hydrocarbons. Ceria has been widely utilized in TWC as the oxygen storage material (OSM) for enhanced oxidation (of CO, H_2, HC) and reduction (of NO and NO_2) since its introduction by Ford [1]. Recently researchers [2] reported that spinel-type mixed metal oxides exhibit excellent oxygen storage capacity (OSC) and can reduce precious group metal (PGM) loading requirements. We studied in a bench scale reactor the abatement of CO and C_3H_6 over a series of (PGM)-OSM catalysts with near-stoichiometric feed under steady-state and lean-rich modulation conditions. $Mn_{0.5}Fe_{2.5}O_4$ was used as the model spinel and conventional zirconia stabilized ceria (CZO) was used as a reference. The results show that the addition of spinel leads to the enhanced CO and C_3H_6 conversion, especially for C_3H_6. The dual-layer design (top PGM layer and bottom spinel layer) results in the best performance while the single layer design with direct deposited PGM onto OSM is the best for CZO. The study provides insight into the beneficial impact of oxygen storage material and provides guidance for optimization of TWC catalyst formulation.

Monolith samples were provided by CDTi Advanced Materials Inc., including PGM-only, OSM-only and PGM-OSM catalysts. The studied OSMs include spinel ($Mn_{0.5}Fe_{2.5}O_4$) and CZO. Two monolith different architecture PGM-OSM monolith samples were used: the first is a single layer with PGM dispersed on OSM; the second is a dual-layer with a top PGM layer and a bottom OSM layer. For brevity, the PGM-only sample was denoted as "PA" and the spinel-only sample was denoted as "SA". The PGM-spinel samples with single or dual layer design were denoted as "PSS" or "PSD" respectively. All catalysts were pre-aged in static air at 700 °C for 33 hrs before testing.

A series simple stoichiometric feed was used to study the light-off behavior of CO and C_3H_6. Also, another complex stoichiometric feed was used to mimic the realistic exhaust conditions.

The complex feed had the composition of 500 ppm NO, 0.17% H_2, 0.52% CO, 400 ppm C_3H_6, 0.5% O_2, 7% CO_2, 7% H_2O. Both feeds had a space velocity of 70 k hr^{-1}. A FTIR (Thermo Scientific, Nicolet 6700) measured the effluent concentrations in the bench reactor system.

Figure 1 shows the C_3H_6 light-off curve over either PGM-only (PA) or spinel-only (SA) catalysts in a simple stoichiometric feed with a ramp rate as 10°C/min. Both SA and PA result in similar T_{50} value (temperature giving 50% conversion) at ~ 400 °C. In comparison with SA, a ~ 40 °C lower T_{80} value (temperature giving 50% conversion) is achieved for PA. Although not shown here, the oxidation of C_3H_6 over spinel is unaffected by the presence of CO while the oxidation of C_3H_6 over PA is inhibited by CO.

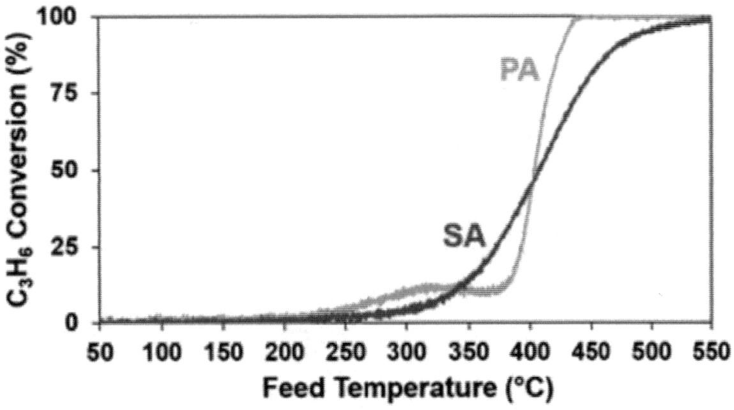

Figure 1. Propylene conversion as a function of feed temperature on spinel-only and PGM-only catalysts [Conditions: ramp rate: 10 C/min; feed: 400ppm C_3H_6, 0.18% O_2, balance Ar].

Figure 2 shows the cycle-averaged CO and C_3H_6 conversion over PGM-only (PA) and PGM-spinel (PSD and PSS) samples using the stoichiometric steady feed from 200°C to 500°C. The results obtained on PA is used as a reference. The addition of spinel (PSD and PSS) promotes CO and C_3H_6 light-off, confirming the promotional impact from spinel. In comparison to PA, a ~ 50 °C lower T_{50} value is achieved for both CO and C_3H_6 conversion with the single-layer design (PSS). The promotional impact from spinel is more exaggerated with the dual-layer design, resulting in a ~ 70 °C lower T_{50} value for both CO and C_3H_6 conversion. Although not shown here, CO conversion is enhanced by addition of spinel and the dual layer design (PSD) provides the largest promotional impacts.

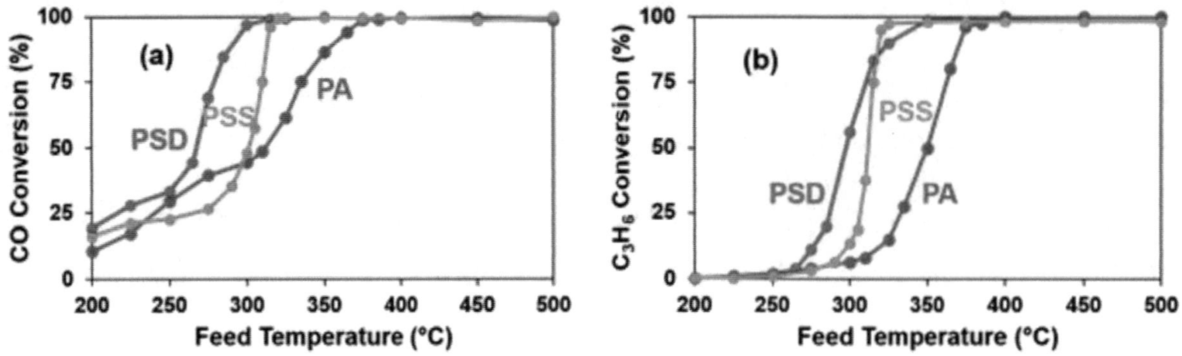

Figure 2. Cycle-averaged CO and C_3H_6 conversion as a function of feed temperature on spinel contained samples with PGM-only sample as reference. [Conditions: steady-state feed: 500ppm NO, 0.17% H_2, 0.52% CO, 400 ppm C_3H_6, 7% CO_2, 7% H_2O, balance Ar].

The reactor data will be supplemented with measurements of the cumulative and dynamic oxygen storage capacity of the materials. The collective results will be interpreted with a phenomenological mechanism.

Reference

1. H.C. Yao, Y.F. Yu Yao, *Journal of Catalysis*, 86 (1984) 254-265

2. S. Golden, Z. Nazarpoor, M. Launois, R-F. Liu, P. Maram, *Society of Automotive Engineering*. SAE 2016-01-0933 (2016)

565191 One-Pot Synthesis of Gold Embedded Ceria Nanoshapes for Catalytic NO Reduction By CO

Growing environmental concern in recent years has resulted in the introduction of more rigorous and stringent environmental laws and regulations. NOx is considered as one of the primary pollutants of the atmosphere because it is responsible for the formation of photochemical smog, acid rain, and the greenhouse effect. The catalytic reduction of NOx by a series of reducing agent (ammonia, H_2, CO and hydrocarbons) is one of the most widespread method and effective technology to decrease the impact of NOx on the environment. The Au-based materials possess high activity at low temperatures, which makes it suitable for catalytic application during the cold start of the vehicles. Oxides like ceria are commonly used as supports due to its outstanding redox ability and high oxygen storage capacity (OSC) associated with the formation of oxygen vacancies. Particularly, the introduction of metal cations into the ceria system will promote its redox behavior and catalytic activity significantly due to the formation of more oxygen vacancies. In this work, differently shaped ceria with embedded gold (Au-CeO_2-rod and Au-CeO_2-cube) were synthesized by one-pot hydrothermal method. XRD measurements show the distinct fluorite CeO_2 patterns with no detected Au related peaks, suggesting that Au atoms are well dispersed into CeO_2 matrix. Their catalytic property was probed by the model reaction: reduction of nitric oxide (NO) with carbon monoxide (CO) under both oxidizing and reducing pre-treatment conditions. Preliminary results present that Au embedded into rod-shaped CeO_2 exhibits higher NO conversion and N_2 selectivity than Au-CeO_2-cube at all reaction temperatures. The catalysts have been further characterized by multiple techniques (TPR, TEM, Raman, etc.) to investigate the support morphology, noble metal dispersion, oxygen defects effect on their catalytic behaviors in NO-CO reaction. *In-situ* FT-IR combined with mass spectrometer is also employed to study the surface chemistry and reaction mechanism of NO reduction by CO over the different Au-CeO_2 catalysts.

ACKNOWLEDGEMENTS: This work is supported by the Center for Understanding and Control of Acid Gas-Induced Evolution of Materials for Energy (UNCAGE-ME), an Energy Frontier Research Center funded by U.S. Department of Energy, Office of Science, Basic Energy Sciences. Part of the work including the IR and Raman was conducted at the Center for Nanophase Materials Sciences, which is a DOE Office of Science User Facility.

568775 Designing of a New Prototype to Detoxify Carcinogenic Cr(VI) Metal from Drinking Water

Porphyrin dye has absorption bands in visible region which resembles with solar light spectrum. So porphyrin dye functionalized nanomaterial can be very much useful in visible light photocatalysis application. Depends on the central metal of the dye, the property of the porphyrin can be modulated heavily. In our present work, we have synthesized Cu-protoporphyrin IX, by attaching Cr (II) within protoporphyrin. In the next step, a porous microsphere TiO_2 (~1.5 μm diameter) was sensitized with Cu-protoporphyrin. The porosity of the structure was confirmed by FESEM and HRTEM. The attachment of the dye with semiconductor was endorsed by Fourier transformed infrared (FTIR) and picosecond resolved fluorescence studies. Hexavalent chromium (Cr (VI)) in the water environment is identified as carcinogenic product, whereas Cr (III) is nontoxic in nature. Our newly developed nanohybrid efficiently converts the toxic metal ion Cr (VI) to non-toxic Cr (III) in aqueous medium under solar light illumination. Cu impregnation in the PP dye keeps the efficacy of the photocatalytic process intact. The interference of other dissolved metal is also nullified after Cu impregnation. We have developed a prototype for the real life application purpose. The nanohybrid has been deposited on the surface of a stainless steel mesh (size 2 cm × 2 cm, pore size 150 μm × 200 μm) which acts as an active filter. While nanohybrid acts as chemical filter, the mesh works as a physical filter to clean the water. The developed prototype exhibits outstanding physical and chemical filtering of toxic Cr (VI) ions (photoreduction to Cr (III)) and physical netting of suspended particulates present in water.

559372 Chemical Warfare Agent Simulant Degradation over Metal Nanoparticles Supported on TiO_2 Aerogels

Chemical Warfare Agent Simulant Degradation over Metal Nanoparticles Supported on TiO_2 Aerogels

Ashley M. Pennington,[1] Paul A. DeSario,[2] Catherine L. Pitman,[1] Debra R. Rolison,[2] Jeremy J. Pietron[2]

[1] National Research Council Postdoctoral Associate

[2] Chemistry Division, U.S. Naval Research Laboratory, Washington, DC 20375, USA

The need to develop and implement materials that can degrade and decontaminate Chemical Warfare Agents (CWAs) has spurred research in the area of both sorbent and catalytic materials. We use titanium dioxide (TiO_2) aerogels, which offer a mesoporous network co-continuous with a covalently bonded solid network of TiO_2 nanoparticles, as catalytic supports for the degradation of chemical warfare agent simulants (CWSs). Mesoporous expressions of metal-oxide catalysts are advantageous as catalytic supports because their pore sizes allow uninhibited mass transport of the CWA to the catalytic active sites, unlike the many microporous structures currently under study. Adding metal (Cu or Au) nanoparticles (NPs) to the mesoporous oxide support adds thermocatalytic and photocatalytic functionality.

The aerogels are characterized via X-ray diffraction (XRD) for crystallinity, N_2 porosimetry for surface area and pore size distributions, and UV–visible spectroscopy for surface plasmon resonance (SPR) response of the aerogel-supported metal NPs. The degree and reversibility of agent binding to the catalyst is investigated via X-ray Photoelectron Spectroscopy (XPS) conducted before and after simulant exposure. Four classes of aerogels are evaluated for their ability to catalytically degrade CWA simulants: TiO_2 aerogels, Cu/TiO_2 with supported copper NPs, Au/TiO_2 with supported gold NPs, and 3D Au–TiO_2 with gold NPs entrained in the TiO_2 network. We measure catalytic and photocatalytic activity of these modified aerogels for degradation of two mimics for nerve agents such as (RS)-propan-2-yl methylphosphonofluoridate (Sarin, GB): dimethyl methylphosphonate (DMMP) and dimethyl chlorophosphate (DMCP). Catalytic activity is monitored via in situ DRIFTS (diffuse reflectance infrared fourier-transform spectroscopy) in anaerobic (50 mL min^{-1} He) and aerobic (40 mL min^{-1} He, 10 mL min^{-1} O_2) flow, under either white light irradiation or dark conditions for a reaction time of four hours at various relative humidities.

The pathways, rates, and product distributions of DMMP degradation differ tremendously based on the conditions to which the catalysts are exposed. Under atmospheric pressure, dry Cu/TiO_2 in the dark under both aerobic and anaerobic conditions exhibits minimal activity for DMMP degradation. Under white-light irradiation, Cu/TiO_2 is catalytically active for the degradation of DMMP under both aerobic and anaerobic conditions. DMMP fragments consistent with hydrolysis products are evident via DRIFTS on the catalyst surface under anaerobic conditions while oxidation products are evident under aerobic conditions. The activity and product distribution on Au–TiO_2 and Au/TiO_2 differ from those observed on Cu/TiO_2 aerogels. Preliminary results from in situ DRIFTS during DMMP degradation over 3D Au–TiO_2 indicate that water plays an important role in DMMP degradation in the dark, which is likely due to proton mobility along water wires that form on the surface of the networked oxide support. These results show the importance of conducting a detailed analysis of the effect of humidity ranges and illumination on the catalyst's efficacy as a CWA decontaminant.

A fundamental understanding of the ability of metal/TiO_2 aerogels to bind and catalytically degrade CWSs under various atmospheric-relevant conditions will forge a path to test these materials using live CWAs, leading to the development and implementation of a highly active, inexpensive, catalytic alternative for field-decontamination.

560994 Fabrication of Natural Clay Based Nanocomposite Materials for Advanced Catalytic Treatment of Organic Pollutants in Wastewater

Natural clays are used in the fabrication of several materials for various industrial applications. This study investigates the fabrication of natural clay-based nanocomposite materials for catalytic removal of dye pollutants in wastewater. The materials were characterized for physical and mineralogical properties using various characterization techniques. X-ray spectroscopy (XRD and XRF) confirmed that the chemical composition was SiO_2, Al_2O_3, Fe_2O_3, K_2O and TiO_2 in the form of quartz, kaolinite, magnetite and illite. The BET showed that the fabricated materials are mesoporous indicating their potential application in wastewater. The presence of TiO_2 and added multi-walled carbon nanotubes (MWCNTs) contributed to the photoresponse of the materials. The natural nanocomposites revealed absorption in the visible range implying that these materials can be applied under the visible light irradiation. The energy band gap values were also relatively low 1.5 and 2.9 eV for the natural nanocomposite materials. Photodegradation was achieved with dyes used as pollutants model. The photodegradation was enhanced (95%) when MWCNTs were added to the natural clays. The addition of CNTs also led to the reduction of the holes-electrons pair recombination which improved the degradation activity of dye molecules from the increased presence of oxidizing species in aqueous solution. The total organic content removal value was quite high 85% and it confirmed the mineralization of the dye molecules. The findings demonstrate that natural clay and MWCNTs can be used successfully for the treatment of dyes wastewater. The implication of these data indicates that the natural nanocomposite materials can be optimized and deploy in the rural area, where natural clays are available and enough solar light can be harvest, for the treatment of wastewater to the benefit of the rural community.

563740 Photocatalytic Metal Organic Framework Induced Ag Nanoparticles Composites for Phenol Detection

Photocatalytic Metal Organic Framework induced Ag Nanoparticles composites for Phenol Detection

Qian Liu and Cerasela Zoica Dinu

Department of Chemical and Biomedical Engineering, West Virginia University, Morgantown, West Virginia 26506, USA

Abstract

Metal-Organic Frameworks (MOFs), novel porous crystal materials constructed by linking metal node (metal ions or clusters) and organic linkers together through bond coordination have been integrated in a variety of applications from membrane separation to catalysis, and from biomedical devices to sensors. Their endowed diversity, geometry, topology and functionality, controllable pore size and large surface area have also shown their advantages in suppling either confined spaces or large amounts of active cites or channels for reactants/product diffusion or synergetic catalytic effects thus ensuring MOFs integration in photo-electron conversion platforms.

In our study, we proposed to create the next generation of hybrid composites based on MOFs capable to be integrated in sensorial platforms for contaminant detection. For this, we used a hydrophilic MOF named MIL-160 which was lab-synthesized to have controllable morphology and ultrahigh stability in water and acidic aqueous solutions as a template for composite formation upon reduction of Ag^{+}. Composite physical and chemical properties were investigated by scanning and transmission electron microscopy, Fourier transform infrared and X-ray photoelectron spectroscopy, and X-ray diffraction. Electrochemical characterization using cyclic voltammetry (CV) confirmed the formation of Ag-based composites as well as its capability to detect *p*-NP, a model contaminant known to be a byproduct of wide usage for manufacturing of drugs and fungicides or degradation of insecticides, with toxic effects on receiving bodies and demonstrated carcinogenicity and mutagenicity, for both liver and kidney. Our CV analysis also showed that electrochemical and sensorial behavior were controlled by the electrode surface characteristics with the sensitivity and detection limit of the fabricated Ag/MIL-160 electrode being evaluated at 8.852 ± 0.528 uA $\mu M^{-1}$and 0.041 μM respectively. Considering the wide usage of *p*-NP and its increased discharge as well as its possible effects on environment and human bodies, we foresee that this new method for this contaminant detection could potentially lead to effective control and regulation of *p*-NP release. Further, we envision that the ability to detect low concentrations of *p*-NP contaminant in solutions could be extended to other contaminants of interest.

575782 Enhancement of Heterogeneous Silver and Copper Catalysts Stability for Catalytic Ozonation in Water

Catalytic ozonation represents a promising method to degrade and mineralize refractory organic pollutants in water. Heterogeneous catalysts facilitate easy separation from water and thus have attracted increasing interests in the field of advanced water treatment.

However, the stability of metal/metal oxides based catalysts under ozonation in aqueous solutions represents a challenge.[1] On one hand, metallic ion leaching from the solid catalysts can degrade the quality of treated water and shorten the catalyst lifetime. On the other hand, transition metal ions have been demonstrated as active homogeneous catalysts for catalytic ozonation. The leaching of these ions from their metal or metal oxide counterparts in heterogeneous catalytic ozonation studies will likely to cause questions and controversies in determining the reaction rate expression and mechanisms.

Two strategies to synthesize stable heterogeneous silver- and copper-based catalysts will be discussed in this talk, namely 1) to coat the catalysts with a thin, anti-leaching protecting layer using atomic layer deposition (ALD), and 2) to doped the active sites in the oxide spinel structure. The catalyst structure was characterized using X-ray absorption spectroscopy, XRD, TEM, XPS, and BET surface area measurements before and after the catalytic ozonation in water treatment. Both strategies showed improving catalysts stability in terms of preventing leaching while the catalytic performance was maintained or improved. The advantages and disadvantages of both methods will be discussed and compared.

Reference:

1. Enhancement of Copper Catalyst Stability for Catalytic Ozonation in Water Treatment, W. Yang, Z. Lu, B. Vogler, T. Wu, Y. Lei, *ACS Applied Materials & Interfaces*, 10 (2018) 43323-43326

577090 Impact of Defects on the Decomposition of Chemical Warfare Agent Simulants in Zr-Based Metal Organic Frameworks

Given the threat associated with chemical warfare agents (CWAs), there is a need to develop new materials that can effectively and rapidly detoxify CWAs. It has been shown experimentally that metal organic frameworks (MOFs) containing zirconium secondary building units, such as $Zr_6O_4(OH)_4$ found in UiO-67 and related MOFs, are highly active for neutralizing the CWA simulant, dimethyl methylphosphonate (DMMP), as well as actual CWAs. It has been assumed that missing linker defects are required for the reaction to proceed in UiO-67. However, it is not known if a single isolated defect is sufficient or if multiple defects are required for facile decomposition of DMMP and CWAs in this material. In this work we use density functional theory calculations to explore the reaction pathways of DMMP on $Zr_6O_4(OH)_4$–based MOFs. We use full periodic and cluster models to study reaction pathways on UiO-67 multiple defects per secondary building unit. We find that multiple defects can dramatically lower the reaction barrier for DMMP decomposition.

558947 Fabrication and Performance Evaluation of Photocatalytic Reactor for the Removal of Malathion from Waste Water

A photo catalytic reactor operates on catalyst suspended in the solution or immobilized on various supports. Photo catalytic reactor with suspended catalyst gives greater contact between the catalyst and dissolved impurities compared to reactors with immobilized catalyst. The aim of the present study is to design an annular UV photo reactor and analyze its performance for the degradation of malathion from water with Titanium dioxide as catalyst and ZnO doped carbon as an alternative catalyst and compare their performance. Various parameters were studied to observe the action of designed reactor like variation in the initial feed concentration of malathion by keeping the mass of catalyst steady and aeration of the system. The reactor performance was assessed by varying the concentration with time. The performance of reactor was studied in fluidized state for a known feed concentration of malathion. The designed reactor has given a degradation of 97.54% within three hours of time which is greater than reactors so far developed.

560421 Insights into H_2S Catalytic Oxidation Performance on Activated Carbon Supported Cu-Based Catalyst at Room Temperature

Hydrogen sulfide (H_2S) is a highly odorous, toxic and corrosive pollutant, which existed widely in industrial operations such as sewage treatment, coke ovens, natural gas processing and petroleum refining. The removal of H_2S is very essential for pollution control and operation safety within industrial processes.

Many technologies including amine scrubbing, Claus process, biological methods and catalytic oxidation process have been reported for H_2S removal. Among them, selective catalytic oxidation of H_2S to sulfur is the preferred one due to its environmental friendliness, low capital cost and high operation flexibility. Moreover, this method could thermodynamically reduce H_2S to less than 0.1 ppm even at room temperature. However, the key for H_2S oxidation is the development of highly efficient catalyst.

In this study, activated carbon (AC) supported Cu was prepared by impregnation method and its catalytic performance for H_2S removal was studied. It was found that Cu/AC could remove H_2S effectively at room temperature, while a low adsorption capacity was detected. As such, a second component (Zn and Ni) was selected and added to the Cu/AC catalyst. Meanwhile, the effects of Zn loading on the activity of catalysts were extensively investigated. The results show that the proper amount of zinc doping could enhance the catalytic performance, and thus remove H_2S significantly. Furthermore, based on the characterization (BET, SEM-EDS, TEM, XRD, XPS and TG) of Cu/Zn/AC catalysts, the mechanism of H_2S oxidation was elucidated.

564463 The Synergistic Effect of Copper and Niobium Species on a Novel Ti-Based Ternary Oxide Catalyst for NO_x Removal with NH_3

The synergistic effect of copper and niobium species on a novel Ti-based ternary oxide catalyst for NO_x removal with NH_3

Xiaoxiang Wang, Sujing Li*, Wei Li*, Yao Shi

Key Laboratory of Biomass Chemical Engineering of Ministry of Education, Institute of Industrial Ecology and Environment, College of Chemical and Biological Engineering, Zhejiang University, Hangzhou 310027, China

Abstract: As the traditional VW(Mo)Ti catalyst existed some drawbacks using in NH_3-SCR reaction, it¡¯s necessary to develop new TiO_2-based alternative catalysts for NO_x abatement. 0.8%Cu/Ti_2NbO_x as a novel catalyst was developed and investigated. The NO_x conversion and N_2 selectivity maintained upon 90% and 97% respectively in the temperature range of 250°C-425°C under a high gas hourly space velocity (GHSV) of 177,000h^{-1}. Water vapor or SO_2 would not pose any negative effect on the SCR activity. Also, 0.8%Cu/Ti_2NbO_x exhibited an excellent hydrothermal stability and alkali mental resistance. The performance characterization indicated 0.8%Cu/Ti_2NbO_x was a promising candidate for the NH_3-SCR catalyst in the future practical application. Besides, an array of analytical techniques were employed to elucidate the correlations among the performance-structure-property and the reaction mechanism over 0.8%Cu/Ti_2NbO_x catalyst. The introduction of niobium oxide increased the surface area of TiO_2 and decreased the crystallite size, promoting the high dispersion of copper species. Meanwhile, the total acidity and the amount of surface active oxygen over the catalyst were increased by the addition of niobium oxide. Copper species mainly existing in the state of isolated Cu^{2+} and non-isolated Cu^{+} enhanced the redox capability of the catalyst, resulting in an excellent catalytic performance. There is a promotional synergistic effect between copper and niobium in 0.8%Cu/Ti_2NbO_x catalyst during the catalytic reaction. Furthermore, the reaction pathway over 0.8%Cu/Ti_2NbO_x catalyst followed both Eley-Rideal mechanism and Langmuir-Hinshelwood mechanism at 225°C.

Key words: Cu/Nb-Ti mixed oxide catalyst; Synergistic effect; Reaction pathway

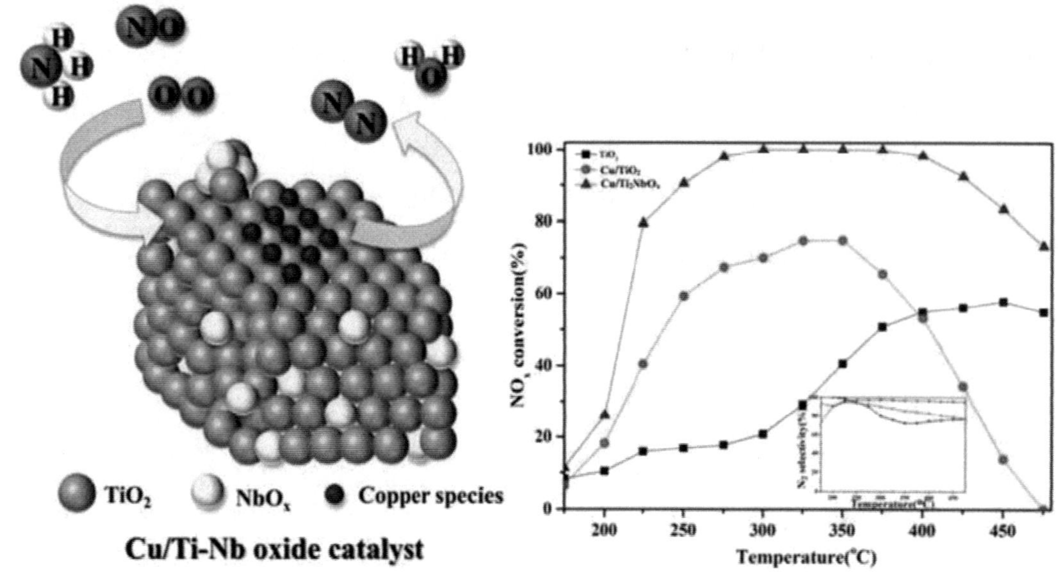

Fig.1. NO_x conversion and N_2 selectivity over TiO_2, Cu/TiO_2 and Cu/Ti_2NbO_x catalysts. Reaction conditions: $[NH_3]$ = [NO] = 500 ppm, $[O_2]$ = 5%, total flow rate = 500 mL min^{-1}, GHSV = 177,000 h^{-1} and N_2 as balance.

References

[1] X.X. Wang, Y. Shi, S. J. Li*, W. Li, Appl. Catal. B: Environ., 220 (2018) 234-250.

[2] X.X. Wang, Q.L. Cong, L. Chen, Y. Shi, Y. Shi, S.J. Li*, W. Li*, Appl. Catal. B: Environ., 246 (2019) 166-179.

565149 Degradation of Gas-Phase O-Xylene Via Non-Thermal Plasma over Fe Doped $LaMnO_3$ Catalysts:the Byproducts Control

Degradation of Gas-phase O-xylene via Non-thermal Plasma over Fe Doped $LaMnO_3$ Catalysts: the Byproducts Control

Voltaile organic compounds (VOCs) are harmful to both environment and human health. Non-thermal plasma (NTP) combined with catalysis is an atlernative thchnology for remediating VOC contaminants, especially aromatic hydrocarbons. During NTP-catalysis degradation process of gas-phase hydrocarbons, the harmless products are CO_2 and H_2O, while the byproducts can be CO, O_3, NO_x, and hydrocarbon fragments. To inhibit the generation of by-products, Fe doped $LaMnO_3$ catalysts were developed. Different concentration of Fe catalysts are synthesized by sol-gel method and supported over Al_2O_3 as follows: $LaMn_{0.95}Fe_{0.05}O_3$, $LaMn_{0.9}Fe_{0.1}O_3$ and $LaMn_{0.8}Fe_{0.2}O_3$. (Marked as Fe0.05, Fe0.1 and Fe0.2 in Figure 1) The catalysts are fixed in the back part of the dielectric barrier discharge (DBD) reactor. The experimental results show that $LaMn_{0.9}Fe_{0.1}O_3$ is the better choice with o-xylene degradation rate of 91.3% when specific input energy (SIE) is 744.46 J/L, 4.2% higher than $LaMn_{0.95}Fe_{0.05}O_3$ and 10.1% higher than $LaMn_{0.8}Fe_{0.2}O_3$. When $LaMn_{0.9}Fe_{0.1}O_3$ is added in NTP system, O_3 emission decreased 84.9% and N_2O decreased 66.7%, while CO_2 selectivity increased from 68.1% to 91.8% and CO_x conversion increase from 51.6% to 80.9%. Various analytical techniques and experiments were employed to explain the efficiency differences. The involvement of Fe results in the lattice deformation of $LaMnO_3$, and further influence the ratio of adsorbed oxygen and lattice oxygen, which mainly explain the degradation rates difference and CO_x conversions difference. Besides, high Mn^{4+} concentration of $LaMn_{0.9}Fe_{0.1}O_3$ contribute to the high inhibition rate of O_3. Moreover, Fe^{2+}/Fe^{3+} redox couple reduce the emission of CO and enhance the CO_2 selectivity. The formation of N_2O is not only resulted from the reaction between N_2 and O_2 directly but also attributed to the oxygen-containing molecules. On the other hand, N_2O can be decomposed in gas-phase or over the catalysts in plasma atmosphere. The joint effects of formation inhibition and N_2O decomposition lead to the less emission of N_2O. As a result, O_3, N_2O, CO and hydrocarbon fragments are all inhibited if $LaMn_xFe_{1-x}O_3$ catalysts are adopted in the NTP system and $LaMn_{0.9}Fe_{0.1}O_3$ is the better choice.

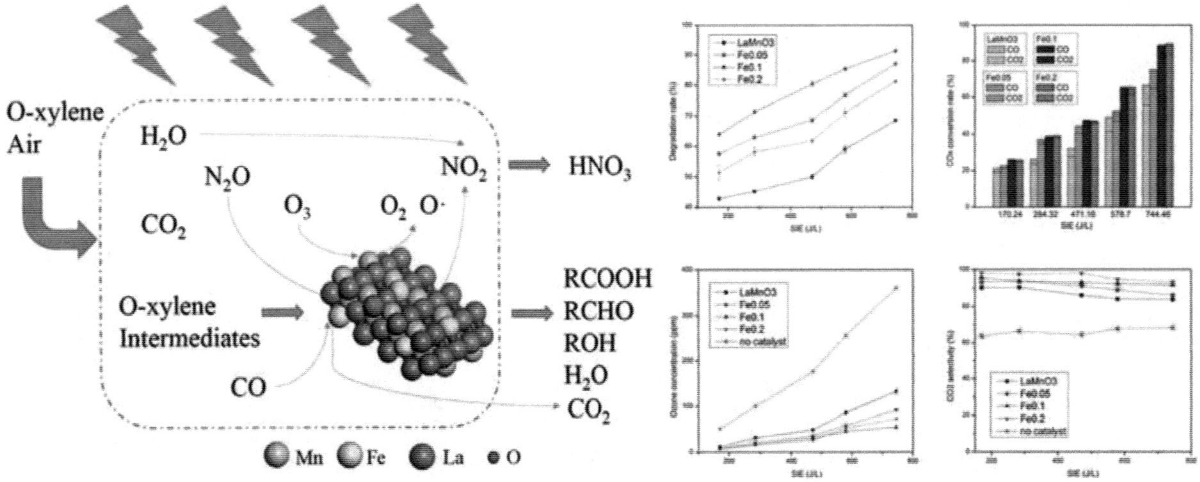

Figure 1. The prevalent reaction with degradation rate, CO_x conversion, ozone concentration and CO_2 selectivity in NTP-catalysis system.

567113 Recent Progresses in Enhancing the Hydrothermal Stability of Cu-SSZ-13 Catalyst for Selective Catalytic Reduction of NO_x with NH_3

Cu-SSZ-13 is a highly active catalyst to remove NO_x by selective catalytic reduction with NH_3 in the emission of diesel-powered engine. However, its stability in high-temperature (HT) hydrothermal conditions is a well-known challenge for applications. The recent progresses in our group in enhancing the HT hydrothermal stability of Cu-SSZ-13 are presented. By mechanically mixing another CHA-type catalyst Cu-SAPO-34 and introducing phosphorus into Cu-SSZ-13 via incipient wetness impregnation, the HT hydrothermal stability is significantly enhanced. The activity of the hydrothermally treated (in 10% H_2O/air at 750 ^oC for 16 h) catalysts increases with the increase of the amount of Cu-SAPO-34 up to 50 wt% in the mixed sample. The delivery of extra-framework P species from Cu-SAPO-34 to Cu-SSZ-13, the migration and coordination of phosphate ions with the framework-bonded Al species are critical in impeding the dealumination process and thus enhancing the HT hydrothermal stability.

567473 Catalytic Oxidation of VOC – Modelling, Reactor Design and Industrial Off Gas Treatment

The treatment of industrial off gas that contains VOC (Volatile organic compounds) before is vented to the atmosphere is mandatory according to the legislation in most of the states worldwide mainly due to undesirable VOC photochemical reactions with NOx forming harmful compounds that can lead in negative climate and public health effects. Some of the individual VOC (e.g. benzene, 1,3-butadiene, formaldehyde, and higher molecular weight aldehydes) are considered harmful on their own due to their negative respiratory, carcinogenic, neurological or other effects.

Frequently the abatement of the VOC in the industry is done by thermal oxidation by simply burning the organic vapours with natural gas in the combustion chambers. In many cases of the industrial off gas treatment containing VOC is possible to change the technology from thermal oxidation into catalytic oxidation and reduce operating costs in terms of difference in the savings because the reduced natural gas consumption and the price of used catalyst. Of course some of the conditions like stable gas flow and no sudden changes in VOC concentration loads must be met to get reasonable reactor dimensions and investment return rate.

This contribution aims to address many aspects of this issue, from reactor design of the pilot scale catalytic reactors to modelling of the deep oxidation of VOC in this type of reactors and treatment of off gas containing either only model VOC pollutant or real industry mixture of different VOC in various concentrations.

As for the modelling of the process in larger scales than laboratory scale, the isothermal reactor can't be used, and instead the adiabatic or even better, the non-isothermal reactor with heat dissipation must be considered. The proposed model (if provided with proper kinetic data) considering non-isothermal plug-flow reactor can predict the achievable conversion and temperature gradient across the reactor, which can help in reactor design, technological design of catalytic bed and scale-up to the full-scale technology. The proposed model is considering changes in gas composition and gas properties due to oxidation reactions, temperature rise due to oxidation reactions and on the other hand temperature decrease due to heat losses and change of pressure due to pressure losses across the bulk catalyst bed.

For the purpose of verification of feasibility in industrial case studies of VOC catalytic oxidation and catalyst testing, the modular design pilot plant was proposed and implemented. This unit is scalable and can be used to test the sprinkled catalysts and catalysts in form of the monoliths both in lab or real industrial conditions.

571450 Improved Catalytic Performance of (RE, TM) Co-Doped Ceria for CO Oxidation

Environmental pollution is becoming a critical issue all over the world, and accordingly, automobile exhaust gas regulations to remove the pollutants such as CO, NO_X, and HC have also been continuously strengthened. Among them, CO is an environmental pollutant that can cause fatal poisoning to humans. The exhaust catalysts working at low temperature to remove it are critical to the automotive industry because they can reduce energy costs and meet regulations. The precious metal catalysts currently used in many applications have high activity, but since they are expensive and limited, it is necessary to design the catalysts using inexpensive metals. As a condition to replace the precious metals, the catalysts using inexpensive metals should have the catalytic activity at least similar to the precious metal catalysts. Furthermore, if those catalysts are proved to have good stability beyond simply having high activity, they would bring great economic advantages. It has been reported from previous studies that a different type of dopant metals have a different effect on metal-doped CeO_2 catalyst systems. Rare earth (RE)-doped CeO_2 catalysts have increased both specific surface area and catalyst stability. Overall, however, they have relatively low CO oxidation activity. On the other hand, transition metal (TM)-doped CeO_2 has the high catalytic activity even at low temperature, but there is a problem of the reduction of active sites due to particle growth.

In this study, we synthesized (RE, TM) co-doped CeO_2 catalysts to take both advantages of TM and RE dopants. (RE, TM) co-doped CeO_2 was characterized by XRD, BET and HR-STEM measurements. XPS, Raman and H_2-TPR were used to confirm the defect (e.g. oxygen vacancies) of the catalyst surfaces, and the catalytic activity test was performed by GC. In addition, CO-TPD, TPR and DRIFTS analysis further confirmed the cause of active site and catalyst stability. Lastly, theoretical analysis including DFT calculations was performed to elucidate the experimental results. It was found that (RE, TM) co-doped CeO_2 catalysts had improved specific surface area and catalytic activity, and furthermore showed good stability to prevent the reduction of active sites by inhibiting the particle aggregation. Therefore, we believe that our approach to combine RE and TM doping into CeO_2 provide the considerable possibility of replacing precious metal catalysts for CO oxidation reaction.

573131 Metal Oxide Catalysts for Low Temperature NO_x Adsorption

The increasing efficiency of automotive engines has resulted in lower exhaust temperatures. Such improvements call for enhanced after-treatment systems to mitigate NO_x emissions, particularly in the cold start period with the exhaust temperature in the range of 100-200 °C. However, the state-of-art commercial deNOx catalysts for diesel vehicles are only effective at temperatures above 200 °C. This work tries to formulate a catalyst which can adsorb and store NO in the form of metal-nitrates below 200 °C and subsequently release NO_2 in 200-400 °C regime, where commercial deNOx catalysts eliminate it effectively. Thermogravimetric analysis results show that nitrates of copper, cobalt and lanthanum supported over alumina are stable at temperatures below 200 °C, but totally decomposed to NO_2 in the 200-400 °C window. This indicates the potential application of these base metal oxides in NO adsorption at low temperatures. In order to accelerate the adsorption of NO and transformation to metal-nitrates at low temperatures, an additional active metal which oxidizes NO to NO_2 will be combined with alumina supported copper oxide, cobalt oxide and lanthanum oxide. The insights drawn from this study will be useful in the design of low-temperature NO adsorption catalysts formulated with base metals.

577932 Surface Redox-Acid Pair Sites Required for Achieving High Performance for the Selective Catalytic Reduction of NO_x with NH_3

Selective catalytic reduction of NO_x with NH_3 (NH_3-SCR) is a well-established technology for NO_x pollution control [1]. Different metal oxides, such as CeO_2, Fe_2O_3 *etc.*, which have appropriate reduction-oxidation (redox) ability, have attracted much attention for NO_x reduction due to their high NH_3-SCR activity and N_2 selectivity. The NH_3-SCR performance of CeO_2-based and Fe_2O_3-based catalysts can be remarkably enhanced by introducing solid acid components such as tungstate and niobate, which is mainly due to the effective integration of excellent redox ability from CeO_x or FeO_x sites and optimal acidity from WO_x or NbO_x components [2-5]. Previous studies have reported that the local mixed oxide structures, *e.g.* surface Fe-O-W and Ce-O-Nb bonds or linkages, were beneficial to superior NH_3-SCR activity and N_2 selectivity [2,4,5]. However, the detailed synergetic effect between redox sites and acid sites has not been systematically studied, namely, the effect of surface redox-acid pair sites for NH_3-SCR reaction.

In order to study the critical role of redox-acid pair sites for NH_3-SCR reaction, a series of binary oxide catalysts (CeO_2-WO_3, CeO_2-Nb_2O_5, Fe_2O_3-WO_3 and Fe_2O_3-Nb_2O_5) were prepared by facile impregnation method, in which the redox components were loaded onto acid supports or the acid components were loaded onto reducible oxide supports. The amount of redox-acid pair sites on the catalyst surface were fine-tuned by adjusting the loading of acid components or redox components. The effect of redox-acid pair sites on NH_3-SCR performance was studied in detail. Catalysts were characterized by XRD, Raman spectra, H_2-TPR, NH_3/NO_x-TPD as well as *in situ* DRIFTS, and the results indicated that the catalysts exhibited the optimal NH_3-SCR activity and N_2 selectivity when the catalysts possessed the most abundant redox-acid pair sites. Based on this research, a comprehensive universal mechanism contributing to high deNO_x performance in NH_3-SCR reaction resulting from highly desired redox-acid pair sites is clearly elucidated, providing theoretical guidance to the design of future highly efficient NH_3-SCR catalytic systems.

References

[1] K Skalska, J S Miller, S Ledakowicz. Trends in NOx abatement: A review. *Sci. Total Environ.*, 408: 3976-3989, **2010**.

[2] R Qu, X Gao, K Cen, et al. Relationship between structure and performance of a novel cerium-niobium binary oxide catalyst for selective catalytic reduction of NO with NH_3. *Appl. Catal. B: Environ.*, 142: 290-297, **2013**.

[3] W Shan, F Liu, H He, et al. Novel cerium-tungsten mixed oxide catalyst for the selective catalytic reduction of NO_x with NH_3. *Chem. Commun.*, 47(28): 8046-8048, **2011**.

[4] F Liu, W Shan, Z Lian, et al. The smart surface modification of Fe_2O_3 by WO_x for significantly promoting the selective catalytic reduction of NO_x with NH_3. *Appl. Catal. B: Environ.*, 230: 165-176, **2018**.

[5] Z Ma, X Wu, Z Si, et al. Impacts of niobia loading on active sites and surface acidity in NbO_x/CeO_2-ZrO_2 NH_3-SCR catalysts. *Appl. Catal. B: Environ.*, 179: 380-394, **2015**.

569100 Descriptor-Based Modeling of CO Oxidation over Alumina-Supported Single Metal Atoms

Single atom catalysts receive extensive attention for reducing noble metal utilization and potential elimination of side reactions. Yet, the rational design of highly reactive and stable single atom catalysts is hampered by the current lack of fundamental insights at the single atom limit. In order to develop a descriptor-based modeling approach for single atom catalysis, the governing reaction mechanisms and scaling relations need to be identified first. To this end, density functional theory calculations are performed in this study for a prototype reaction, namely CO oxidation, over different single metal atoms supported on alumina. As expected for oxidation reactions, a large change in the oxophilicity of the supported single atom can have an important effect on the reaction mechanism. Hence, a novel descriptor-based modeling approach is developed in the present work using several correlations without *a priori* assumptions about rate-determining steps and most abundant reaction intermediates. In addition, the stability of the single atom catalysts is also accounted for in conjunction to their reactivity. In that respect, we find that atomically dispersed iron on alumina is a promising and relatively stable non-platinum group metal (non-PGM) alternative for low temperature CO oxidation.

Acknowledgements: This work is performed in the framework of the PARTIAL-PGMs project – funded by the European Union's Horizon 2020 research and innovation programme under grant agreement No 686086.

573055 Flame-Synthesized Pd-TiO_2 Catalyst for Oxygen Removal from Oxy-Coal Combustion Flue Gas

Oxy-coal combustion has been developed to effectively capture CO_2 in power plants. It uses nearly pure oxygen (O_2) to combust coal, and thereby a concentrated CO_2 stream (~63%) can be achieved in its exhaust [1]. The concentrated CO_2 stream can be utilized for enhanced oil recovery (EOR), however, the O_2 level in the stream should lower to 100 ppmv for satisfying the requirement for EOR [2]. Catalytic O_2 removal system with methane (CH_4) has been recently applied to solve this problem, and a high conversion of O_2 was reported with noble metal (Pd) doped metal oxide catalysts [3, 4]. In this study, Pd-TiO_2 catalysts with different Pd loadings (< 1.5 wt%) were synthesized by using a flame aerosol reactor which is a continuous one-step process [5]. The catalytic properties and kinetic characteristics of the synthesized Pd-TiO_2 catalysts were evaluated for O_2 removal. The crystal structure and the size of the bulk Pd-TiO_2 were not changed by varying the Pd loading, while the size of the Pd subnano clusters/nano particles and the fraction of the three Pd species (metallic Pd, intermediate PdO_x ($0<x<1$), and PdO) were significantly affected by changing the Pd loading. Scanning transmission electron microscopy (STEM) and X-ray photoelectron spectroscopy (XPS) were utilized to determine the size of the Pd subnano clusters/nano particles and the oxidation states of Pd. Based on the analytical results, the total surface area of the different Pd species were calculated, and a correlation to the apparent reaction rate constant was evaluated. Among the three different Pd species, the total surface area of the metallic Pd showed the strong correlation to the apparent reaction rate constant, which reveals that the metallic Pd is a major active phase in the O_2 removal reaction. Our finding emphasizes a pivotal role of metallic Pd for the effective O_2 removal, and a further modification of the catalyst could be achieved based on this finding to enhance the O_2 removal efficiency.

References:

[1] A. Gopan, B.M. Kumfer, J. Phillips, D. Thimsen, R. Smith, R.L. Axelbaum, Process design and performance analysis of a Staged, Pressurized Oxy-Combustion (SPOC) power plant for carbon capture, Appl Energ, 125 (2014) 179-188.

[2] DOE/NETL, Quality guidelines for energy system studies: CO_2 impurity design parameters, (2013).

[3] Q. Zheng, S. Zhou, M. Lail, K. Amato, Oxygen Removal from Oxy-Combustion Flue Gas for CO_2 Purification via Catalytic Methane Oxidation, Ind Eng Chem Res, 57 (2018) 1954-1960.

[4] A. N. Kuhn, Z. Chen, Y. Lu, H. Yang, Sequential oxygen reduction and adsorption for carbon dioxide purification for flue gas applications, Energy Technol-Ger, (2018).

[5] V. Tiwari, J. Jiang, V. Sethi, P. Biswas, One-step synthesis of noble metal-titanium dioxide nanocomposites in a flame aerosol reactor, Appl Catal a-Gen, 345 (2008) 241-246.

563905 In-Situ X-Ray Absorption Spectroscopy (XAS) Study on Bimetallic Iron-Nickel Based Nanoparticles for Oxygen Evolution Reaction (OER) Catalysis

Recent studies have revealed that iron-nickel based nanoparticles are excellent catalysts for oxygen evolution reaction (OER) in alkaline conditions. It is critical in the development and optimization process of catalysts to perform an in-depth characterization. We have centered this work on x-ray absorption spectroscopy (XAS); a dynamic characterization technique. XAS is divided into x-ray absorption near-edge structure (XANES) and extended x-ray absorption fine structure (EXAFS). For this research, we designed an in-situ electrochemical cell that could collect spectroscopy data on the catalysts while an electrical potential is applied. We selected two iron-nickel based nanoparticles ($Fe_{20}Ni_{80}$ and NiO_x – NiO_x/FeO_x) developed from varying synthesis approaches as the experimental samples and performed cyclic step chronoamperometry (CSCA) and cyclic voltammetry (CV) on the nanocatalysts in the operando cell. $Fe_{20}Ni_{80}$ and NiO_x – NiO_x/FeO_x have exhibited great promise as OER catalysts in our previous investigation. Therefore, the work focused on trying to understand the mechanism behind the high OER activity of the two catalysts. The role of synthesis methods on the catalytic behavior of the two nanoparticles was also investigated. XANES and EXAFS analysis also revealed additional electronic and geometric information to better comprehend the complex bimetallic nanoparticles.

574289 Structure-Property Relations of $PtWO_x/C$ Inverse Catalysts

Owing to the strong synergy between Pt and WO_x, the $PtWO_x$ catalyst has been found to be efficient for a number of important reactions such as the oxidation of propane[1], ring-opening of tetrahydrofurfuryl alcohol[2] and hydrodeoxygenation of biomass derived compounds[3,4,5]. Accumulated knowledge underscores that the close proximity of Pt and WO_x is crucial. However, catalysts prepared via conventional methods are heterogeneous in composition and particle size, rendering the establishment of structure-property relations difficult. Recently, atomic layer deposition (ALD) has been employed deposit a near monolayer of WO_x film on the surface of carbon-supported Pt. Comparing to the conventionally prepared $PtWO_x$ catalysts, the inverse catalyst has demonstrated much higher efficiency for ring opening of tetrahydrofurfuryl alcohol.[2] It has been proposed that the partially reduced WO_x acts as a Brønsted acid that activates C-O bonds and stabilizes cationic intermediates.[2] However, the interaction between Pt and WO_x remains elusive. Structure-function correlations of the $PtWO_x/C$ catalysts under working conditions are still lacking.

In this study, we use probe molecules, e.g., alcohols, to investigate the Brønsted acid sites on WO_x (@Pt) under different reaction environments and pretreatment conditions. Those probe molecules undergo acid-catalyzed elimination reactions and are sensitive to the configurational changes of $PtWO_x$ catalyst during reaction. Combined strategies including kinetics, theoretical calculations as well as *operando* characterization reveal an intriguing interplay between Pt and WO_x.

References:

1. Wu, X. *et al.* Total Oxidation of Propane on $Pt/WO_x/Al_2O_3$ Catalysts by Formation of Metastable Pt^{1+} Species Interacted with WO_x Clusters. *J. Hazard. Mater.* **225–226,** 146–154 (2012).

2. Wang, C. *et al.* A Study of Tetrahydrofurfuryl Alcohol to 1,5-Pentanediol Over $Pt-WO_x/C$. *Catal. Letters* **148,** 1047–1054 (2018).

3. Janampelli, S. & Darbha, S. Selective and Reusable $Pt-WO_x/Al_2O_3$ Catalyst for Deoxygenation of Fatty Acids and Their Esters to Diesel-range Hydrocarbons. *Catal. Today* **309,** 219–226 (2018).

4. Wang, J. *et al.* Hydrogenolysis of Glycerol to 1,3-propanediol under Low Hydrogen Pressure over WO_x-Supported Single/Pseudo-Single Atom Pt Catalyst. **116023,** 784–790 (2016).

5. Wang, C. *et al.* Mechanistic Study of the Direct Hydrodeoxygenation of m-Cresol over WO_x-Decorated Pt/C Catalysts. *ACS Catal.* **8,** 7749–7759 (2018).

559242 Annealing Studies of Cobalt Supported on Hydrothermally Synthesised Carbon Spheres for Fischer-Tropsch Synthesis

Carbon spheres (CSs) synthesised by a low temperature (190 °C) hydrothermal synthesis method using sucrose as a substrate are reported. The effects of annealing the synthesised CSs at temperature range of 250-800 °C under N_2 gas was investigated. Annealing the CSs changed the physical and chemical properties of these materials. BET studies showed that annealing these CSs could increase their surface area from 2.1m^2/g to 431m^2/g. The CSs were also characterised by TEM, XRD, TGA and Raman spectroscopy. Raman spectra ratios of I_D/I_G, I_{D2}/I_G, I_{D3}/I_G, I_{D4}/I_G and the S-band are reported for the annealed CSs. Co supported on the annealed CS were characterised and evaluated for Fischer-Tropsch synthesis.

568198 Design and Evaluation of Nanostructured Doped Perovskite Oxygen Carriers

Doped perovskites have demonstrated facile oxygen vacancy formation and stability over many reduction cycles, making them good candidate materials for oxygen carriers in chemical looping processes [1]. Prior work has illustrated that the introduction of dopant in the perovskite lattice can have a dramatic impact on the material's phase stability and reducibility [2,3]. We consider a $BaFe_{1-x}In_xO_3$ doped perovskite that has been shown to release oxygen at lower temperatures upon addition of In in the lattice.

Using density functional theory (DFT)+U method, we have evaluated the oxygen vacancy formation energy for a set of 74 rotationally-distinct motifs spanning a range of dopant concentrations. From these data, we can make simplified structure-function relationships to predict the reducibility of different patterns of dopant in the lattice. Since the placement of dopant is highly combinatorial, it is intractable to enumerate the possible dopant patterns. Instead, we have formalized the search space in a mathematical optimization model that allows an efficient search over the entire space of dopant positions to identify patterns with optimal properties [4].

In this work, we investigate the sensitivity of oxygen excess energy against several parameters that are expected to impact the material performance. Specifically, we first consider the impact of multiple oxygen removals on the trends of oxygen excess energy. We investigated trends with both the number of oxygen removals and the distance between vacancies. We also study the energy of forming different combinations of dopant and vacancy structures in the lattice, improving our predictions for the likelihood of structures forming spontaneously. This allows us to relax some of our previous modelling assumptions and develop more refined models for the performance of our doped perovskite oxygen carrier.

The specific model improvements are demonstrated in the context of a $BaFe_{1-x}In_xO_{3-\delta}$ doped perovskite, but are formulated in a generic way so as to be applicable to a wide range of material systems. Generally, we demonstrate how simplified structure-function relationships can be refined and embedded in an optimization model in a modular way, preserving the overall structure of the optimization problem. The resulting optimization models can be solved via standard optimization approaches and provide solutions that can inform future synthesis efforts and provide theoretical bounds of material performance.

References

[1] Motohashi, T., Ueda, T., Masubuchi, Y., Takiguchi, M., Setoyama, T., Oshima, K., & Kikkawa, S. (2010). Remarkable oxygen intake/release capability of BaYMn2O 5+δ: Applications to oxygen storage technologies. *Chemistry of Materials, 22*(10), 3192–3196.

[2] Baiyee, Z. M., Chen, C., & Ciucci, F. (2015). A DFT+U study of A-site and B-site substitution in $BaFeO_{3-\delta}$. *Phys. Chem. Chem. Phys., 17*(36), 23511–23520.

[3] Ishihara, T., Matsuda, H., & Takita, Y. (1995). Effects of Rare-Earth Cations Doped for La Site on the Oxide Ionic-Conductivity of $LaGaO_3$-Based Perovskite-Type Oxide. *Solid State Ionics*, 79, 147–151.

[4] Hanselman, C.L., Alfonso, D.R., Lekse, J.W., Tafen, D.N., Matranga, C., Miller, D.C., & Gounaris, C.E., Tuning oxygen desorption in a doped $BaFe_{1-x}In_xO_3$ perovskite oxygen carrier. Accepted for Publication.

570663 Synthesis of Zeolite-Encapsulated Metal Nanoparticles Via a Cationic Polymer-Assisted Strategy for Substrate-Selective Catalysis

Metal nanoparticles (NPs) on a solid support have been widely used as heterogeneous catalysts. However, supported metal NPs frequently suffer from the sintering and leaching of NPs. A proven strategy for addressing these challenges is to encapsulate metal NPs into zeolites. We developed a novel cationic polymer-assisted synthesis method for embedding metal NPs within MFI zeolites (denoted as M@MFI, M = Pt, Au and Pd). This method could be extended to other zeolite frameworks such as BEA and CHA. The resultant M@MFI, M@BEA and M@CHA catalysts were investigated for substrate size-selective reactions, and showed 10 times higher substrate size-selectivity than the corresponding zeolite-supported metal NPs, i.e., M/MFI, M/BEA and M/CHA. These results were attributable to the encapsulation feature of the NPs inside zeolite frameworks.

562015 High Performance Enzymatic-Transesterification in an Anhydrous Gas Flux Using Lipase Encapsulated in Graphene Oxide Aerogel

Gaseous enzymatic catalysis that holds great promise to chemical engineering and related fields is hindered mainly by the extremely low activity of enzyme in a dry form. The adding of moisture, however, leads to a poor stability of the enzyme. Here we propose and demonstrate a method to prepare an enzyme encapsulated in graphene oxide (GO) for gaseous enzymatic reactions. A lipase-GO aerogel (LGA) was prepared by mixing Candida Antarctica lipase B (CALB) with the oppositely charged GO at pH 3.3 followed by lyophilization, and examined with gaseous transesterification of geraniol and vinyl acetate. While exhibiting a 5 to 10-fold apparent activity over lyophilized CALB powder independent of the water activity, the LGA maintains its initial activity in an anhydrous gas flux, being 67 % of the peak value obtained in the optimal water content, over the temperature range from 20 to 80°C. The solid-state circular dichroism (ssCD) measurement confirms that the lipase encapsulated in LGA keeps its native conformation, and the thermogravimetric analysis (TGA) shows that the GO sheets replace a significant amount of water molecules essential for the lipase activity in the bulk, possibly. These results indicate that GO stabilizes the lipase activity by forming a water-like microenvironment through multiple hydrogen bonding and eliminating the random attacks by free water molecules. The LGA is able to display its initial activity over 500 hours of in anhydrous gas-phase reactions, highlighting its great potential for industrial synthesis of chemicals by gas-phase enzymatic reactions.

573546 Rational Catalyst Design: Kinetics Put into Action for Small Open Data

Introduction

The development of more sustainable processes as the main source for carbon-based molecules calls for a breakthrough in catalyst design. To do so, the exploration of the chemical space for a new catalyst formulation can be efficiently achieved by computers. Such computational research can be conducted via molecular modelling or via experimental data-based mathematical models. In the latter approach, a set of catalysts featuring highly diverse formulations and properties (e.g. combining metals and supports) is firstly screened experimentally prior to any mathematical model development[1]. Only afterwards, the data will be exploited to extract the underlying chemical information leading finally to the regression and/or training of a mathematical model. In such a manner, a long period of experimental data acquisition is required. Alternatively, the upcoming Open Data might be used.

Up to now, research has been mainly focused, on the one hand, on automating data visualization and statistical inference[2] for Big Data, and, on the other hand, on the development of the models themselves, either "black-box" (e.g. machine learning based) or kinetic ones, for catalytic data. However, to take full advantage of the Open Data revolution, new methodologies for scientific data mining, adapted to the characteristics of catalytic small data, have to be developed.

In Figure 1, a general approach for data-based rational catalyst design is described. Step 1 is the extraction of chemical (preferably kinetic) information from the data. In Step 2, this information, together with the data, serves as input for a mathematical model. Finally, in Step 3, this mathematical model, via relationships with catalyst structure and/or properties, provides guidelines for the synthesis of better catalysts (Step 4). However, suitable methodologies to convert catalytic data into useful kinetic information (as in Step 1) or information into knowledge on optimized catalyst structure, via catalytic data, (as in Step 3) are lacking.

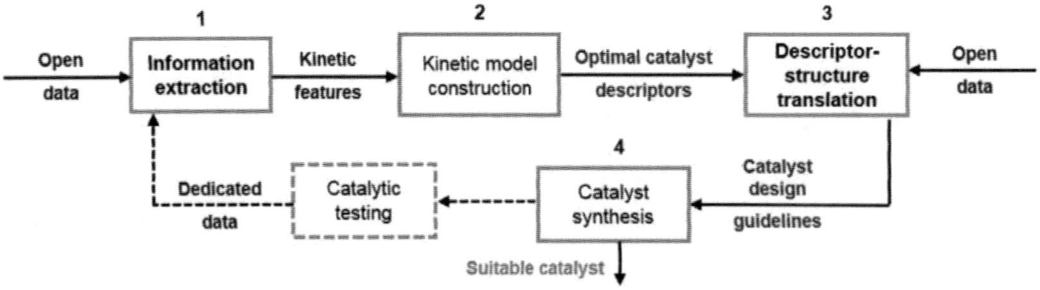

Figure 1: Data-based rational catalyst design cycle. Focus of this work: green.

Hence, this work aims at shedding light at the state-of-the-art of Open Access catalytic data and developing the suitable tools to evaluate the underlying kinetic information and relate it to catalyst structure or properties by making use of these small datasets.

Open Access catalytic data

As a starting point, the current storage of data on catalysis was evaluated. Open Access storage of data has been mainly driven by recent policies from funding bodies and publishers. Hence, to investigate the ongoing potential, the focus should be on a reaction widely investigated in the past few years. Hydrodeoxygenation has been receiving significant attention in the last years, but catalyst design remains an open challenge [3]. By employing a broad search for whole reaction families under "hydrodeoxygenation", well-known repositories and search engines were surveyed (Table 1). The overall number of available datasets is modest as compared to the number of publications in the same field (ca. 2600, according to WoS at the same moment). Particularly, if one excludes Figshare (mainly a repository of article's preprints). In the other cases, not all search results were relevant (e.g. data generated by molecular modelling calculations), and data were often unstructured and flawed.

Table 1: Number of Open Access datasets on hydrodeoxygenation (October 2018).

REPOSITORIES	DATASETS	SEARCH ENGINES	DATASETS
Mendeley Data	3	OpenAIRE	3
Zenodo	0	DataCite	5
Figshare	349	Google Dataset Search	11
ResearchGate	13	SHARE	3

In summary, one cannot presently rely on Open Access data, but more data sharing, better curation and standardized formats, following the FAIR Data Principles [4], will turn it into reality in the near future. Furthermore, literature mining software, already employed in biology, can enable the access to virtually the whole data generated till present. In that sense, the characteristics of (upcoming) Open Access data will result from the combination of individual datasets, as we know them today. For a given reaction, these will result in a significant amount of data, but very far in volume and balance from the massive volume of big data. Conversely, the existing data visualization and inference tools, designated under the umbrella term "machine learning", have been designed for truly big data, i.e. huge amounts of well-balanced data[11]. Hence, tools adapted to the considerable smaller size of catalytic data must be developed.

On the way to automate kinetic information extraction

As of today, the first step of information extraction (Figure 1) relies solely on the researcher's prior knowledge and experience. The methodology under development aims at filling in this gap. In practice, this means that all kinetic features (e.g. variations in conversion) must be identified and, preferably, classified in terms of relevance. This can be achieved via the recognition of patterns and fingerprints [5], e.g. abrupt variations in the performance indicators. The first step is to visualize the general trends in the dataset. Typically, a researcher draws a curve based on his/her intuition, which represents the overall trend in the data, acknowledging experimental error. The tool developed herein is, hence, meant to mimic how the researcher would draw the overall trend in a dataset.

The developed algorithm is based on the class "UnivariateSpline" in "SciPy module" of Python [6]. The latter consists of an iterative procedure in which the number of piecewise polynomials (which altogether constitute a spline) is increased until the residual sum of squares is below the defined tolerance level. The results with fictive data (Figure 2.a and .b) indicated that the algorithm could not adequately reproduce simple shapes. By testing data featuring different trends, it also became clear that the optimal tolerance level depends on the dataset.

A new algorithm featuring lower tolerance was thus developed. To prevent overfitting, the tolerance level is decreased until a maximal number of piecewise polynomials is reached. In some cases, particularly for small datasets, the piecewise polynomials can still be superfluous. Therefore, the spline is replaced by a lower-degree polynomial if the generated trends are not physically realistic (e.g. excessive variability) or if that polynomial yields a goodness-of-fit sufficiently similar to the higher-degree one. In addition, a feature classification function in terms of shape and variability was also introduced. The results are shown for a fictive dataset with an S-shape (Figure 2.c) and two real anisole hydrodeoxygenation datasets by Otyuskaya et al. [7] (Figure 2.e and .f). For the conversion, the trend could be described by two shapes, while the selectivity could be described by a linear increase. The developed algorithm is hence able to follow the intuitive trends in data for variable number of points, shape, and variability.

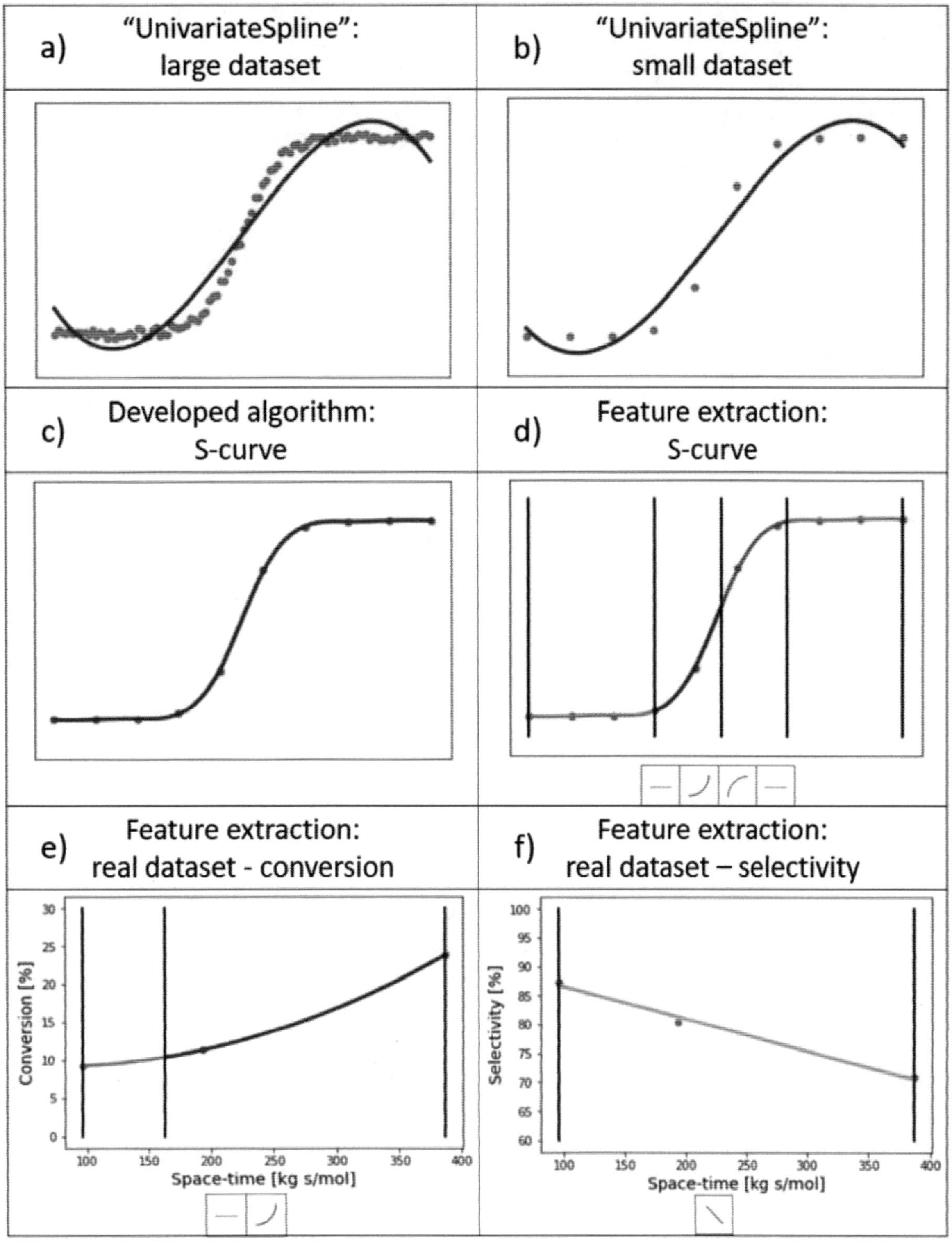

Figure 2: Performance of state-of-the-art and developed algorithms.

From descriptors to structure: a case study

In order to establish relationships between kinetically-relevant catalyst features and parameters which can be tuned during the synthesis procedure, data diverse in catalyst performance is required. Fortunately, in the field of Oxidative Coupling of Methane, studies involving more than a few catalysts have been carried out paving the way for its utilization in catalyst design even before the advent of Open Access data. The most prominent case comprises forty-four catalysts tested at similar operating conditions[8], i.e. small catalytic data.

To match experimentally observed performances with simulated ones and potentially draw significant relationships, microkinetic simulations were carefully combined with statistical tools. The microkinetic simulations were carried out using a state-of-the-art model [9] and varying the catalyst descriptors by the means of Design of Experiments. By combining the simulated catalyst with ones from the referred dataset, this resulted in the identification of four clusters of catalysts holding distinct performances (Figure 3). Interestingly enough, a cluster (the blue) with optimal performing catalyst could be identified, but no tested catalysts were included. More importantly, by comparing the descriptors of different clusters, relationships between the composition and properties of the catalysts tested by Kondratenko et al. [8] and the simulated catalyst descriptors are being drawn.

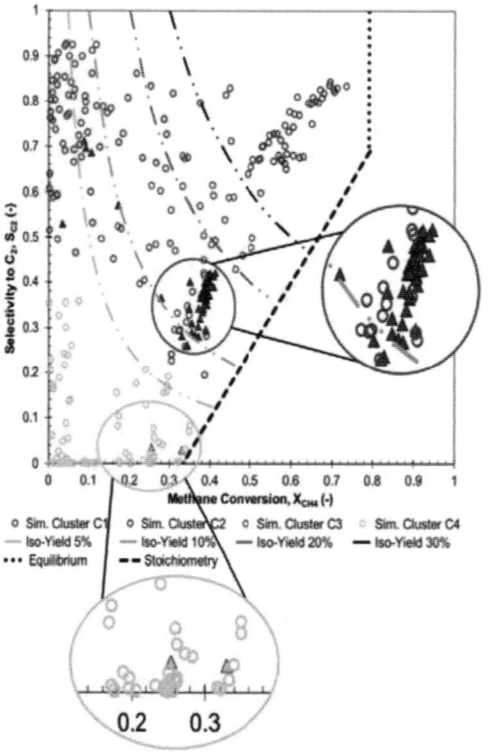

Figure 3. Comparison of experimental[8] (closed) and simulated (open symbols) data for OCM catalysts at iso-operating conditions. The color code distinguishes clusters of catalyst with comparable performance.

Summary

The lack of curated and standardized Open Access data on catalysis precludes its use at present, but the ongoing policies will overcome this obstacle. This will generate small catalytic data which cannot for which machine learning techniques are not adapted. To efficiently make use of such data, a tool for automated kinetic information extraction is under development which can, as of today, recognize the relevant patterns in small data, mimicking the intuition of a researcher. Finally, a methodology able to extract knowledge for catalyst design for typical catalytic data has been also developed.

Acknowledgments

Funding was granted by Ghent University BOF (BOF18/PDO/093) and EU commission (ERC Grant No. 615456).

References

1. Van der Borght, K. et al., *Catalysts* **2015,** 5, 1948.

2. Chiang, L. et al., *Annu. Rev. Chem. Biomol. Eng.* **2017,** 8, 63-85.

3. Chen, S. et al., *Renewable and Sustainable Energy Rev.* **2019,** 101, 568-589.

4. Wilkinson, M. D. et al. , *Scientific Data* **2016,** 3, 160018.

5. Caruthers, J. M. et al., *J. Catal.* **2003,** 216, 98-109.

6. Bellussi, G. et al., *Catal. Sci. Technol.* **2013,** 3, 833-857.

7. Otyuskaya, D. et al., *Energy Fuels* **2017,** 31, 7082-7092.

8. Kondratenko, E. V. et al., *Catal. Sci. Technol.* **2015,** 5, 1668-1677.

9. Pirro, L. et al., *Ind. Eng. Chem. Res.* **2018,** 57, 16295-16307.

563679 Novel Nanostructured Transition Metal Oxides for Catalysis

Transition metal oxides (TMOs) have been the subject of growing interest in the catalysis community in the past decades owing to their low cost as a great alternative for precious metals and excellent catalytic performance. Their unique electronic structure of the outer d-electrons has offered them versatile abilities for optical, electrical and catalytical property engineering. The high oxygen mobility in TMOs also gives rise to easy stoichiometry tuning, heteroatoms doping as well as modifications of electronic band structures, making property control more feasible. In addition, TMOs possess high structural tunability thus allowing for the synthesis of various nanostructures with different dimensionalities ranging from 0D to 3D. In this work, a variety of facile wet synthetic approaches, such as soft templating, hard templating and template free, will be exploited to controllably synthesize novel TMO nanomaterials including copper oxide, iron oxides and manganese oxides with different nanodimensionalities. These nanomaterials and their composites were well characterized and understood via an assortment of tools including SEM/TEM/AFM, XRD/XPS/XAFS/XRF, FTIR/UV-VIS/UV-DRS/NMR/EPR and BET physisorption/chemisorption etc. Temperature programmed reactions such as TPR (reduction) /TPO (oxidation) /TPD (desorption) were used to study the reductive/oxidative/desorptive behaviors of these catalysts. To investigate their catalytic performance, a combination of straight gas phase catalysis, electrocatalysis and photocatalysis will be explored including CO_2 hydrogenation via reverse water gas shift route (rWGS) for iron oxides and manganese oxides, CO_2 electroreduction for CuO, photocatalytic water splitting for iron oxides and CuO, etc. In addition, first principle computations will be highly coupled with our experimental results in order to guide the catalyst designs and mechanistic understandings. The core of this work will be hypothesis-driven, enlightening from TMO structural and computational studies, and application-oriented in a sense of seeking fascinating properties.

564645 Defect Engineering of in-Situ Exsolved Nanoparticle Catalysts on Perovskite Supports

2019 AIChE Annual Meeting

Abstract for Poster Presentation

Defect Engineering of In-Situ Exsolved Nanoparticle Catalysts on Perovskite Supports

Soham Shah*[1], Samuel Sayono[2] and Kandis Leslie Abdul-Aziz[1,2]

Department of Chemical and Environmental Engineering, University of California, Riverside.

Materials Science and Engineering Program, University of California, Riverside

sshah034@ucr.edu

Perovskites are a class of mixed metal oxides with general stoichiometry of ABO_3 and have drawn the interest of researchers for their versatility in applications from solar cells to fuel cells and catalysts. An interesting phenomenon occurs in some defective perovskites – dopant metals leave their positions in the lattice, when subjected to high temperatures in a reductive atmosphere. These atoms diffuse or "exsolve" through the perovskite matrix and form nanoparticles on the surface of the parent perovskite. Exsolved nanoparticles display strong adherence to the perovskite support from which it emerged. Consequently, their enhanced resilience and activity makes them stand out as candidates for catalysis, fuel cells and other emerging energy conversion technologies.

With the objective to gain insight about the driving forces behind exsolution, a series of lanthanum ferrite perovskites were synthesized, with nickel and cobalt as dopant metals. These perovskites either had deficiencies in La (A site), Fe (B site) or neither site (Over-Stoichiometric). The perovskites were reduced and characterized with XRD, SEM, TPR, XPS, STEM, EDS and these methods confirmed the exsolution of Ni and Co oxide nanoparticles in samples reduced at 600°C. However, samples reduced at 700°C had nanoparticles that also contained iron while what remained of the parent support material was largely lanthanum oxide. We conclude that at higher temperatures as the diffusion of B site atoms is accelerated, the Fe iron atoms could also exsolve along with the dopant atoms and form alloyed nanoparticles. This could open new avenues in the study of in-situ exsolution and their application in devising multi-metallic particles for catalysts.

The objective is to understand the role of defects in perovskites in driving exsolution of nanoparticles. "Defect Engineering" can then be exploited to direct perovskite materials to have optimized properties relevant to specific applications.

565014 Characterization of Cyclodextrin-Based Metal-Organic Frameworks with Catalyst Molecules

Cyclodextrin-based metal organic framework (CD-MOF) consisting of g-cyclodextrin (g-CD) and potassium hydroxide has a porous cubic crystal structure in which the primary units of $(g\text{-}CD)_6$ were packed into a body-centered cubic array. CD-MOF is easily prepared by the methanol vapor diffusion method at room temperature. The spherical nanopores with a diameter of 1.7 nm are formed at the center of $(g\text{-}CD)_6$ units and show a hydrophilic nature by the free primary 6-OH groups in the $(g\text{-}CD)_6$ unit. Meanwhile, a pair of g-CD tori in CD-MOF forms hydrophobic nanopores (size of 1 nm). Thus the channel in CD-MOF is a pearlnecklace-like channel connected by hydrophilic and hydrophobic nanopores. Once larger guest molecules compared to the aperture size of g-CD tori are encapsulated into the CD-MOF, they cannot escape from CD-MOF without the collapse of the crystal. CD-MOF has high crystallinity and is a stable crystal at temperatures below 473 K in organic solvents. CD-MOF encapsulating catalytically active molecules would be expected as an advanced catalyst having a high specific surface area.

In the previous paper (Nagai et al., 2018), water soluble porphyrin (TCPP) was found to be introduced into the spherical hydrophilic nanopores during the crystallization process. Fullerene C60 was successfully encapsulated in the hydrophobic nanopores of CD-MOF by the methanol vapor diffusion using C60/g-CD complex in KOH solution.

In this research, Co(II)TCPP was synthesized from cobalt(II) acetate tetrahydrate and encapsulated into CD-MOF (Co(II)TCPP/CD-MOF). The catalytic activity of Co(II)TCPP/CD-MOF was investigated in the coupling reaction of 2-methoxy-4-methylphenol. In addition, we investigated hydroxylation of phenylboronic acid over C60/CD-MOF under visible light irradiation to confirm the generation of singlet oxygen from C60/CD-MOF.

The crystallization yield of CD-MOF was 74% and decreased to 28% for C60/CD-MOF and 57% for Co(II)TCPP/CD-MOF, respectively. However, an XRD analysis proved that there was no change in crystal structure of CD-MOF after introducing the C60 and Co(II)TCPP molecule.

The oxidative coupling of 2-methoxy-4-methylphenol (cresol) was carried out in methanol solution of Co(II)TCPP and Na_2CO_3 under oxygen atmosphere at 333 K for 15 hours. Conversion of cresol was as high as 88%. Meanwhile, Co(II)TCPP/CD-MOF behaves as a heterogeneous catalyst in methanol. Under the same conditions as Co(II)TCPP, the conversion of 90% for Co (II) TCPP/CD-MOF revealed the almost same catalytic activity as the homogeneous catalyst. These results indicate that CD-MOF is effective as a catalyst support of organic complex catalysts and organocatalysts.

Fullerene C60 is known to generate singlet oxygen under visible-light irradiation and therefore attracting attention as a photodynamic therapy drug that attacks cancer cells. In addition C60 capable of generating singlet oxygen can be used as a metal-free photocatalyst. In this study, photocatalytic hydroxylation of phenylboronic acid in chloroform was performed using C60/CD-MOF. After 6-h reaction, the conversion for the agglomerated C60 catalyst was 65.3%, while the conversion for the C60/CD-MOF catalyst greatly decreased to 6.0 %. This low conversion is considered to be influenced by pore diffusion rate of phenylboronic acid in CD-MOF. Effect of the pore diffusion rate on the reaction rate in this system will be investigated in the future.

565687 Surface Phase Diagrams of Strontium Titanium Oxide Using an Automated Ab-Initio Grand Canonical Monte Carlo Method

Solid state fuel cells are an alternative energy technology with great potential to impact the energy sector but require improvements to their cost and efficiency to achieve real-world viability. Currently, platinum electrodes are used as catalysts for the oxygen evolution reaction (OER) and oxygen reduction reaction (ORR) in regenerative fuel cells. However, Pt resources are limited, and the electrodes are limited by chemisorption scaling relations which prevents optimization of OER and ORR simultaneously. Perovskite materials have been studied for a variety of catalytic applications, including the OER and ORR. Transition metal oxide perovskites of the structure ABO_3 are potentially good replacements for the anode and cathode catalysts due to their stability under OER and ORR and their ability to be altered structurally, allowing for a greater optimization of the reaction for a fraction of the cost of noble-metal electrodes. Also, these materials are comprised of earth-abundant elements.

Reconstructions under different chemical environments of the material surfaces influences the surface chemistry and effectiveness of the ability to catalyze the reaction. Surface reconstructions of the perovskite catalysts was explored using an automated ab initio Grand Canonical Monte Carlo (GCMC) method to sample across thermodynamically relevant surface configurations. In this work, this method was utilized for the generation of surface phase diagrams of strontium titanium oxide (ST) which is important in predicting and understanding the surface chemistry behavior of the materials under differing conditions.

Reconstructing the surfaces of perovskite catalysts can circumvent factors that limit the OER. Theoretical and experimental studies on STO3 have shown that a TiO_2 rich surface can reduce the overpotential of the catalyst for this reaction.

Regeneration of $PtSOxZrTi/SiO_2$ By Hydrogen Treatment. Part I. Spent Catalyst Characterization

Roberto Galiasso Tailleur,**, Carlos Farina** and Rough P Ronaldi***
*Simon Bolivar university Caracas Venezuela and **Hypro Consultants Orlando, FL USA

Abstract

$PtSO4TiZr/SiO_2$ spent catalysts used for alkylation of isobutane with 2-butene in presence of hydrogen were produced in a pilot plant and then extracted with CS_2 at different operating conditions. The spent, and extracted catalysts were obtained were characterized, and alkylation activity measured in a continuous stirred reactor. Solid were characterized by nitrogen adsorption, FTIR, XPS, TEM, MASS ^{13}CNMR, CS2 extraction and chemical analyses to determine the characteristic of the coke. It was found that there are three different types of coke (hydrocarbons) present on spent catalysts produced during alkylation of isobutane by olefins in pilot plant; The characterization determined the average species present in these three types of cokes. The characteristic and reactivity of coke in presence of hydrogen depend of alkylation operating conditions and the number of passes through alkylation-regeneration process. The mechanism of coke formation in three active sites and the role of hydrogen is discuss based on solid characterization and DFT simulation.

Keyword spent alkylation catalyst characterization

1) Introduction.

The definition of coke is arbitrary and, by convention, related to their origin. Coke is produced by decomposition or condensation of hydrocarbons on catalyst surfaces and typically consists of polymerized heavy hydrocarbons. Nevertheless, coke forms may vary from high molecular weight paraffinic adsorbed hydrocarbons to primarily carbons such as graphite, depending upon the conditions under which the coke was formed and aged. Books and reviews treat the coke on catalysts and the attendant deactivation of the catalysts (see for example Melon et al. review [1]). Catalyst deactivation the type of coke formed have been analyzed in different type of catalysts [2,3], but there is no information about the carbonaceous deposits on SO4Zr, SO4Ti and $PtZTiSO_x/SiO_2$, other than our own previous papers [4,6]. The $PtZTiSO_x/SiO_2$ catalyst was used in a combine pilot plant alkylation-regeneration to alkylate light olefins [5,6]. Therefore, we proceed to perform the characterization of the alkylation spent catalyst obtained in a continuous alkylation pilot plant to understand the mechanism of deactivation.

The above-mentioned catalyst was then regenerated by hydrogen at different operating conditions to recover their activity [7]. The current study is directed to improve this regeneration stage. The following section present the characterization of carbonaceous deposited on metal and acid active sites.

1) Experimental

2.1 Catalyst preparation. $PtSO_xZrTi/SiO_2$ catalyst was prepared by a gel co-precipitation of beta-diketone (2,4-pentane-dione) tetra-ethyl-orthotitanate, and a tetra-ethyl-orthozirconate (Aldrich) in ethanol. Then C22-N4 -C22, under a sodium-free condition, and Tetra-ethoxy-silane (TEOS) were added under intense stirring. The precipitate was

washed with ammoniac solution at 298 K and then dried under vacuum at 350 K for 8 hours. The porous structure of as-synthetized solid was Silicalite-type zeolite, hard spheres with a particle size between 60-80 microns. Then, this solid was treated first in steam ammonia for 2 h, and then by oleum; the solid described in detail reference [8] was dried, impregnated with H_2PtCl_6 in water, dried again at 393 K and calcined in air at 623 K for 4 hours. This solid was activates in hydrogen at 363 K for 10 minutes before use in alkylation. The catalyst particle size is controlled by the synthesis between 60-80 microns.
A large sample of spent $PtZrTiSO_x/SiO_2$ catalyst was produced in a three liters CSTR pilot plant in continuous operation for 5 days on stream. Pilot plant was operated at 363 K, 1.4 MPa, 0.4 h of liquid residence time, 0.2 h of hydrogen residence 0.4 h solid residence time. Steady state operation was achieved by adding and withdrawing the catalyst each five minutes using an automatic rotating valves system. Catalyst activity was kept constant at different level of 2-butene conversion. The reactants have 0.3 molar ratio of $C_4^{=}/iC_4$, and the reactor was operated at 0.6 MPa of hydrogen partial pressure. The solid withdraw was vacuumed at 1 kPa, 303 K for 0.4 h (D^1 spent catalyst). D^1 sample was characterized by different techniques and activity measured for alkylation at 353 K, 0.6 MPa of hydrogen and 1.4 MPa of total pressures.

2.2. Spent catalyst characterization

Spent and extracted catalysts were characterized by BET, MAS ^{29}SiNMR, MAS ^{13}CNMR, IR, XPS and Pyridine (Py) and hexyl-cyclo-hexane adsorptions.

2.2.1 Chemical analysis. Catalyst bulk composition was determined by atomic adsorption spectroscopy (**Table 1**). Carbon analysis was performed by Mac Bain micro-combustion thermo-balance coupled with a GC-MS. Sulfur concentration was determined by combustion in a Leco CS-244 analyzer and platinum concentration by using a Perkin-Elmer ICP/5500 plasma spectrometer (ICPAES).

2.2.2 Analyses of liquid extracted by CS_2-Benzene. Hydrocarbon obtained by CS_2 plus benzene extraction (mention as extract) in a CSTR at different operating conditions (1.4 MPa, 363 K, for 0.2 h) were analyzed by GC-MS. The samples of catalyst obtained from pilot plant (D^1) was extracted in a pressurized vessel to produce the sample D^1_{CS2}. The solution (estract) was filtered to remove the insoluble carbonaceous solids and analyzed by GC-MS.
The gas chromatograph is an HP 6890 equipped with 10 m, 0.53 mm (i.d.) Poropak column. Gas used as carrier was helium. Inlet split-ratio was set at 10:1. Mass Spectrometer is an (HP 5972) with scan range of 20-200 amu. Interface temperature: 250°C Ion source temperature: 200°C Solvent elution time: 0.50 min, Data sampling time: 1 min to 50 min, measurement mode: Scan, Mass range: *m/z* 35 to 500, event time: 0.3 sec. Mass spectra provide the C_{10}-C_{16} carbon distribution and HNMR the average di and tri-branches distribution.

The mass spectrometer interface temperature was 450 K the ion source temperature: 473 K, solvent elution time of 0.50 min, data sampling time of 1 min to 50 min. Mass spectra provide the C_{10}-C_{16} carbon distribution and the HNMR analyses the average di and tri-branches distribution.

The Pyrolysis-GC-MS was performed using in-house built pyrolizer coupled with above mentioned GC-MS. The carbonaceous solid obtained by dissolving the catalyst matrix by HF is place in the solids-injector and introduced into the furnace pyrolizer at 700 K. The pressure of helium carrier gas at the inlet to the furnace was 102 kPa. The pyrolizer was connected to the GC- that use a silica capillary columns 100 m long with Turex-400 stationary phase. The spectra of mass fractions obtained was based on three main pics. The temperature of the split/splitless injector was 523 K and the split ratio was 20:1. The transfer line temperature was 550 K. The ionization occurred with a kinetic energy of the impacting electrons of 70 eV. The quadrupole temperature was 498 K. Mass spectra and reconstructed chromatograms were obtained by automatic scanning in the mass range m/z 35 – 750 u. GC/MS data were processed with the "Aditus" program.

2.2.3 Surface area and pore volume. The pore structure of the for spent catalysts samples were measured using nitrogen sorption Micromeritics ASAP 2010 apparatus at 70 K; then BET or BJH analyses of the data were performed by a dedicated program. Adsorption-desorption procedure were automatically performed with 5s equilibration time for each point. The results are reported in **Table 1**. Nitrogen adsorption isotherms were evaluated in the low-pressure region of $P/P_0 < 0.1$, where micropores are located, and in medium-pressure region of $0.4 < P/P_0 < 0.5$ to measure mesopores (**Fig 1**). The former region could be interpreted as due to the N_2 filling in the micropores, and the latter to capillary condensation in the mesopores, according to the nonlocal density functional theory.
The relative accessible volume of pore was measured by pore filling and adsorption with hexyl-cyclo-hexane (HCH) at 303 K, 1.4 MPa of nitrogen partial pressure, for 2 h. This large molecule cannot access the micropores. Dried catalysts, with empty pores, were contacted by HCH and the amount adsorbed measured by mass balance and desorption. The value of mmol of HCH /g are reported in **Table 1**.

2.2.4 ^{29}Si MAS NMR spectra. The spent catalyst D^1 was analyzed siliceous^{29} (MAS) NMRspectra using a Bruker DSX-300 spectrometer equipped with a 6 mm MAS probe with resonance frequencies of 59.4 MHz. Typically, a pulse width of 3.6 and a recycle delay of 10 s were used. Chemical shifts were externally referenced to TMS (tetra-methyl-silane) at 0.0 ppm. The spectra with overlapping bands were deconvoluted by in-house software.

2.2.5 Mass ^{13}CNMR. A Mass ^{13}CNMR (Bruker 300 apparatus) was used to analyze spent and extracted by CS_2 catalysts. The spectra provide information on the different type of Carbon structure in the support referenced to the resonance of $Al(H_2O)_6^{3+}$ set to 0 ppm (**Fig. 2**).

The HF-extracted solid (insoluble deposit) was analyzed by PYGC-MS equip (mentioned in 2.2.2) and by the MAS ^{13}CNMR equip mentioned above; the spectrometer analyses were performed at 303 K at frequency of 300 MHz and a rotation frequency of 4 kHz. Free induction decays (FID) were recorded with 45° flip angle preparation pulses of 3.5 μs duration, 25 μs ring down delay, 4 s recycle delay and 38 scans for accumulation.

2.2.6 FTIR. A Perkin Elmer 1800, Fourier-transform infrared (FTIR) spectroscopy, was used to determine the most characteristic carbonaceous frequencies in the infrared region. The samples were introduced in the IR cell, dried in argon, vacuumed and then the spectrum recorded with a fully computerized data storage and data handling capability. Measurements were carried out at room temperature using a Raflex type optical interferometer and a source of light covering the range from 4500 to 400 cm^{-1}. 50 particles of the four spent catalyst was powdered compressed in a self-supported ultra-thin film (about 18 mg cm^{-2}) and place in a special cell to measure the 1000-3000 cm^{-1} region bands.

2.2.7 Acid sites measured by Pyridine adsorption in hexane.
Then these samples were contacted with pyridine in hexane at 303 K 1.4 MPa of nitrogen pressure for 0.2 h; then the solid was vacuumed at 303 K for 0.2 h and the amount of pyridine calculated by mass balances and FTIR analyses. The acid sites evaluation of samples before and after extraction were performed in the same ways. FTIR of remaining adsorbed pyridine at 343, 353 363 K were performed in the range 1400-1600 cm^{-1}. The extinction coefficient ratio of coordinately bonded pyridine (LPy) and protonated pyridine (BPy) was determined in situ for different concentration of Py in the solid. The spectra were repeated three times with 64-scan data accumulation and a resolution of 4 cm^{-1}

2.2.8 XPS metal dispersion. Samples of 10 particles of spent catalyst D^1 and extracted catalyst D^1_{CS2} (60-80 microns diameter) were retained into copper foil and placed in a XPS reaction chamber of a Laybold Hereus XPS-340 apparatus. Samples were evacuated at10^{-9} Torr and introduced into the X-Rays spectrometer. The XPS instrument employed monochromatic Al Ka X-rays and the pass energy of the analyzer was 58.7 eV. The C 1s spectra was fits with two and three symmetric components.

2.2.9. Catalyst activity measured by 2-butene and isobutene.
The activity of spent and extracted catalysts with different amounts of deposit on surface was obtained in the alkylation micro-reactor reactor mentioned in reference [8]; operating conditions were 363 K, 0.4 h of contact time at 0.6 MPa of hydrogen and 1.4 MPa of total pressure. H_2/HC used was 0.3 molar and Cat/HC ratio of 0.08 by weigh. Liquid sample were analyzed by GC-MS and spent catalyst by micro-combustion and other techniques mentioned above.

3. Results
3.1 Physical properties of spent and extracted catalyst. Surface area and pore volume of the catalyst D, D^1 and D^1_{CS2} are shown in **Table 1**. The spent catalyst had accumulated 0.75 wt. % of carbonaceous deposits during the alkylation of isobutane with 2-butene to produce alkylate (C_6- C_9). The vacuum treatment (303 K, 1 kPa) that produce D^1 sample reduces the amount of deposit in 5% due to desorption of hydrocarbon weakly adsorbed in pores. D^1 catalyst do not desorb additional hydrocarbon under 10^{-8} Torr at 297 K (according to the XPS analyses of carbon signal intensities). Thus, the value of surface area and pore volume reported in **Table 1** are reproducible (±0.8%) base on three successive measures. The surface area of D^1 catalyst is reduce in 89%, mesopore volume in 19% and micropore volume in 91% respect to fresh catalyst (D) due to the accumulation of hydrocarbons deposit in pores. The extraction recovered surface area and mesopore volume, but most of

the micropores remain blocked by insoluble coke (**Table1**); the treatment had extracted 89% of deposits at 303 K, 1.4 MPa of nitrogen, and vacuum treatment at 303 K, 1 kPa. The pore distribution analyses (**Fig. 1**) indicated that there is 4% shift in the average mesopores diameter and 12 % in micropore diameter; the latter happen by the solvent deblocking of some micropores access.

3.2 Metals in the solid (bulk). The bulk metal contents of the catalyst are also depicted in **Table 1**. This type of catalyst has a relative low amount of SO_x (S is less than 0.85% by mol of Zr plus Ti) Pt content is also low in comparison $PtSO_xZr$ catalysts used for isomerization of paraffins. Maximum calcination temperature is set to 653 K based on TPO results, because above this value the catalyst loss sulfur. The Zr to Ti ratio in the bulk is less than 1.

3.3 Structure of the sample. The XRD spectra does not show any crystalline signal associated to silica or carbon. The presence of an incipient Silicalite-I structure was observed in a sample of and as-synthesized aged for 25 days in the mother liquor, but after activation in steam ammonia all the incipient silica structure disappeared.

Table 1 Physical properties of D^1 catalyst

Properties	D^1	D^1_{CS2}	D
Carbon content wt. %	0.75	0.16	-
Molecular weight	165.37	-	
PtO wt%	0.15		
ZrO_2 wt %	3.8		
TiO_2 wt%	4.2		
SO_4 wt%	5.5		
Area m^2/g by (N_2)	28	59	180
Micropores vol. cm^3/g	0.008	0.12	0.06
Mesopores vol. cm^3/g	0.23	0.26	0.29
HCH mmol/g	0.16	0.17	0.21

Table 2. XPS Metal dispersion

Cat/disp.	D^1_{CS2}	D
Pt	0.02	0.04
S	0.53	0.63
Zr	0.86	0.88
Ti	0.21	0.23

Table 3 acid sites mmol Py/g

Cat./temp.	D^1_{CS2}	D
343 K	0.23	0.24
353 K	0.11	0.15
363 K	0.07	0.09
B/L	<0.01	<0.01

The analyses of the as synthetized by ^{39}SiNMR shows two broad resonances with chemical shifts at 91 and 98 ppm were observed. These bands could be attributed to polymerized Si(OSi)x(OH)4x framework units with x ¼ 3 (Q3), and x ¼ 4 (Q4).The high ratio Q3/Q4 is due to the method of adding surfactant that form strong bonding between the silanol groups and the surfactant, which stabilized the Q3 environment. The steam ammonia treatment and posterior use of catalyst in alkylation produce a complete amorphization of the silica matrix.

Microphotography (x125000 image) of the spent catalyst D^1 (**Fig 2**) obtained by TEM analyses shows nano-spot of insoluble carbon deposits across the particle. The images confirm the characteristic of a structured meso-microporous system with quasi-hexagonal pore packing. The software calculated from HRTEM images the average mesopores diameter in this sample, but the analyses present a high dependency on sample thickness and imaging focus. The meso-pore structure seems to have an average diameter of 8 nm (±0.5 nm).

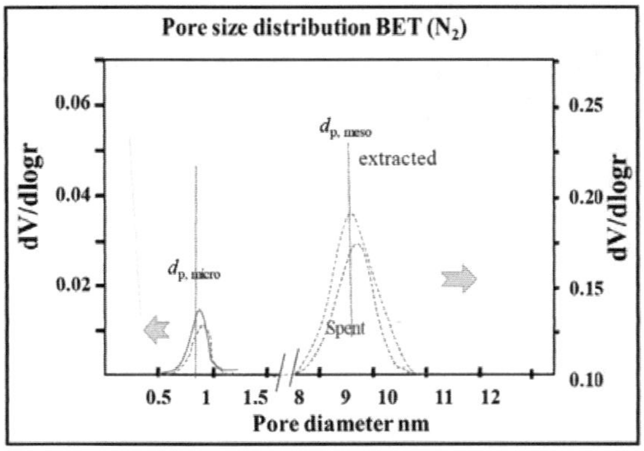

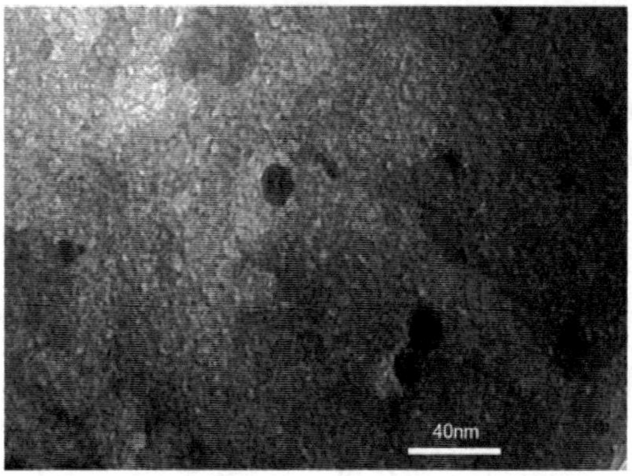

Figure 1 pore size distribution. **Figure 2**. TEM x 12500 Structure of the solid

3.4 Metal on surface. Metals dispersions were measured by XPS using the area of the 2p signal of sulfur, zirconium and titanium. **Table 2** shows the ratio Me to total metal on surface for D and extracted D^1_{CS2} catalysts. D^1 catalyst cannot be analyzed because the desorption of adsorbed species during the measures. After extraction the vacuum is stabilized and can be analyzed. Notice that the ratio Zr to Ti is different on surface than in the bulk. D^1_{CS2} shows different ratio Zr and Ti, and sulfur to Zr plus Ti on surface than D catalyst. In D^1_{CS2} mesopores contribute with 11 % of total surface area sample because most of the micro-porosity is blocked by insoluble deposit. Thus, it can be said that above-mentioned metals ratios are different in mesopores than in micropores (see the effect on acid sites, activity and selectivity in the next points).

3.5 Volume of accessible pores measured by adsorption of HCH. On spent catalysts, the accessible surface depends on minimum dimension of the adsorbates (critical diameter) respect to micropores diameter due to their accessibility reduction by coke deposition. **Table 1** shows the amount of HCH present in porous system of D, D^1 and D^1_{CS2} catalysts. The relative pore volume calculated based on dimension of the molecule decrease 22% from fresh to spent catalyst due to the deposit formed during alkylation, but slightly increase after extraction. The adsorption of HCH confirm that there is small modification of accessible mesopores volume by soluble deposits.

3.6 Acid sites measured in pyridine/hexane. Previous studies demonstrated that the adsorption of pyridine produces desorption of hydrocarbons adsorbed on spent catalyst, thus the method does not measure the real number of acid sites. Pyridine, being a stronger base than CS_2 and benzene, displaces all hydrocarbon strong adsorbed (iso- and cyclo-paraffins) but not those hydrocarbons polymerized. The analyses of the surface area by nitrogen adsorption of samples used to adsorb Py/hex (after vacuuming) show that micropores are still blocked by insoluble material. Py/hex practically did not desorbed hydrocarbons from D^1_{CS2} samples. Acid sites present on D and D^1_{CS2} samples were measured by adsorption of Py/Hex (303 K 1.4 MPa, 0.2 h contact time); then Py was desorbed at either 343, 353 or 363 K and the value of Py adsorbed per g reported in **Table 3**. Notice that the method only measures the acid sites present in mesopores because hex fill the micropores of the two samples. The comparison shows that extraction recovered a small fraction of acid sites located in mesopores. FTIR spectra of pyridine adsorbed on regenerated and spent catalyst present the characteristic bands in the range of 1400-1650 cm^{-1}. There are a very small bands at 1545 -1640, assigned to B sites, and the strong bands at 1455-1607 cm^{-1}, attributed to L acid sites; band at 1495 cm^{-1} is normally a combination of both B and L acid sites. The analyses bands obtained by FTIR confirm that D and D^1_{CS2} catalysts contain mainly L sites (**Table 3**).

3.7 Carbon characterization by MAS ^{13}CNMR on spent catalyst. Fig. 3 shows an example of the solid MAS ^{13}CNMR analyses. Spent D^1 and extracted D^1_{CS2} catalysts present the characteristic strong signals of paraffinic-naphthenic deposits and small signals due aromatics deposit (**Table 4**).

Notice in **Fig. 3** that paraffinic and naphthenic signals are reduced by extraction with CS_2 of these compounds from the surface. The MAS ^{13}CNMR analyses depicts important differences in the bands between (D^1) and (D^1_{CS2}.) The main constituents of soluble deposit are iso-paraffinic-naphthenic compounds, whose are soluble in CS_2. The spectra of catalysts contain small signals due to aromatic bands at ~130.0 ppm accompanied by spinning side bands at 65 and 195 ppm (not depicted). The aliphatic region of the spectra shows broad bands attributed to inhomogeneous distribution of carbonaceous species. There is strong overlap in these signals, especially those from methyl group of aromatics. Some of these aliphatic carbons are bound to naphthenic and aromatic nuclei of insoluble coke and the dominant signals in the aliphatic region come from dimer and trimer of isobutane and alkyl-cyclo-hexanes with minor amount of isopentanes and cyclo-pentanes in agreement with the analyses of the extracted material depict in **Table 4.** The small amount of aromatic species seems to be of poly-alkyl-toluene.

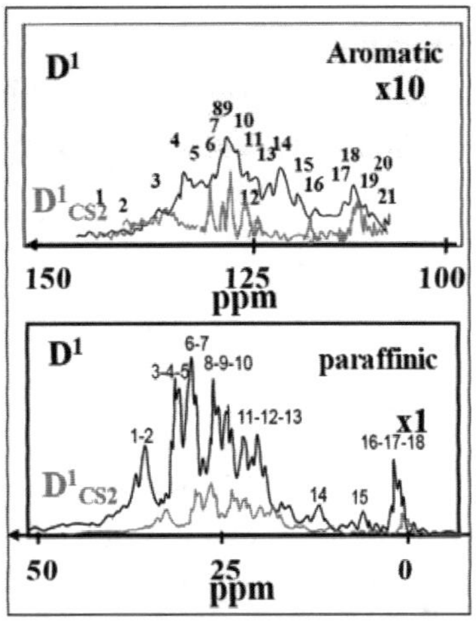

Figure 3 MAS ^{13}CNMR

Table 4 Carbon distribution in the CS_2-extract

Carbon N	%	P	N	Ar
C_{10}	20	98	1	1
C_{11}	13	97	2	1
C_{12}	37	99	1	-
C_{13}	10	100	-	-
C_{14}	7	100	-	-
C_{15}	5	100	-	-
C_{16}+	8	100	-	-

Table 5 Pyrolysis-GC-MS (700 K) insoluble

Peak	RT min	Hydrocarbon	%
4	9.27	dodecane	12
12	11.2	hexadecane	32
18	17.8	Heavy alkanes	13

The MS spectra of extract exhibit a molecular weight distribution between 140 and 224 Da as it is shown in **Table 4**. These ions formed in the fragmentation, in particularly those of secondary carbocations, are often prominent in the spectra of branched alkanes and can be diagnostic of alkyl substituent location. They are typically characterized by a pair of odd mass (even electron)/even mass (odd electron) ions but can be inhibited when methyl or other alkyl groups are in proximity (even up to seven carbons apart.) Methyl and tetramethyl group are not present in this C_{10}-C_{16} deposit, while di, and trimethyl are the most abundant. The fragmentation data obtained agrees with the literature. Molecular ion, albeit of low abundance due to facile loss of alkyl substituents, was evident in each of the mass spectra in MS spectra, allowing carbon number assignment. The classical odd mass [CnH2n+1] fragmentation pattern typical of n-alkanes (i.e., m/z 57, 71, 85, etc.) were not prominent in the MS spectra since naphthenic compound are mainly present (CnH2n). Presence of substituted benzene was stablished by the mass of fragments. Molecular ion, albeit of low abundance due to facile loss of alkyl substituents, was evident in each of the mass spectra in MS spectra, allowing carbon number assignment. The classical odd mass [CnH2n+1] fragmentation pattern typical of n-alkanes (i.e., m/z 57, 71, 85, etc.) were not prominent in the MS spectra since naphthenic compound are mainly present (CnH2n). Presence of substituted benzene was stablished by the mass of fragments 77, 91 and 105. There is primary and secondary carbocations formed by cleavage adjacent to a tertiary carbon.

The main constituents of the insoluble coke are mentioned in **Table 5** obtained by Py-GC-MS at 700 K. They are heavy iso-paraffins with two or three alkyl groups and insoluble in CS_2 plus benzene. The siliceous matrix was dissolved by HF and then the solid (insoluble hydrocarbons) was Pyrolyzed at 700 K. The product was analyzed on-line by a GCMS. Main constituent of this insoluble polymer are families of dodecane, hexadecane and heavy iso-paraffins; MAS ^{13}CNMR analyses of insoluble material indicates that the polymer contains between three to four alkyl groups. Clearly, they are formed by polymerization of isobutane in a restrain environment.

3.6 Adsorbed hydrocarbons characterization by IR. The FTIR of spent cat D^1 present the typical bands for adsorbed hydrocarbons. The asymmetric stretching vibrations band of CH_3 at 2963 cm^{-1}, those of CH_2 at 2938 cm^{-1} and the poorly defined C=C at 1640 cm^{-1} decrease with the extraction. The ratio of intensities of these three bands also decrease; that is indicative of the change species from paraffinic to naphthenic and then to aromatic type of coke. There are other symmetric and vibration of CH_3 bands at 1460 and 1350 cm^{-1} as well two poorly defined alkenyl bands at 1531 and 1505 cm^{-1} present the IR spectra that are reduced by the extraction. In addition, there is an increase in small bands due to silanol at 3740 and this at 3605 cm^{-1} due acidic hydroxyls. After the extraction appear additional weak-bands between 3740 and 3500 cm^{-1}assignee to hydroxyls groups and the band due to SO (symmetric and asymmetric vibrations) that were cleansed by the extraction. The insoluble coke presents the characteristic FTIR of asymmetric bands of CH_2 and CH at 2925 and 2852 cm^{-1} and the (C=C) three aromatics bands in the region at 1750-1500 cm^{-1} that confirm the presence of alkyl-aromatics. In addition, it shows poorly defined wagging.

3.7 XPS of spent and extracted catalyst. Fig. 4 shows the XPS results of carbon in Cat D^1 and $D^1_{CS2.}$ The C1s spectra display peaks at 284.5 and 286.5 eV due to aromatics

carbonaceous deposits; these catalysts show in addition a shoulder at 289.0 eV which is characteristic of a sulfonated carbon formed by interaction of naphthenic polymer with the SO_x groups on surface. There is another shoulder that was not identified properly but it seen to be carbon associated to the sulfate sites since it disappears by treating with hydrogen at high temperature. The 284.5 eV peak
are complex, it was not reported previously on spent catalyst, and looks likes those observed by us on $PtSO_xZr$ spent catalyst.

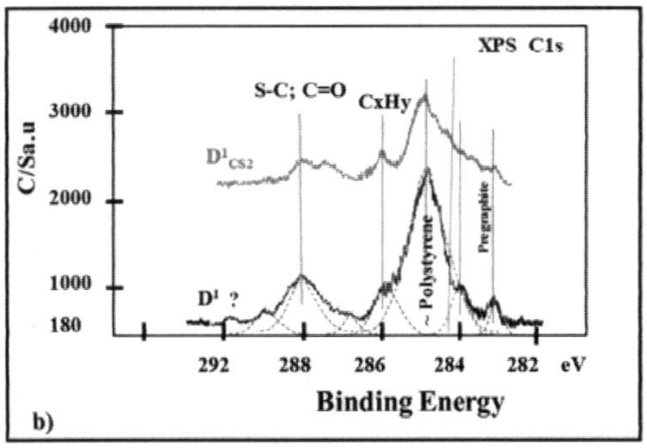

Fig 4 XPS of C signals

Table 6 Yields and activity for alkylation

Yield wt%	D^1	D^1_{CS2}
Alkylate	0.2	0.3
Dimers	0.12	0.18
Other	0.18	0.23
Activity (-)	0.74	0.82

3.8 Catalysts activity. The activity of the D^1 and D^1_{CS2} samples were tested for the alkylation of isobutane with 2-butene. **Table 6** depicts the results obtained at 353 K, 1.4 MPa 0.5 h o contact time using a ratio $C_4^=/iC_4$ of 0.25 (mol) and HC/Cat of 0.08 (wt.). The Yields of alkylate, dimers and other hydrocarbons increase by the extraction and 45% of the catalyst activity was recovered.

3.9 Deposits structure. Previous characterization of D^1 catalyst demonstrated that carbonaceous deposit present on the surface, after 2-butene alkylation, are formed by three type of adsorbed products with different degree solubility and molecular weight; the amount of deposit depend upon the time on stream, mass transfer and temperature used during the alkylation reaction. The above characterization of spent $PtSO_xTiZr/SiO_2$ catalyst has shown the presence of:

(1) **Reversible adsorbed reactant and products (high H/C ratio)**. They are at equilibrium with the species present in in solution (pores filled with liquid). This deposit is formed by isoparaffinic and cyclo-paraffins species. Some of them fill in the mesopores and they are not strongly adsorbed since are reduced by vacuum at 303 K and 1kPa.
(2) **Irreversible adsorbed reactant and products (medium H/C ratio)**. They remain on surface in spite that liquid is evaporated from the pores. They can be extracted by CS_2-benzene and by pyridine. The affinity for the surface is lower than those for the solvent. They are composed by 98% iso-paraffins and 2 % cyclo-paraffins.
(3) **Irreversible deposits formed polymerization of isobutane (low in H/C ratio).** They are formed in micropores with strong diffusion limitation. They are insoluble in CS_2-benzene and formed by C_{12}^+ hydrocarbons with high degree of branching. Microscopic analyses indicated that they are formed by amorphous nanoparticles.

4. Conclusion

Spent $PtZrTiSOx/SiO_2$ catalyst was characterized by different techniques. The sample was extracted with CS_2 - benzene and characterized to determine the structure of the carbonaceous deposits. Results confirm that:

- there is two type of hydrocarbon present on catalyst that deactivate the catalyst: adsorbed alkylate, dimer and trimer soluble hydrocarbons, and insoluble polymer;
- Soluble deposit can be extracted, and it is formed by 98% of adsorbed C_{10}-C_{16} isoparaffins and 2% alkyl-cyclo-hexanes;
- Insoluble deposit is formed by polymer of iC_4 with high molecular weight (>C_{12}) and branching degree (two to four). They have amorphous isoparaffinic structure;
- The extraction restores most of the acid sites, sulfur and Pt dispersion present in mesopores by freeing sites cover by soluble hydrocarbons;
- Coke seems to be multiple adsorbed on several nearby clusters of Ti-O-Si and Zr-O-Si that are used to grow the deposited based on initial alkyl-butonium adsorption.

5. Acknowledgment

The authors want to acknowledge the experimental work of Eng. P Zanabria and Eng J. Costa who perform the CS_2 extractions, CNMR and FTIR study, the valuable discussion with Prof R. Defounce and the economic support of the HyPro Consultants.

References

1. Menon, P.G. Coke on catalysts—harmful, harmless, invisible and beneficial types. *J. Mol. Catal.* **1990**, *59*, 207–220.
2. Forzatti P., Lietti L. "Catalyst deactivation Catalysis", Cat Today 52 (1999) 165-181
3. Argyle M.D., Bartholomew C.H., "Heterogeneous Catalyst Deactivation and Regeneration: A Review" Catalysts, 5, (2015), 145-269.
4. Roberto GaliassoTailleur and Jose V. Andretti Salva "1-butene alkylation reaction on $PtSO_x/TiZrO_y$ catalyst" Preprint Aiche Meeting Houston (2008)
5. Galiasso Tailleur R, Rodríguez S., Farina C. and Derjani-Bayeh S. "Simulation of C_2-C_3 olefins solid alkylation in the presence of hydrogen using a slurry transport reactor-hydrocyclone-regenerator system and $PtSO_4TiZr/SiO_2$ catalyst" Part I, Alkylation; *Energy Fuels*, 2018, 32 (2), 2527–2548
6. Galiasso Tailleur R, Rodríguez S., Farina C. and Derjani-Bayeh S. "Simulation of C_2-C_3 olefins solid alkylation in the presence of hydrogen using a slurry transport reactor-hydrocyclone-regenerator system and $PtSO_4TiZr/SiO_2$ catalyst" Part II, catalyst regeneration. E*nergy Fuels,* **2018 *32*** (2), 2549-2570.
7. Galiasso Tailleur R. and Farina C. Regeneration of $PtSOxTiZr/SiO_2$ by Hydrogen treatment. Part II Catalyst regeneration in a semibatch fluidized bed reactor - catalyst reactivation. Preprint 560ha, Aiche Meeting, Orlando, November (2019)
8. Peretti A.G., Álvarez K., Galiasso Tailleur R. and Derjani-Bayeh S. Comparative study of 2-Butene dimerization and alkylation at low temperature medium pressure in presence of hexane or decalin, using $CeNaX/SiO_2$ and $SOxZrTi/SiO_2$ structured-catalysts. Sent to Applied Catalysis (2020)
9. Galiasso Tailleur R. and Ronaldi R. "Effective Diffusion Coefficient of Isobutane, 2,3 dimethyl-Butane and hexyl-cyclo-hexane in Ternary Blends of Adsorbates (liquid phase) in Spent-regenerated alkylation Catalysts**.** Preprint 560ha, Aiche Meeting, Orlando, November (2019).

564789 In Situ Synthesis of Zn-Doped Ordered Mesoporous Carbon: Structure Characterization and Catalytic Performance Evaluation

Zinc acetate/activated carbon as a traditional catalyst was used for the production of vinyl acetate (VAc). However, the yield of VAc is low due to the rapid decline of activity and poor stability of zinc acetate/activated carbon catalyst. In the present study, to improve the reaction performance and the service life of catalyst, Zn-doped ordered mesoporous carbon (Zn-OMC) and HAc modified Zn-OMC were separately synthesized by a new in situ method. The crystal structure, pore properties, thermal stability and surface morphology of catalysts were characterized by XRD, N_2 adsorption-desorption isotherms analysis, TG, TEM, and SEM. The catalytic performance of Zn-OMC and HAc modified Zn-OMC for the vapor-phase synthesis of VAc from acetylene was tested using a fixed-bed reactor. The effects of temperature and zinc content of catalysts on the reaction activity were examined in detail. The results showed that HAc modified Zn-OMC catalysts have mesoporous channels, large specific surface area, high thermal stability, and high selectivity to VAc.

567354 Development of an Accelerated Deactivation Protocol for Vacuum Gasoil Hydrocracking Catalysts

INTRODUCTION

Hydrocracking is one of the most flexible refining processes for converting heavy, high-boiling feedstock molecules to smaller, lower boiling products. This process is carried out using hydrogen at high pressure conditions in presence of heterogeneous bifunctional catalysts, combining an acid and a metallic function. This catalyst is continuously deactivated in the presence of hydrocarbon feedstocks at hydrocracking conditions. Deactivation at normal operational conditions occurs slowly, depending on the feed and the operation severity; cycle lengths can vary between a few months and a couple of years. The heavier the feedstock the more severe the deactivation. The loss of activity determines the time of an operation cycle and consequently the profitability of the process. This is due to the fact that the increase of cycle length reduces the number of catalyst replacements and consequently the downtime of the unit. Deactivation is inevitable, but it can be slowed down or prevented and some of its consequences can be avoided or delayed. In that sense an obvious action is the purification of the feed to prevent poisoning. One common industrial practice is to gradually increase the operating temperature to compensate the activity loss. A different option could be to reduce the deactivation rate through catalyst stability improvement. The knowledge of the chemical and physical aspects of catalyst deactivation is fundamental for the design of more stable catalysts. For this reason, one of the main concerns in the hydrocarbons processing is the study of catalyst deactivation, for which different approaches exist. Some authors have evaluated the effects of coke or metal deposition through the characterization of industrial spent catalysts. However, this approach was not proved to give enough insight on deactivation phenomena due to the changing-character of the industrial operation. As mentioned above, deactivation in an industrial plant takes place over a period of time that makes impractical to do absolutely equivalent experiments in the laboratory. Additionally they are expensive and high time-consuming and, depending on the sampling program and the analyzes carried out, would not provide much more information than a full scale operation. The accelerated deactivation appears as an instrument capable of providing relevant information on the deactivation phenomenon in reduced time. Despite of the importance of this subject, it is hardly discussed in the literature for hydrocracking catalysts. Most of deactivation studies are focused on catalysts used for residues hydrotreatment, where the main sources of poisoning are the deposition of asphaltenes and metals contained in these heavy fractions. In the case of vacuum distillates, these two poisons are much less abundant than in residues; therefore, they may not be the main sources of deactivation. Some other studies have been carried out with model feedstocks, whose behavior is not representative of real feedstocks.

OBJECTIVES AND METHODOLOGY

The purpose of this work is to develop an accelerated deactivation experimental procedure for a vacuum gasoil hydrocracking catalyst, that allows to assess this phenomenon in a short period of time, compared to the one in an industrial unit. The experiments were carried out in a fixed bed pilot unit. The feed was a vacuum gas oil for studying the effect of real compounds that are typically in contact with the industrial catalyst. The catalyst is constituted of a nickel-molybdenum metallic phase dispersed on Y zeolite. The methodology consists of two main steps: the first one is a preliminary screening of the variables that impact the deactivation rate. Based on the literature these variables are: temperature, space velocity, hydrogen/hydrocarbon ratio, and organic and total nitrogen content. This step consist on submitting the catalyst to the following conditions: 403°C, 140 bar, H_2/HC: 1500 NL/L, LHSV: $2h^{-1}$, total nitrogen content 2200 ppm and an organic nitrogen content of 150 ppm. This set of conditions constitutes the reference case. Subsequently each one of this variables is changed once at a time and catalyst is under each new condition for 10 days. After each variable change, conditions goes back to the reference case. Measurement of deactivation is carried out by tracking the conversion comparing to the conversion of the last point at reference conditions. According to these results we can determine what variables have the bigger impact on the deactivation rate.

Once the preliminary screening was finished, the operating conditions for a standard protocol were selected. In order to validate this protocol, the catalyst was submitted to these conditions until the conversion decreases between 30 to 50%, in a maximum period of one month. This loss of conversion is representative of the typical increase in temperature in a commercial unit between the start and the end of the run, which is around 20 °C. At the end of the test temperature was increased until the initial conversion is achieved. In that way, it was possible to determine the deactivation in terms of a delta temperature, the common practice in industrial scenarios.

During the test, catalyst samples were taken at different bed locations and in the middle and at the end of the test run, in order to assess the evolution of the catalytic properties with time on stream and reactor location. An experimental procedure for catalyst sampling without catalyst modification and catalyst reloading by keeping the same conversion as the one immediately obtained before catalysts unloading for sampling purposes was developed. In the best of our knowledge there are no publications on accelerated deactivation experimental procedures for hydrocracking catalyst with real feedstocks nor do they consider the evolution of the catalyst properties with time and location inside the reactor.

Respect to the laboratory analysis, concerning the catalyst these are mercury porosimetry (to evaluate changes in porosity, establishing which pores are more sensitive to the variables that are going to be modified), nitrogen isotherms (to establish the loss of active area after catalyst deactivation), elemental analysis (to quantify the H/C ratio which gives an idea about the aromaticity of the sample, type of coke deposited), TGA (to measure number and strength of acid sites), HTA to evaluate the changes in the metallic phase and eventually mass spectroscopy to quantify known materials, to identify compounds and its structure, infrared spectroscopy and eventually STEM (to characterize the morphology and surface of solid materials, occurrence and location of zeolites embedded in the matrix, size of supported metal particles and changes in their size, shape, and location during catalyst use). Respect to the reactor effluents gases are online analyzed by gas chromatography meanwhile the feed stock and liquid products are characterized by density, refraction index, nitrogen content, simulated distillation, type of carbon, type of aromatics and viscosity.

RESULTS

The methodology described above allows first of all to determine the effect of the variables on deactivation rate which should follow the subsequent behavior:

- Increase temperature to 410°C: conversion increases, coke precursors formation increases.
- Decrease H_2/HC ratio to 1000 LN/L: conversion decreases, increase of dehydrogenation reactions involved in coke formation.
- Increase LHSV to 3 h^{-1}: conversion decreases, quantity of contaminant compounds (nitrogen compounds and coke precursors) flow from the feedstock increases.
- Increase organic nitrogen content to 300 ppm: conversion decreases, increase of neutralized acid sites

Based on the results of the conversion obtained when operating conditions return to the base case conditions, the impact of each variable change on deactivation was determined and the specific value for each variable were set for the establishment of the final accelerated deactivation experimental procedure.

Taking into account the information from the samples of the accelerated deactivation test, the set of operating conditions was validated in terms of the loss of activity reached during one month. As mentioned in the methodology description, this experiment was carried out until obtaining a loss in conversion representative of 20°C of delta temperature. Besides, information about the differences on catalyst properties such as porosity, surface area, pore volume, coke deposition (quantity, location and type), and the effect on the metallic function for three different locations of the catalyst inside of the reactor and at two different times were obtained.

577575 Investigation on Control of Surface Reactivity Towards Carbon, Oxygen, and Hydrogen of Intermetallic Compounds in Wet Reforming of Hydrocarbons and Oxygenates

The heterogeneous catalysis community has long been married to late transition metal catalysts for hydrocarbon reforming despite significant issues with coke formation, catalyst deactivation, and the need for high temperatures and water-rich feed. It is proposed that the focus on late transition metal catalysts has polarized our understanding of the mechanism and that materials that present higher surface reactivity towards O and H and lower towards C may be beneficial in the reaction. It has already been shown by others that the addition of a secondary catalytic element with elevated reactivity towards O can limit coke formation and enhance overall activity. However, the catalyst formulations often produced active portions of the catalyst that could not be effectively characterized, e.g., a surface alloy or off-stoichiometric intermetallic compound.

Our approach is to utilize supported intermetallic compound solid nanoparticles (Ni+Ga/SiO_2) with well-defined bulk and surface compositions to build in the favorable properties already demonstrated but also benefit from more stable particles and quite tunable surface chemistry towards C, O, and H. Quantum chemical surface reaction modeling indicated that surface reactivity towards carbon, oxygen, and hydrogen could be tuned via p-block element selection as well as adjustment of the bulk and surface compositions. These features allowed for control of the dissociation of propane and H_2O and surface reactions of H-transfer, H_2 evolution, and carbon oxidation steps. Experimentally, suits of IMCs with controllable bulk and surface composition as well as particle size were successfully used in the reaction. Significantly improved H_2 selectivity and catalyst stability were achieved as well as comparable overall rates of reaction. The selectivity of CO vs. CO_2 could also be controlled from a 1:1 H_2:CO synthesis gas ratio to a CO-lean H_2/CO_2 composition. These studies and the new understanding of IMC surface chemistry developed can be extended to aid in the development of small-oxygenate reforming, biomass deoxygenation, and controlled hydrogenation catalysts.

573460 Dual Role of Surfactants in Zeolite Synthesis and Catalyst Optimization

Zeolites are crystalline microporous aluminosilicates with unique properties, such as tunable acidity and exceptional (hydro)thermal stability, which are utilized in commercial processes ranging from ion-exchange and adsorption to catalysis and separations. Surfactant templating has emerged as one of the most effective and versatile strategies for the construction of well-defined porous architectures in zeolites.[1] Despite tremendous efforts to elucidate the mechanisms of surfactant templating in zeolite synthesis, these pathways are not well understood. Here, we will discuss the dual roles of the cationic surfactant cetyltrimethylammonium (CTA) as an organic structure-directing agent (OSDA) and as a mesostructuring agent in the rational design of zeolites.

One of the most common zeolites is ZSM-5 (MFI type), which is used as a catalyst and sorbent in a wide range of industrial applications.[2] In ZSM-5 synthesis it is challenging to identify new and inexpensive OSDAs that can tailor the physicochemical properties of the final product. The most frequently used OSDA in ZSM-5 synthesis is tetrapropylammonium (TPA); however, recent studies have shown that CTA can be used as an alternative.[3,4] Here, we will present the effects of using CTA and TPA as OSDAs in combination with a variety of alkali metals as inorganic structure-directing agents. Our findings reveal that the selection of inorganic/organic combinations has a significant impact on the kinetics of ZSM-5 crystallization, as well as the properties of the resulting crystals. Notably, we show that TPA/Na and CTA/K are optimal combinations of structure-directing agents that can markedly alter the size, morphology, and aluminum distribution in ZSM-5. Using a combination of experiments and molecular modeling, we explore the use of CTA as an alternative OSDA for zeolite MFI and show that we can achieve smaller crystals (ca. 600 nm) in similar time (< 24 h) as syntheses employing TPA.

In addition to OSDA design, the development of mesoporosity in zeolites has been a long-standing goal in catalysis to alleviate the diffusion limitations imposed by micropores.[5,6] One area of research that has garnered considerable interest, yet is not fully understood, is the rearrangement of zeolite crystals post-synthesis to accommodate mesoporosity. Here, we will present in situ observations of intracrystalline mesoporosity in USY zeolite (FAU type) assisted by CTA using atomic force microscopy. Our findings capture in real time the structural, morphological, and textural evolution of initially rough crystals to smooth crystals with a uniform distribution of mesopores. Collectively, these findings provide new and unequivocal insights that confirm surfactant templating is a facile and efficient method to induce reconstruction of the crystal and the healing of defects towards the generation of optimal zeolites for catalytic applications.

(1) Choi, M et al. *Nature* 461 **(2009)** 246–249

(2) Chawla, A et al. *Mol. Syst. Des. Eng.* 3 **(2018)** 159-170

(3) Meng, L.; Mezari, B.; Goesten, M. G.; Hensen, E. J. M.; *Chem. Mater.* 29 **(2017)** 4091–4096

(4) Moteki, T.; Keoh, S. H.; Okubo, T.; *Chem. Commun.* 50 **(2014)** 1330–1333

(5) Sun, J.; C. Bonneau, et al. *Nature* 458 **(2009)** 1154

(6) Chawla, A et al. *Chem. Mater.* **(2019)** In Press DOI: 10.1021/acs.chemmater.9b00435

573516 Elucidating the Factors Governing the Organic-Free Interzeolite Transformation

Elucidating the Factors Governing Organic-Free Interzeolite Transformation

Rishabh Jain[1] and Jeffrey D. Rimer[1]

[1] *Department of Chemical and Biomolecular Engineering, University of Houston, Houston, Texas*

There is increased interest to identify new commercially viable routes to produce zeolites with well-defined physicochemical properties. One such approach is seeded growth, which can expedite zeolite crystallization by reducing the induction time and impeding the formation of unwanted impurities. Seeding also provides a route to eliminate expensive organic structure-directing agents (OSDAs) from the synthesis; however, the theory behind seeded growth is still unclear due to the inherent complexity of zeolite crystallization and the vast number of species in synthesis mixtures. Interzeolite transformation involves the use of a parent zeolite with a different framework as that of the seed, often used as a sole source of silica and alumina, which transforms into a desired daughter structure. Okubo and co-workers have hypothesized these transitions are facilitated by a common composite building unit (CBU) shared between the parent and daughter zeolite.[1]

In this presentation, we will discuss the role of different physiochemical parameters such as temperature, time, seed properties (Si/Al ratio, crystal structure), and growth solution composition, to elucidate the mechanistic details of seeded growth and interzeolite transformation. We will show the role of kinetics and thermodynamics in such interzeolite conversions. To this end, we utilize ternary (kinetic) phase diagrams to select the compositions for the seeds and growth solutions to avoid the formation of polymorphs.[2] From these studies we have found that the seed phase is metastable, and can facilitate the formation of the same phase at short times; however, longer synthesis time often leads to secondary transformation to a putatively more thermodynamically stable isostructure.

Here, we will also present our study on the synthesis of ZSM-5 (MFI) via direct transformation of USY (FAU) in the absence of organics and seeds.[3] To the best of our knowledge, this is the first report of such transformation involving two zeolites lacking a common CBU. This process, which was previously deemed improbable, is achieved using a growth medium that favors OSDA-free ZSM-5 crystallization. We observed that the crystalline product obtained from this transformation has fewer surface defects (e.g. intergrowths) compared to ZSM-5 crystals obtained by direct synthesis (i.e. non-zeolitic reagents).

(1) Itabashi, K., Kamimura, Y., Iyoki, K., Shimojima, A., Okubo, T.; *J. Am. Chem. Soc.* 134 **(2012)** 11542-11549.

(2) Maldonado, M., Oleksiak, M.D., Chinta, S., Rimer, J.D.; *J. Am. Chem. Soc.* 135 **(2013)** 2641-2652.

(3) Qin, W., Jain, R., Hernandez, F.R., Rimer, J.D.; *Chem. Eur. J.* 25 **(2019)** 1-7.

559320 In Situ Analysis and Dynamic Response Surface Modeling of Energetic Material Hydrolysis

In an effort to improve the demilitarization of propellant materials and provide alternative methodologies to open burning disposal, a process involving the conservation of propellant nitrogen and subsequent assimilation into algae for production of biofuel is investigated. Optimized caustic hydrolysis of a double base propellant is presented as an efficient process to remove the nitrogen from nitrocellulose and nitroglycerine in the form of nitrites and nitrates for further downstream processing. Factors affecting the hydrolysis nitrogen yield and reaction kinetics are analyzed using a Design of Experiments methodology to develop Dynamic Response Surface Models. Reactions are executed in an EasyMax reactor utilizing in-situ FTIR, FBRM, and calorimetry to provide continuous quantification of ion concentration, particle size, and heat flow.

575588 Understanding Molecular Structural Changes in Sulfated Mixed Metal Oxides

Sulfated metal oxides, a class of solid acid catalysts, have received considerable attention over the past three decades for their promise in replacing reactions traditionally catalyzed by homogenous acids. These homogenous acids present a number of problems ranging from the difficulty of separation to corrosivity that do not exist for heterogenous acid such as sulfated metal oxides, which provide equal or greater activity while benefitting from generally mild reaction conditions and relatively facile methods of separation from reaction solutions1,2. The strength and availability of both Lewis and Brønsted acid sites on the surfaces of sulfated metal oxides provides the opportunity to catalyze a range of reactions including isomerization, alkylation, acylation, and oxidative dehydrogenation reactions. However, much work towards characterization and elucidation of the pertinent structure-reactivity relationships is still required.

This work seeks to characterize the behavior of sulfate species on the surface of mixed metal oxides containing tin and zirconium at varying molar ratios. It is hypothesized that these materials could optimize the selectivity and yield of a given reaction by effecting the distribution of catalytically active sites in order to avoid undesirable side products. *In-situ* Raman spectroscopy is coupled with ATR-FTIR, XRD, and BET analysis in order to probe the effects of support composition and environment on the predominant sulfate chelation structure. Our findings suggest that a tridentate configuration of the dispersed sulfate species is preferred at dehydrated conditions for the entire range of catalysts, while a large number of different hydrated species are present for the various catalysts tested.

References:

[1] Tanabe, K.; Hölderich, W.F. App. Cat. A. 181 (1999) 399-434.

[2] Corma, A. Chem. Rev. 95 (1995) 559-614."

575783 Fin-like Zeolite Catalysts: A New Class of Hierarchical Materials

Zeolites are microporous crystals that have been widely used in catalysis, separation, and adsorption processes owing to their unique properties, such as high surface area, controllable acidity, ion exchange capacity, and shape selectivity. However, the small pore sizes (4 - 7 Å) of typical zeolites can impose severe mass-transfer limitations for internal diffusion. Our group and others have shown that reduced crystal size lessens the rate of catalyst deactivation and can alter product selectivity, e.g. promote the olefin cycle in methanol-to-hydrocarbon (MTH) reactions.[1] Reported methods to reduce diffusion limitations include the preparation of nano-sized zeolites[2] as well as 2-dimensional materials,[3] which often involve expensive multi-step processes. Inspired by conventional fins which enhance interfacial heat transfer, we report the synthesis of fin-like zeolite crystals with improved catalyst performance in MTH reactions.

In this presentation, we will discuss the preparation of hierarchical ZSM-5 and ZSM-11 catalysts using both one-pot and seeded growth methods. Catalytic testing revealed a nearly three-fold decrease in the rate of catalyst deactivation during MTH reaction relative to conventional zeolite crystals. The fin-like zeolite catalysts also promoted the olefins cycle of the hydrocarbon pool (HCP) mechanism. HRTEM and tomography measurements confirmed the rough surface protrusions were well-aligned with the seed crystal, without potential pore blockage. Operando UV-vis spectroscopy of these fin-like crystals revealed the slower evolution of heavy coking species. Collectively, our findings indicate that this novel synthesis method enhanced the performance of ZSM-5 and ZSM-11 catalysts, thus offering an efficient and versatile platform to synthesize optimal zeolite catalysts with reduced diffusion limitations for diverse applications in the (petro) chemical industry.

1. Shen, Y., et al. *ACS Catalysis.*8 **(2018)** 11042-11053.
2. Tosheva, L. and V.P. Valtchev; *Chem. Mater.* 17 **(2005)** 2494-2513.
3. Jeon, M.Y., et al. 543 **(2017)** 690.

576924 Exploration of Si-Based Precursor Pyrolysis with Theoretical and Experimental Studies of Semiconducting Nanomaterial Properties

Silicon-Germanium alloy materials (SiGe) are one of the principal semiconductors used for high-frequency wireless communication chips, high-speed information communication processors, and modems. With the advent of not only the "Internet of Things" but also 5th generation cellular mobile communications, demand of SiGe semiconductor materials is expected to increase progressively. Silicon alloy materials are also attractive for batteries and energy storage applications. These alloy materials can be synthesized by vapor deposition processes such as Chemical Vapor Deposition (CVD) or Atomic Layer Deposition (ALD) methods to make continuous films, as well as by homogeneous chemical nucleation through pyrolysis to make nanoparticles which can then be tempered to make films. Our study here will focus upon the latter materials synthesis route. However, understanding semiconducting nanomaterials formation from the pyrolysis of mixtures of silane (SiH_4) and germane (GeH_4) under the mildest conditions is still incomplete. It is because homogeneous gas-phase nanomaterials formation is a complex phenomenon in which hundreds to thousands of species undergo simultaneous reactions. The methodology used in our previous work for thermodynamic properties[1] and the kinetics[2] of Si nanoparticles was robust enough to validate the experimental results of pyrolysis of monosilane by Wyller et al.[3] Wyller et al. was able to detect higher order silanes up to pentasilane isomers from the monosilane pyrolysis by developing an advanced gas chromatography/mass spectrometry (GC/MS) method. Further computational modeling can play a significant role in narrowing the gap between controlled experimental studies and practical operating conditions. Automated network generation techniques is one way to describe the kinetics of inorganic nanoparticle formation and investigate all the possible reactions under certain conditions. Rate coefficients are to be estimated for every elementary step comprising the mechanistic model, and kinetic correlations are used to make this tractable. One common method for predicting activation barriers (E_a) is the Evans-Polanyi correlation; however, these structure-activity correlations require detailed thermochemical information for each reacting species. Unfortunately, there are limited studies available that predict the thermochemical properties of SiGe clusters. For this purpose, we conducted a computational study of hydrogenated silicon-germanium alloy clusters ($Si_xGe_yH_z$, $1<X+Y\leq 6$) to predict the thermodynamic and electronic properties.[4] The optimized geometries of the $Si_xGe_yH_z$ clusters were investigated systematically using quantum chemical calculations and statistical thermodynamics. All electronic energies for the clusters were calculated using Gaussian-n methods, in order to validate our approach, we compared our methodology to other popular composite methods such as the CBS-QB3 method and available observed data. Our studies have established trends in thermodynamic properties (standard enthalpy of formation ($\Delta H^o{}_f$), standard entropy (S^o), and heat capacity (C_p)), as a function of cluster composition and structure. Furthermore, we compared HOMO−LUMO energy gaps, and HOMO and LUMO electron distributions in order to gain insight into the electronic stability of the hydrogenated Si, Ge, and SiGe clusters. These properties will be discussed in the context of tailored nanomaterials design and generalized using a machine learning approach.

1. Adamczyk, A. J.; Broadbelt, L. J., Thermochemical Property Estimation of Hydrogenated Silicon Clusters. *J. Phys. Chem. A* **2011,** *115*(32), 8969-8982.
2. Adamczyk, A. J.; Reyniers, M. F.; Marin, G. B.; Broadbelt, L. J., Kinetics of Substituted Silylene Addition and Elimination in Silicon Nanocluster Growth Captured by Group Additivity. *ChemPhysChem* **2010,** *11*(9), 1978-1994.
3. Wyller G. M., Preston T. J., Skare M. O., Klette H., Anjitha S. G, Marstein E. S., Exploring thermal pyrolysis of monosilane through gas chromatography-mass spectrometry measurements of higher order silanes. Silicon for the Chemical and Solar Industry XIV, 2018
4. Choi, Y.; Adamczyk, A.J., Tuning Hydrogenated Silicon, Germanium, and SiGe Nanocluster Properties Using Theoretical Calculations and a Machine Learning Approach. J. Phys. Chem. A

2018, 122, 9851−9868

577998 Raman-Spectrokinetics to Gain Insights on Support Effects of Supported Vanadium Oxide Catalysts

Bridging the gap between observed structure and measured reactivity of supported metal oxide catalysts continues to be a challenging goal for the complex catalytic systems typically used in oxidative and non-oxidative dehydrogenation of hydrocarbons. Supported ternary metal oxides are no exception with reactivity being influenced by catalyst coverage, relative amounts of species on the support, and the kind of support used. In this study, by combining operando Raman spectroscopy with conventional time-resolved reaction kinetics, we measure the redox reaction rates of surface species directly from the Raman spectra. This approach, coined Raman-spectrokinetics, exceeds one major limitation of measuring kinetics by mass spectrometry (MS) and gas chromatography (GC). It allows us to observe reaction rates at a specific catalytic site while conventional MS and GC analysis only provide overall reaction rates (average rates from the entire samples' sites). This study is a continuation of an earlier study which explored the oxidation of $(VO_x)_m/(NbO_y)_n/SiO_2$catalysts. Herein, we aim to determine support effects using our Raman-spectrokinetics approach to gain new insights on the oxidation kinetics of supported vanadium oxide catalysts by following the evolution/depletion of the vandyl (V=O) vibration as a function of O_2/H_2concentration. It is known that turnover frequency in catalysts remains constant as a function of coverage, and we are observing the same phenomena via Raman-spectrokinetics. It is our expectation that this study will launch our use of the Raman-spectrokinetics approach in studying the reactivity of metal oxides at different MO_xcoverage (2D and 3D species), on different supports (SiO_2, Al_2O_3, TiO_2, ZrO_2), with different promoters ($(M_1O_x)_m$-$(M_2O_y)_n$/support multi-component systems), and with different surface structures. These insights will further our goal of preparing catalysts with rational design choices in order to achieve unprecedented catalytic performances with supported metal oxide catalysts.

578001 Nitrogen-Doped Carbon: A Support to Synthesize Ultra-Small and Stable Pt-Nanoparticles

Abstract

Strong electrostatic adsorption (SEA), a simple and rational approach, has been used to synthesize ultra-small, high metal loading (~20wt%) and stable Pt-nanoparticles supported on nitrogen doped carbon (NC). It is well-known that high metal dispersion accompanied by stable nanoparticles are often desired in supported metal catalysts to maximize active sites [1]. Carbon supported Pt-nanoparticles (Pt/C) is one of the examples of such highly disperse catalysts that typically form 2-4 nm particles for high metal loading (>15wt%). These catalysts are widely used in many electrochemical (i.e. oxygen reduction temperature, methanol oxidation) applications, however, activity degradation of Pt/C catalysts are still an issue [2]. Platinum metal sintering and metal oxide formation are assumed to be the causes for catalyst deactivation. This work has produced ~1.5 nm particles by SEA for 20wt% of Pt supported on NC. Additionally, particles are much stable and do not sinter even exposing at 500^0C in inter condition. Furthermore, nitrogen atoms stabilize metallic platinum by preventing the formation of platinum oxide, which is confirmed by X-ray diffraction.

References

[1] R. Banerjee, Q. Liu, J. Meynard, and M. Tengco, *Catal. Letters*, vol. 147, no. 7, pp. 1754–1764, 2017.

[2] R. Banerjee, D. A. Chen, S. Karakalos, M. C. Piedboeuf, N. Job, and J. R. Regalbuto, *ACS Appl. Nano Mater.*, vol. 1, pp. 5876–5884, 2018.

574298 Reactions of Methyl and Ethyl Fragments on α-$Cr_2O_3(10\bar{1}2)$

The reaction of methyl and ethyl fragments formed by the dissociative adsorption of methyl iodide and ethyl chloride, respectively, has been examined on the non-reducible α-$Cr_2O_3(10\bar{1}2)$ surface. XPS indicates that ethyl fragments bind at surface cation sites while methyl fragments can bind at either a surface cation or anion site. TPD results show that the rate-limiting step for ethyl decomposition is β-hydrogen elimination to produce ethylene directly and ethane via a subsequent ethyl hydrogenation step. Methyl fragments decompose to methylene via a rate-limiting α-hydrogen elimination step. Methane and ethylene are the primary products, with methane formed by hydrogenation of surface methyl, and ethylene formed both by methylene coupling and methylene insertion into the methyl surface bond to form an ethyl fragment that reacts via β-hydrogen elimination. Density functional theory (DFT) and DFT+U simulations were also conducted to examine the adsorption sites and calculate these reaction barriers. DFT gives a better representation of the adsorption site and the activation barriers to reaction for the rate limiting steps in contrast to the prevailing narrative that DFT+U performs better for transition metal oxide surfaces with highly-correlated electronic structures.

577920 Colloidal Gold Nanoparticles for Catalytic Applications

The distinct optical, electronic and catalytic properties of colloidal gold nanoparticles (AuNPs) which are not seen in bulk gold have made them an attractive research subject in the past few decades. Oxidation of alcohols, reduction of nitroarenes, hydrogenation of quinolone and carbon - carbon coupling reactions are the reactions oxidized by colloidal (AuNPs). The preference for supported catalysts in industries over colloidal catalysts is due to their usability in continuous reactors and simplicity in recovery through simple separation methods like centrifugation, sedimentation and filtration for reuse. In comparison, recovery of colloidal catalysts is a bit complicated, employing methods like temperature or pH induced aggregation and redispersion, solvent extraction etc. However, the prospective advantages of colloidal (AuNPs) over supported catalysts lie in improved catalytic activity, selectivity and milder reaction conditions with applications in chiral catalysis. The colloidal (AuNPs) need to be stabilized with ligands to remain dispersed and be functional as catalysts. This also causes surface passivation of the nanoparticles thereby reducing their catalytic activity. We are exploring the effects of the organo thiol stabilizing ligands on the catalytic activity of colloidal (AuNPs). Reduction of 4-nitrophenol to 4 -aminophenol with sodium borohydride ($NaBH_4$) as a reducing agent catalyzed by colloidal (AuNPs) is our model reaction. Effects of the chain length of ligands and surface coverage of nanoparticles on catalytic activity are studied. Studying the relationship between the molecular structure of the stabilizing agents and the surface coverage of the nanoparticles is vital for understanding the mechanism of the catalytic rate of colloidal (AuNPs) for redox reactions. The use of pH sensitive ligands for pH triggered recovery and reusability of colloidal (AuNPs) catalysts is also studied.

562827 Coupling the Magnetic and Catalytic Properties of Fe_3O_4 Via Shape-Controlled Routes and Cr Doping

Iron catalysts play a vital role in the leading chemicals industry, favoring the production of ammonia, hydrogen, and liquid hydrocarbons through the Haber-Bosch, Water-Gas Shift, and Fischer-Tropsch processes, respectively. These catalysts are benefited by an increase in surface area, which can be achieved through reduced dimensionality. To produce efficient catalysts, thermal decomposition routes capable of controlling size and shape of nanoparticles have been reported, enabling fine tuning of the catalytic properties of a given material. Consequently, for iron oxide (Fe_3O_4) nanoparticles, for example, structural control also enables improved magnetic properties. In the context of catalysis, Fe_3O_4 is a promising material due to its ability to generate localized heat by the conversion of electromagnetic radiation into heat, as seen in magnetic hyperthermia. Therefore, combining the properties of iron catalysts with synthetic control and doping have the potential to make an impact in the chemicals industry.

In this work, thermal decomposition is used to produce monodisperse, spherical, cubic and octahedral shaped, $Fe_{(3-x)}Cr_xO_4$ nanoparticles as a strategy to improve the activity of the iron oxide (Fe_3O_4) host nanoparticles via Cr^{+3} doping, while maintaining the magnetic properties of the host lattice. Cr^{+3} replaces Fe^{+3} in the octahedral sites of Fe_3O_4 inverse spinel structure at low doping concentrations ($x < 15$ mol%), and tetrahedral and octahedral sites are substituted at higher concentrations. Low doping concentration improves activity and redox performance due to coupling of the Fe^{+3}/Fe^{+2} with Cr^{+2}/Cr^{+3}, and regeneration of Fe^{+2} on the surface. To confirm this regeneration effect and the site occupancy of Cr in Fe_3O_4, L-edge X-ray Absorption Spectra of Fe and Cr is collected as well as X-ray Photoelectron Spectroscopy (XPS), to determine the formation of oxygen vacancies and to probe the valence state of Cr and Fe. In addition, Cr doping changes the pore structure from mesopores to micropores, thereby increasing the surface area, demonstrated through BET measurements. Finally, magnetic properties are characterized via Superconducting Quantum Interference Device (SQUID) measurements to show how the saturation magnetization, and coercivity are preserved. Induction heating is characterized by determining the Specific Loss Power of these materials under various magnetic field strength.

565329 Synthesis of Supported Metal Nanoparticles By Controlling the Interaction between Colloidal Metal Nanoparticles and Supports

Supported metal nanoparticles (NPs) represent a large category of heterogeneous catalysts for the fuel and chemical industry. Considerable amount work has been devoted to the synthesis and characterization of supported metal NPs. Industrial methods for the synthesis of supported metal NPs include impregnation and precipitation of molecular precursors followed by calcination and reduction, which often yield polydispersed NPs. Deposition-precipitation, ion exchange, electrostatic adsorption, photodeposition, surface reduction, surface organometallic chemistry, and atomic layer deposition have been developed for the synthesis of supported metal NPs with improved uniformity. These methods heavily hinge on the surface properties of catalyst supports, such as charges, defects, and reactivity of surface hydroxyls. Therefore, they cannot be simply extended to different types of supports.

In this presentation, we will discuss a new protocol to deposit colloidal metal NPs onto various catalyst supports by tuning the interaction between colloid metal NPs and supports. This approach is applicable to many types of metal NPs and supports, and can be readily scaled up without affecting the dispersion of NPs. The metal loading can be adjusted by changing the ratio between metal NPs and supports. Furthermore, the adsorption of colloidal NPs mostly occurs on the external surface of catalyst supports, thus eliminating the diffusion control in catalytic reactions. The Pd catalysts synthesized via our strategy show much lower alkane selectivity in competitive acetylene hydrogenation.

572199 Synthesis of Well-Defined Heterogeneous Catalysts Using Atomic Layer Deposition

Using Atomic Layer Deposition (ALD) to synthesize heterogeneous catalysts with atomic level precision has gained increasing interest in recent years. In our laboratory we have synthesized well-defined heterogeneous catalysts by depositing different noble metals onto various well-defined nanoparticle oxide shapes using ALD. The benefit of using well-defined oxide shapes is that the surface facets of each nanoparticle are well defined, as the CeO_2 octahedra have (111) surface terminations, the CeO_2 cubes expose (100) surface facets, while the rutile TiO_2 rods consist (110) surface facets. Therefore, this allows us to determine the influence of the support oxide surface structure on the active metal and ultimately on the probe reactions, which facilitates identifying structure-activity relationships for these catalyst materials. By using ALD during catalyst synthesis and tuning the ALD conditions, such as precursor temperature, deposition temperature and dose length, we have shown that we can carefully control the active metal loading on the nanoshapes. The resulting catalysts have been tested using CO oxidation and partial oxidation of alkanes as probe reactions.

569804 Initial Stages of Chloride Enhanced Depassivation of Chromium Protected Layer, a Density Functional Theory Study

Corrosion is one of the major problems to the service life of concrete structures and energy infrastructure. Stainless steels are widely used as primary piping material because of their high corrosion resistance largely due to the formation of a protective passive film of metal oxides, such as chromium (e.g. Cr_2O_3). Understanding the surface chemistry of these oxide surfaces, in different conditions, provides an essential insight into the stability and functionality of the protective layers. Aggressive ions, such as chlorides, are often part of the environment surrounding stainless steel, but these ions can lead to a breakdown of the oxide film resulting in the loss of protection at the metal surface. Here we used density functional theory to investigate the interactions of Cl and adsorbed water with a-Cr_2O_3 (0001) passive film to better understand the breakdown mechanism of the passive film. Previous studies have suggested a mixed hydroxyls and adsorbed water termination of the Cr_2O_3 passive film in aqueous solution, so these are included in our study. The initial step of Chloride interactions with the partially hydroxylated surface was modeled by the Cl substitution with hydroxyls/water. The effect of Cl surface concentration on substitution and insertion was also calculated from 1/6ML to a monolayer coverage. The Cl insertion via exchange with subsurface oxygen (O) and Cl induced chromium (Cr) vacancy formation on the surface were examined at various coverages of Cl. These have been proposed to be the initial steps of Cl induced breakdown of the passive film. Competitive adsorption of Cl by substitution with OH and H_2O is more favorable at low Cl coverages but becomes less favorable at higher coverages while the insertion of Cl is unfavorable at lower coverages but becomes favorable at monolayer coverage. Cr vacancy formation was found to be favorable at full coverage where Cr atoms on the surface are saturated by three Cl atoms. The result of these calculations implies that the initial stages of Cl induced breakdown of stainless steel passive film is concentration dependent in agreement with experimental observations.

572813 Engineering Dopant Position in Core-Shell CeO_2-ZrO_2 Nanoparticles to Control Catalytic Activity

Tailoring active sites position and concentration on the catalyst surface has been shown to effect activity and stability of a catalyst. Specifically, the surface coverage of transition metals on CeO2-ZrO2 (CZO) nanoparticles (NPs) is known to directly affect the redox property and ability to utilize lattice oxygens. Recently, Ni incorporation in the core of a CZO NPs has caused a two-fold increase in catalytic activity per surface area due to high oxygen storage capacity. However, the activity of the particle is dependent on the Ni surface concentration and is traditionally deposited via the insipid wetness impregnation technique, forming Ni on the surface of the particle. In this work, the role of Ni position on catalytic activity is probed to develop a two-step synthesis process which allows for spatially controlled dopant distribution for improved catalytic activity.

A two-step process, co-precipitation/molten salt synthesis (MSS), has been explored for the catalyst core, and urea deposition synthesis for the shell of CZO:Ni NPs. The concentration of oxygen vacancies in these NPs was studied using Raman spectroscopy, which shows an order of magnitude increase in oxygen vacancies based on the change in intensity of defect-induced (D) band at 600 cm-1 as the position of the active site is modified. High resolution transmission electron microscopy (HRTEM) and electron loss energy spectroscopy (EELS) identify that a core-shell structure is obtained, and incorporation of nickel changes the d-spacing of the CZO crystal structure. UV-Vis spectroscopy proved presence of Ni in the lattice whereas for the standard catalyst Ni is positioned on the surface. To further correlate the dopant position and oxygen vacancies, local bond structure and geometry of the active site (Ni) was studied by performing L edge X-ray Absorption Spectroscopy (XAS) measurements. Geometric distortions, eg and t2g orbitals, is dependent on the distribution of Ni within the NPs, therefore determines the Ni bonding and position. These preliminary results on CZO|Ni core-shell NPs demonstrate that controlling the position of the active site (Ni) modifies the oxygen vacancies of the NPs which is the first step to understand the diffusion of the active site during a reaction and deactivation pattern of the catalyst.

573161 Role of Support Reducibility in the Catalytic Activity of CO Oxidizing Reactions on Pd Loaded Cu-Doped Ceria

Ceria (CeO_2) has been spotlighted as one of the essential components for automobile exhaust-control catalyst due to its facile redoxability or oxygen storage capacity (OSC). In particular, Pd/CeO_2 catalyst has attracted remarkable attention due to the low-temperature CO oxidation activity and the high Pd dispersion owing to strong metal-support interactions (SMSI), which is also important for water-gas shift (WGS) reaction activity at low temperatures.

In this study, Pd-loaded ceria support was doped with Cu to enhance the reducibility. We found that among WGS and CO oxidation reactions, only the activity of WGS reaction was significantly improved on Pd/Cu-doped ceria, indicating that the active oxygen from different source might have been utilized in each reaction. To elucidate this observation, density functional theory (DFT) calculations were performed. The different mechanisms between CO oxidation and WGS on the Pd/Cu-doped ceria catalyst provided the reason why doping the ceria support with Cu resulted in a significant effect on the activity of WGS reaction, not CO oxidation

578584 Sequential H_2-CO Pulse Chemisorption for Estimating in-Situ Ni Dispersion : Application on ALD-Coated Catalyst

Measuring catalytic properties in-situ provides much more valuable information about catalyst deactivation mechanisms than ex-situ measurements. In this work, a binary in-situ H_2-CO pulse chemisorption technique was used to study the changes in metallic Ni dispersion. Catalysts coated through Atomic Layer Deposition (ALD) technique are known to provide coking and sintering resistance. A commercial 20% Ni on alumina catalyst was coated with 5,10 and 20 cycles of alumina using ALD process. The uncoated commercial catalyst was then compared with the ALD catalysts for changes in dispersion, which was measured periodically in 40 h of Dry Reforming of Methane (DRM) reaction at 650 °C and 1 atm. Plotting the dispersion for the uncoated and the 5-ALD catalyst shows that the rate of decrease of dispersion is much faster in the uncoated catalyst than for the 5-ALD catalyst. For the 5-ALD catalyst, the dispersion dropped at the rate of ~0.02% per hour. For the uncoated catalyst, it was about 5 times higher at ~0.11% per hour. After 40 h, the uncoated commercial catalyst did not show any chemisorption results whereas the 5-ALD catalyst still showed measurable dispersion, indicating exposed Ni active sites. TEM imaging before and after the reaction provide a more direct measurement of the crystallite size. The average crystallite size for the uncoated commercial catalyst increased from 8.5 nm to 24.5 nm during the 40 h of DRM, indicating sintering phenomenon. Under similar conditions, the 5-ALD catalyst retained the initial crystallite size, and the post-reaction crystallite size was ~8.7 nm. Carbon formation was measured by doing a Temperature-Programmed-Oxidation (TPO) after the 40 h of DRM, and the 5-ALD catalyst showed a lesser rate of coke formation than the uncoated catalyst, which could be due to the lesser crystallite size of the 5-ALD catalyst. The chemisorption technique described herein can be quite helpful for core-shell systems where in-situ chemisorption is to be performed, and this is a much more faster and cheaper alternative to in-situ imaging of catalysts.

567836 Kinetic Modelling of the Ammonia Temperature Programmed Desorption of ZSM-5

Kinetic Modelling of the Ammonia Temperature Programmed Desorption of ZSM-5

Rebecca L. Gibson[a,b], Prof Mark Simmons[c], Prof Athanasios Tsolakis[d], Prof Hugh Stitt[a], Dr Stephen Schuyten[b], Robert Gallen[a]

a. Johnson Matthey, Belasis Avenue, Billingham, Stockton-on-Tees, TS23 1LB, UK.

b. Johnson Matthey, Eli Whitney Blvd, Savannah, GA, 31408, USA.

c. School of Chemical Engineering, University of Birmingham, Edgbaston, B15 2TT, UK.

d. School of Engineering, University of Birmingham, Edgbaston, B15 2TT, UK.

Temperature programmed desorption (TPD) is a commonly used thermal analysis technique for investigating the strength of adsorption sites [1]. The adsorption energies of these sites are however currently rarely quantified from TPD data [2]. These adsorption sites are important features in the performance of a sorbent or catalyst and can be influenced by formulation and chemical structure. Characterising these sites would offer improved understanding and enable predictive modelling of performance.

It is common for TPD results to feature more than one peak, which makes the resolution of these peaks an important part of processing the results. Historically, the shapes of these peaks, as well as their overall contribution to the extent of reaction would be estimated before mechanistic modelling was carried out [3]. The inherent assumptions can be a source of error. In the approach used in this study, no assumptions are made about the peak shapes and contributions, rather, the Sestak Berggren model (Eq. (1)) [4] is used to fit the data directly,

$$\frac{d\alpha}{dt} = \sum_{i=1}^{n_{events}} E_{v,i} \cdot A_i \cdot \exp\left(\frac{Ea_i}{R.T_{b,i}}\left(1 - \frac{T_{b,i}}{T}\right)\right) \cdot (1-\alpha_i)^{n_i} \alpha_i^{m_i} \qquad \text{Eq. (1)}$$

where

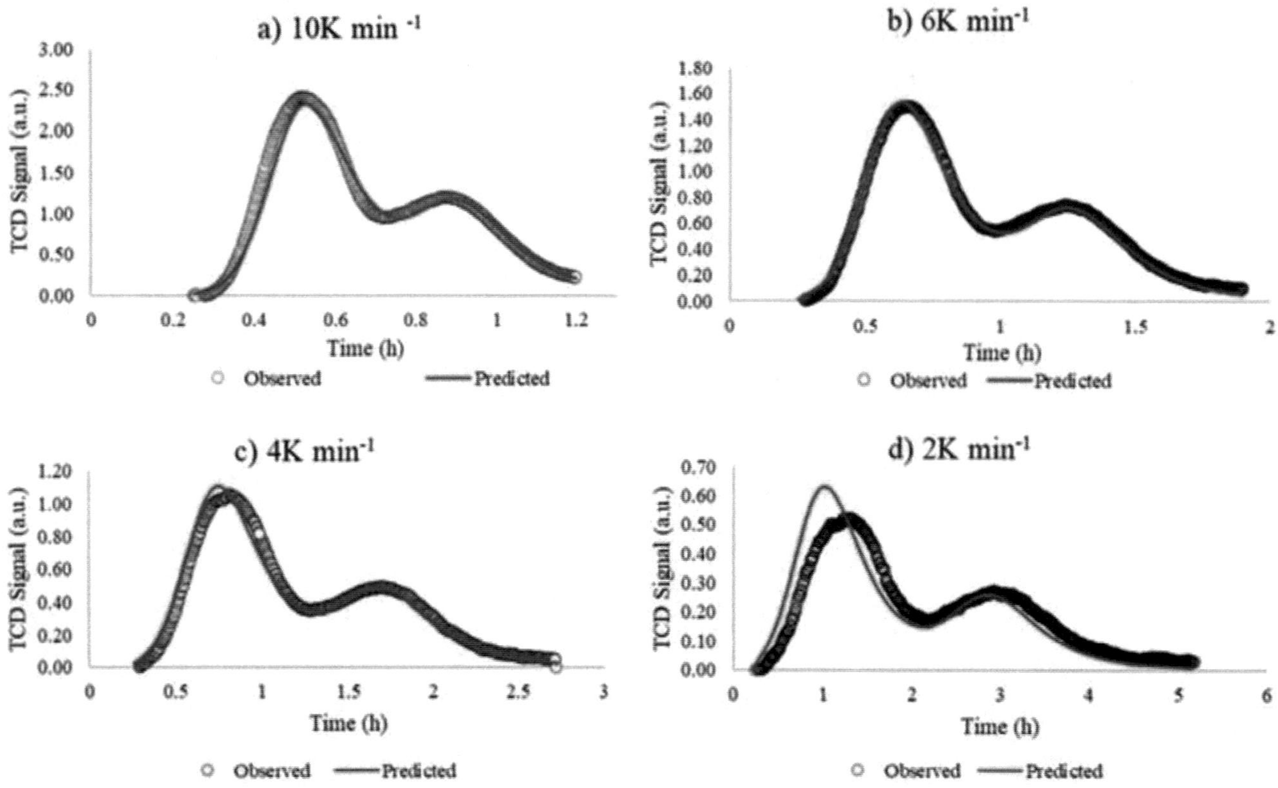

Figure 1: Sestak-Berggren Model fit for various ramp rate experiments

is the extent of reaction, A is the pre-exponential factor, E_a is the adsorption energy, R is the universal gas constant, T_b is the base temperature of the peak, T is the temperature, n and m are exponent parameters and E_v is a contribution term.

This model is a numerical fit and the exponent parameters (n and m) indicate which mechanisms could be plausible for the system under investigation [4]. These are analysed, and model discrimination carried out, allowing the statistically adequate fitting of the kinetic parameters, such as adsorption energy. Additionally, this gives information about the rate limiting step occurring during the process [2].

This Sestak-Berggren equation, along with the mechanistic models it represents, captures the behavior of kinetically-limited, irreversible chemical reactions. Transport phenomena and reversible reactions are not represented by this modelling scheme.

In silico data has been previously used for model validation, allowing rigorous methods to be developed [5]. This methodology has then been applied to a commercially available ZSM-5; testing the method on experimental data. Although this is a widely studied class of material, extracting kinetic information using ammonia TPD is still uncommon.

The results of the Sestak-Berggren modelling provide a good statistical fit for the two overlapped thermal events present for these experiments. The statistical metrics used to access the quality of fit for this parameter estimation study are: R-squared, 95% confidence intervals and residual sum of squares. This Sestak-Berggren fitting then allows for the prediction of a statistically relevant kinetic triplet (pre-exponential factor, activation energy and kinetic mechanism), as is the intention of the method.

In order to discriminate between the many mechanistic models possible for this desorption, multiple ramp rate experiments must be used [6]. In this case, five experiments were conducted at ramp rates of 2, 4, 6, 8 and 10

K min^{-1}. The parameter estimation was carried out using all datasets.

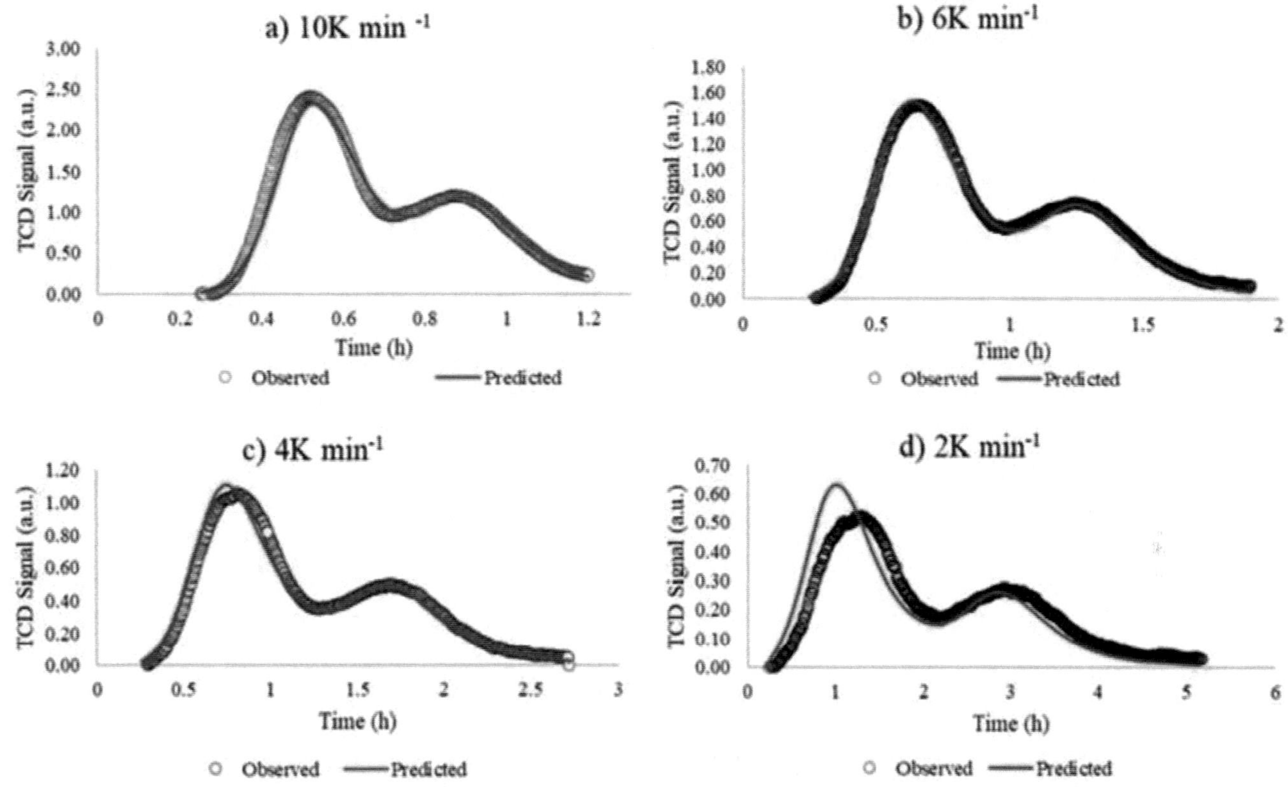

Figure 1: Sestak-Berggren Model fit for various ramp rate experiments

Although the Sestak-Berggren modelling provides a suitable statistical fit for these experiments, a shift in the fitting has been observed which varies with ramp rate as shown in Figure 1. This shift is not consistent with mass transport effects and has been attributed to uncaptured reversibility in the system [7].

This study concludes that the Sestak-Berggren model can successfully indicate mechanisms for multiple overlapped thermal events, such as those observed in the ammonia TPD of ZSM-5. However, the fit of each data set should be analysed individually to check for trends. In this case it is believed that the drift in model fit with ramp rate is a result of the adsorption process reversibility. Further model development to capture reversibility is underway.

Acknowledgements

Rebecca Gibson has been funded by the EPSRC Centre for Doctoral Training in Formulation Engineering at the University of Birmingham (EPSRC grant number EP/L015153/1) and Johnson Matthey.

References

[1] P. J. Barrie, 'Analysis of temperature programmed desorption (TPD) data for the characterisation of catalysts containing a distribution of adsorption sites', *Physical Chemistry Chemical Physics*, vol. 10, no. 12, p.

1688, 2008.

[2] M. Niwa, N. Katada, M. Sawa, and Y. Murakami, 'Temperature-Programmed Desorption of Ammonia with Readsorption Based on the Derived Theoretical Equation', *The Journal of Physical Chemistry*, vol. 99, no. 21, pp. 8812–8816, May 1995.

[3] Q.-L. Yan *et al.*, 'Decomposition kinetics and thermolysis products analyses of energetic diaminotriazole-substituted tetrazine structures', *Thermochimica Acta*, vol. 667, pp. 19–26, Sep. 2018.

[4] J. Sestak and G. Berggren, 'Study of the kinetics of the mechanism of solid-state reactions at increasing temperatures', *Thermochimica Acta*, vol. 3, pp. 1–12, 1971.

[5] R. Gibson, M. Simmons, A. Tsolakis, H. Stitt, J. West, and R. Gallen, 'Kinetic Modelling of Ammonia Temperature Programmed Desorption Using the Sestak-Berggren Equation: An In Silico Study.', *Industrial & Engineering Chemistry Research*, vol. Under Review.

[6] M. Maciejewski, 'Computational aspects of kinetic analysis. Part B: The ICTAC Kinetics Project the decomposition kinetics of calcium carbonate revisited, or some tips on survival in the kinetic minefield', *Thermochimica Acta*, p. 10, 2000.

[7] L. Lietti, I. Nova, S. Camurri, E. Tronconi, and P. Forzatti, 'Dynamics of the SCR-DeNOx reaction by the transient-response method', *AIChE Journal*, vol. 43, no. 10, pp. 2559–2570, Oct. 1997.

556794 To Leach or Not to Leach? Combined Data Analysis to Elucidate the Leaching Behavior of Strongly Acid Resin Supported Pd and Pdco NP Catalyst in Suzuki Coupling

To leach or not to leach? Combined data analysis to elucidate the leaching behavior of strongly acid resin supported Pd and PdCo NP catalysts in Suzuki coupling

Beau Van Vaerenbergh [1], Jeroen Lauwaert[1], Joris W. Thybaut[2], Pieter Vermeir[3] and Jeriffa De Clercq[1]

(1) Department of Materials, Textiles and Chemical Engineering, Ghent University, Ghent, Belgium.

(2) Laboratory for Chemical Technology, Ghent University, Ghent, Belgium.

(3) Department of Green Chemistry and Technology, Ghent University, Ghent, Belgium.

Introduction

Suzuki reactions are frequently investigated due to their relevance in pharmaceutical and fine chemical industries and are often carried out with Pd complexes. However, present research focuses on replacing these homogeneous catalysts by suitable heterogeneous counterparts due to the well-known challenges in product separation and catalyst recyclability in the area of homogeneous catalysis [1]. In heterogeneous catalysis, however, metal leaching undeniable needs to be assessed [2]. Ion exchange resin supported Pd nanoparticle (NP) catalysts have attracted a lot of attention in this respect because of their excellent performance in terms of NP stabilization (i.e., electrostatic as well as steric NP stabilizers), resulting in highly active NP catalysts with low leaching [3,4]. Moreover, the introduction of non-noble metals (e.g., Co [5]) induces synergistic effects between Pd and the former enhancing the catalytic performance in terms of activity, selectivity as well as stability and significantly reduces the catalyst cost. However, to date, literature still lacks of in-depth mechanistic leaching studies with combined data analysis for these NP catalysts, causing great ambiguity about the actual leaching behavior in Suzuki reactions.

Methodology

The leaching behavior and corresponding mechanism(s) of strong acid resin (Lewatit K2629, i.e., a macroporous, cross-linked polystyrene based resin with sulfonic acid groups), supported Pd and 1:1 Pd:Co (mol:mol) alloyed NP catalysts were investigated in the model Suzuki reaction between iodobenzene and phenylboronic acid with K_2CO_3 as base in 1:1 (v/v) DMF/H_2O at 40°C. The Pd content in the solution and iodobenzene conversion were monitored as function of time. Each point on the reaction/leaching curve represents one batch reactor experiment with a specific reaction time.

A hot filtration test was performed to unambiguously identify the type of catalysis that is occurring, i.e., heterogeneous versus (pseudo-)homogeneous.

Leaching tests (i.e., single and two component leaching tests), Transmission Electron Microscopy (TEM) analyses (i.e., NP size and distribution analysis) and recycle reaction and leaching kinetic experiments were performed and combined to elucidate their leaching mechanism in detail.

Type of catalysis

For Pd-NP/Lewatit K2629, the Pd leaching exhibited a maximum (35.1%) after 60 minutes (i.e., 0.23 min · mmol Pd) at 30% conversion and significantly decreased to 5.6% after 3h (i.e., 0.68 min · mmol Pd) at

75% conversion (Figure 1) [6]. The measured reaction and leaching profiles were not directly correlated, i.e., when the Pd concentration in the liquid phase rapidly decreased, the reaction rate did not change significantly. This implies a heterogeneous contribution to the overall catalytic activity.

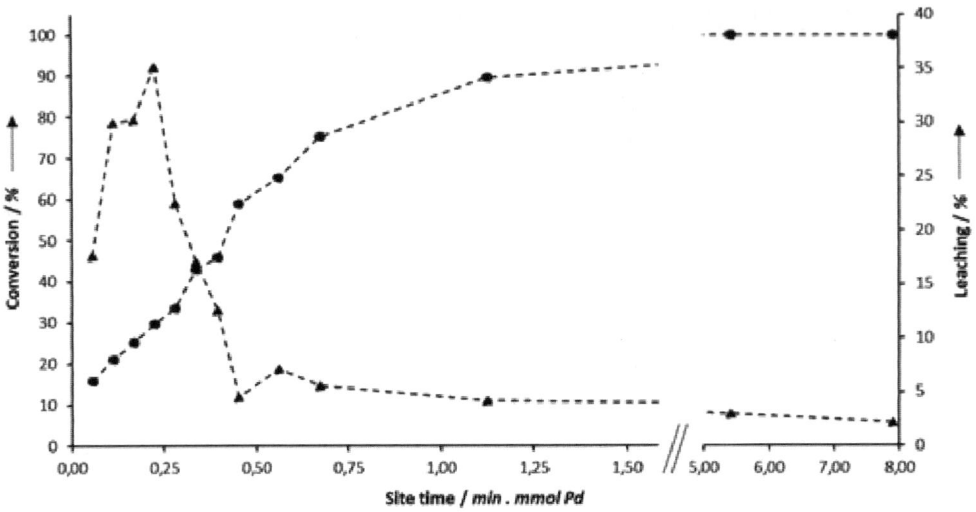

Figure 1. Iodobenzene conversion (•) and Pd leaching (△) as function of time for Pd-NP/Lewatit K2629 [6]. Lines are added to guide the eye.

Moreover, a hot filtration test pointed out that the leached Pd species were also catalytically active (i.e., from 30% after 1h reaction to 84% conversion after another 2h upon filtration). As a result, a homogeneous-heterogeneous catalytic system is proposed.

For PdCo-NP/Lewatit K2629, the leached Pd species in solution sharply decreased from 20.0% after 2h at 40% conversion to 1.4% after 3h at 100% conversion. Remarkably, the reaction rate significantly increased as function of time which is in strong contrast to traditional reaction profiles for which the reaction rate decreases as function of time due to decreased reactant concentrations. This indicates a significant homogeneous contribution of leached Pd to the overall activity. Besides Pd, Co also leached (i.e., 5.5% and 9.5% after 2h and 3h respectively). In order to more thoroughly explain its catalytic nature, further research is focused on completing the reaction and leaching kinetic profiles as well as performing a hot filtration test.

Leaching inducing character of the involved chemicals

For Pd-NP Lewatit K2629, single and two component leaching tests showed that the leaching behavior occurs via two pathways. On the one hand, Pd-NP leaching is induced by dissociated K_2CO_3 which causes a disturbance of the electrostatic NP stabilization (i.e., 45.8% and 44.4% after 60 minutes and 24h respectively). On the other hand, iodobenzene in the presence of the base induced molecular Pd^{2+} leaching as phenylpalladium(II)hydroxide complexes (i.e., 13.6% and 100% after 60 minutes and 24h respectively). In that respect, the re-deposition of leached Pd species at high conversions is due to the decreased amount of these leaching inducing reagents. This caused enlarged NPs in the resin (from an initial NP size of 2.4 nm to 4.1 nm after the 2nd run) as observed by TEM. In addition, the decreased catalytic activity (i.e., 75% versus 47% conversion after 3h) as well as the decreased and shifted leaching peak to higher site times (i.e., 35.1% Pd leaching after 60 minutes versus 25.2% after 90 minutes) in a second run confirms the leaching/re-deposition behavior of Pd-NP Lewatit K2629. For PdCo-NP/Lewatit K2629, research concerning its leaching behaviour and corresponding mechanism is ongoing.

Conclusions

For Pd-NP/Lewatit K2629, the reaction and leaching kinetic analysis in combination with the hot filtration test unambiguously point towards heterogeneous-homogeneous catalysis. Single and two component leaching tests showed that both Pd-NP and molecular Pd^{2+} leaching (as phenylpalladium(II)hydroxide complexes) occurs. For PdCo-NP/Lewatit K2629, the increased reaction rates as function of time indicate a more significant homogeneous contribution of the leached Pd species to the overall activity than for Pd-NP/Lewatit K2629. For both NP catalysts, the leached Pd species re-deposit at high conversion.

Overall, it is clear that well-combined data analysis is crucial to unambiguously elucidate the leaching behavior and corresponding mechanism of supported metal nanoparticle catalysts in Suzuki reaction.

References

[1] Torborg, C., and Beller, M., Recent Applications of Palladium-Catalyzed Coupling Reactions in the Pharmaceutical, Agrochemical, and Fine Chemical Industries. *Advanced Synthesis & Catalysis*, **351** (2009) 3027-3043.

[2] Biffis, A., Centomo, P., Del Zotto, A., and Zecca, M., Pd Metal Catalysts for Cross-Couplings and Related Reactions in the 21st Century: A Critical Review. *Chemical Reviews*, **118** (2018) 2249-2295.

[3] Van Vaerenbergh, B., Lauwaert, J., Bert, W., Thybaut, J.W., De Clercq, J., and Vermeir, P., Effect of Ion Exchange Resin Functionality on Catalytic Activity and Leaching of Palladium Nanoparticles in Suzuki Cross-Coupling Reactions. *ChemCatChem*, **9** (2017) 451-457.

[4] Liguori, F., Moreno-Marrodan, C., and Barbaro, P., Metal nanoparticles immobilized on ion-exchange resins: A versatile and effective catalyst platform for sustainable chemistry. *Chinese Journal of Catalysis*, **36** (2015) 1157-1169.

[5] Alonso, A., Macanás, J., Shafir, A., Muñoz, M.,Vallribera, A., Prodius, D., Melnic, S., Turtab, C., and Muraviev, D.N., Donnan-exclusion-driven distribution of catalytic ferromagnetic nanoparticles synthesized in polymeric fibers. *Dalton Transactions*, **39** (2010) 2579-2586.

[6] Van Vaerenbergh, B., Lauwaert, J., Thybaut, J.W., Vermeir, P., and De Clercq, J., Pd nanoparticle and molecular Pd^{2+} leaching pathways for a strongly acid versus strongly basic resin supported Pd nanoparticle catalyst in Suzuki coupling. *Chemical Engineering Journal*, **xx** (2019) xx-xx. Manuscript submitted for publication.

576720 An Ambient Pressure XPS Study of Methane Dissociation and Oxidation on $IrO_2(110)$

Recent studies demonstrate that the $IrO_2(110)$ surface is highly active in promoting methane dissociation, with C-H bond cleavage occurring at temperatures as low as 100 K in ultrahigh vacuum. This finding stimulates interest in advancing the fundamental understanding of $IrO_2(110)$ surface chemistry, and exploring the possibility of developing IrO_2-based catalysts for applications in the partial and complete oxidation of methane. In this talk, I will discuss recent AP-XPS (Ambient Pressure X-ray Photoelectron Spectroscopy) investigations of CH_4 oxidation on $IrO_2(110)$ that we performed at methane partial pressures up to 1 Torr. We find that methane efficiently reduces an ~10 layer $IrO_2(110)$ film on Ir(100) at temperatures above 500 K, with the rate of reduction depending on both temperature and CH_4 partial pressure. Complete reduction of the oxide film demonstrates that lattice oxygen migrates rapidly to the surface to sustain methane oxidation under the conditions studied. I will also discuss results obtained during the steady-state oxidation of CH_4 over $IrO_2(110)$ and the nature of surface species observed during reaction for varying $CH_4:O_2$ mixtures and temperatures.

583781 Effect of Pd Precursors on Pd/TiO_2 Catalysts Prepared by Different Methods and Their Catalytic Activity in Hydrogenation of Maleic Acid

The size, shape, and distribution of metal clusters on TiO_2 with strong metal-support interaction are generally controlled by the preparation methods, the nature of palladium precursors, pH values, etc.[1,2] Herein, a series of 5 wt% Pd/TiO_2 catalysts were prepared by varying two parameters; the types of palladium precursors and preparation methods. The catalytic performances were evaluated via hydrogenation of maleic acid (MA). The reaction was carried out under temperature=90 °C, pressure of H_2=3 bar, reaction time=60 min, and stirring rate=700 rpm. We prepared the Pd/TiO_2 catalysts using four different precursors such as H_2PdCl_4 (H), K_2PdCl_4 (K), Na_2PdCl_4 (N), and $Pd(NH_4)Cl_2 \cdot H_2O$ (PA) by different methods. We also investigated the effect of preparation methods (i.e. chemical reduction (CR), incipient wetness impregnation (IWI), and deposition-precipitation (DP)) on the Pd particles of the Pd/TiO_2 catalysts. The prepared catalysts were denoted as follows: Pd/TiO_2_A_*B* (A: Pd precursor, B: preparation method). The XRD, FT-IR, N_2 physisorption, CO chemisorption, UV-vis spectroscopy, and XPS were employed to characterize the physicochemical properties of prepared catalysts. The CO chemisorption results showed that Pd/TiO_2_K_*DP* had the highest Pd dispersion of 12.4%. K_2PdCl_4 (K) in the aqueous solution formed rapidly $[Pd(OH)_2]_n$, which could be uniformly precipitated the hydroxyl groups on TiO_2 support.[3] However, Pd/TiO_2_PA_*CR* had the lowest Pd dispersion due to the self-reduction and agglomeration of Pd by interaction with the NH_3 ligands, causing lower Pd dispersion.[4] From the XPS results, Pd/Ti atomic ratio of Pd/TiO_2_K_*DP* was the highest value of 0.079, while Pd/Ti atomic ratio of Pd/TiO_2_PA_*CR* was lower than 1.97 times. It was found to be in accordance with the CO chemisorption results. The catalytic activities were evaluated via MA hydrogenation over prepared Pd/TiO_2 catalysts. The MA conversion of almost 40% was achieved over Pd/TiO_2_K_*CR*. Although Pd/TiO_2_K_*CR* had similar Pd dispersion to Pd/TiO_2_K_*DP*, the maleic acid conversion was different. From above XPS results, we suggested that the Pd^{2+}/Pd^0 ratio influenced the catalytic activity of maleic acid hydrogenation.[5]

Keywords: Hydrogenation, Maleic acid, Palladium catalyst, Pd/TiO_2

References

1. J. Schoell et al., AIChE J., **2019**, 65, e16560
2. M. Y. Byun et al., J. Nanosci. Nanotechnol. **2018**, 18, 6283-6287
3. F. Kettemann et al., CrystEngComm, **2015**, 17, 1865-1870
4. Z. Paal et al., J. Catal., **1997**, 168, 164-175
5. 5. N. S. Babu et al., Catalysis Today, **2009**, 141, 120-124

560441 Demonstrating the Power of Custom Modeling Integrated into Flowsheet Simulation

Various custom modeling tools are available to model chemical reactors using experimental kinetic data and governing equations. Integration of these powerful models into useful flowsheet simulations can be challenging to implement, but the user-friendly interface offered by gPROMS provided the opportunity to utilize an integrated modeling approach in one platform. We set out to demonstrate the level of accuracy and illustrative power that can be achieved when a custom reactor model is integrated with a flowsheet simulation.

This presentation will present some examples of utilizing this approach in heterogeneous hydrosilylation chemistries within Dow Performance Silicones. Reaction kinetic data was experimentally generated and used in a scalable high-fidelity custom reactor model which numerically solves the governing equations for the fluid and solid phases. The reactor model was then connected to other unit operations within an integrated flowsheet using gPROMS. Complex scenarios of interest can be modeled to help mitigate risk and build the business case for process changes. This integrated approach also requires a lesser investment of time and resources compared to a non-integrated approach. Examples to be presented include capacity improvements through debottlenecking and process changes to maximize catalyst life. This application of an integrated modeling approach demonstrates its power to unlock improvements in process capacity, cost, product quality, and process safety.

569999 Generalized Adsorption Models on Metal Nanoparticles

Computational catalyst design has attracted tremendous interest over the last years as a means of identifying active and selective catalysts without the need of trial-and-error experimentation in the lab. However, there is still a scarcity of generalizable, physically-grounded models capable of both rapidly and accurately capturing the adsorption behavior of reactants and intermediates on the catalyst surface. With adsorption being central to catalyst design efforts, it is crucial to develop models that capture adsorption as a function of metal nanoparticle size, shape and metal composition. Using a combination of symbolic regression and statistical methods, we develop a novel first-principles-based adsorption model which is parameterized by both experimental literature data, and DFT calculations on very simple systems. Our model accurately captures a wealth of reported adsorption data and predicts adsorption energies of many intermediates onto nanoparticles of any size, shape and metal composition, in agreement with first principles calculations.

574061 Enhancing Ab Initio Microkinetic Models with Machine Learning

Mean-filed microkinetic models (MF-MKM) based on energies derived from density functional theory (DFT) are widely used to study heterogenous catalytic system, in terms of elucidation of mechanisms and quantification of reaction fluxes of catalytic systems. However, the model usually deviates from kinetic experiments, due to the intrinsic error of DFT calculation, and over-simplified mean-field approximation, etc.

Here, we present two examples of the application of statistical learning in kinetic modeling. (I) We demonstrate the improvement of a microkinetic model by statistically calibration the intrinsic error of DFT energies, and further improvement by incorporating less experimental data. (II) After accounting the influence of the adsorbate-adsorbate interaction with Monte Carlo simulation and statistical learning, MF-MKM captures missing phenomenon (bistability of oxidation system) missing from mean-field approximation.

We show how this approach systematically improve the prediction of MF-MKM combining machine learning model. This work demonstrates the promising performance of statistical learning in the application of kinetic modeling of heterogenous catalytic system.

566940 Catalyst Design for Glycerol Valorization through Data Science, First Principles Modeling, and Experimental Verification

Renewable carbon resources such as biomass and carbon dioxide are perceived to play an increasing role for the industrial production of fuels and chemicals. However, great challenges are faced in the valorization of these renewable carbon resources. For biomass valorization, the challenge is to efficiently utilize its highly functional form in the production of the target fuel or chemical. Despite the significant advances in the past few decades, there remains a great need for the discovery of more efficient catalysts and reaction processes for biomass valorization. Building on our recent mechanistic and experimental investigations on glucose conversion via pyrolysis and catalysis (*1-3*), here we apply data science analysis, first principles simulations, and catalyst synthesis, characterization, evaluation and optimization to design a Pt/WO_x-based heterogeneous catalyst for the efficient hydrodeoxygenation (HDO) of glycerol into 1,3-propane diol (1,3-PDO). Data science analysis shows the importance of both oxophilicity and reducibility; mechanistic studies reveal the importance of the Pt/WO_x interface in the selective cleavage of the ß-C-OH bond and in achieving the kinetic control of the HDO reaction; and experiments identify a promising zeolite-supported Pt/WO_x catalyst for the selective conversion of glycerol to 1,3-PDO. Our integrated studies show the promise of combining data science, modeling and experiment in industrial catalyst design for biomass valorization.

1. Bao, L. et al, ChemSusChem 2017, 10, 3040-3043.

2. Tan, Z. et al, Appl. Catal. A 2018, 560, 28-36.

3. Sun, Y. et al, unpublished work.

574052 Coupling Experimental Kinetics and Thermodynamic Modeling with IR Spectroscopy and Machine Learning for Fundamental Studies and Fast Product Quantification

The chemo-catalytic conversion of non-edible lignocellulosic biomass into furanic compounds such as 5-hydroxymethyl furfural (HMF) and furfural represent critical steps in biomass upgrading to fuels and chemicals. Metal salts enhance the yield and selectivity in the biphasic dehydration of fructose to HMF by improving the extraction of the product from the aqueous phase reducing the formation of byproducts. Furthermore, the dehydration reaction carried out in microreactors enhances the mass and heat transfer rates resulting in further increase in the HMF extraction at short reaction times. However, our understanding of the effect of the salts on the rate of dehydration and selectivity is limited, and the quantification of the products is slow compared to the fast throughput of microreactors.

In the present work, we couple thermodynamic modeling with experimental kinetics to understand the effect of different metal salts and Brønsted acid species on the dehydration of fructose to HMF. We focus on different alkali and alkaline earth metal salts on the acidity and speciation. Furthermore, we develop a method for fast online product quantification by utilizing machine learning for the systematic analysis of IR spectra.

565703 Elucidating Reaction Pathways through Combined Insights from Experimental and Computational Hammett Analysis

Density functional theory (DFT) has been extraordinarily successful in advancing mechanistic studies and catalysis. However, application of DFT to transition metal chemistry can be challenging owing to high sensitivity to the choice of theory coupled with limitations associated with describing multireference characteristics, spin-state energetics, and relativistic effects. This issue is illustrated in the case of bio-inspired oxo-complexes of copper, for which prior DFT studies have led to conflicting conclusions regarding the long-standing problem of identifying the preferred mechanism of CH activation. Our work goes beyond the traditional approach of contrasting proposed mechanisms through the calculation of potential energy surfaces for a single catalyst-substrate system. Instead, we systematically vary the identity of N-donors bound to the active center to achieve twin goals of (a) examining the barrier response of the two proposed mechanisms – oxo-insertion and radical recombination – and contrasting them with experiment, and (b) identifying characteristics or descriptors that enable the design of active candidates. Our approach combines DFT with energy decomposition analysis (EDA) and the activation-strain model. By varying the electrophilicity of amine- and imidazole-based N-donors bound to a dioxo-dicopper center and contrasting the resulting (inverted) Hammett plots with experimental Hammett studies, we demonstrate that methane activation most likely occurs via the single-step oxo-insertion pathway. Radical barriers are insensitive to catalyst variations by virtue of a very early transition state. Further, oxo-insertion barriers vary linearly with EDA-based charge transfer contributions, which are in turn dependent on electrophilicity of the ligand motif. We are now combining these insights with a DFT+EDA investigation into the role of strain and steric factors on catalytic activity, through systematic variation of the N-Cu-N bite angle in the dioxo-dicopper complex.

571263 Reaction Mechanism Modeling of By-Products in Semiconductor Manufacturing Process Using Reaction Mechanism Generator (RMG)

The semiconductor manufacturing process is one of the biggest industries, and researchers have studied a lot of research to improve it, but there is not much research on the by-products formed from the process. However, with increasing interest in the environment and safety, it is necessary to study the kinds of the by-products and the environmental and safety problems that each product may cause. For this reason, in this study, we investigate the mechanism of by-products generated by semiconductor process and compute the kind and amount of the products produced. Preferentially, construct database of the chemical reactant used in the semiconductor process, find the main reaction path using the open source program RMG, and calculate the sort and amount of the by-products formed by the main mechanism using the program such as Chemkin. In this way, the main reaction mechanism for the semiconductor process is modeled, and the type and amount of the by-products coming out of the reaction are calculated and predicted so that it is possible to know whether toxic substances are present in advance before the new process and handle it. Furthermore, in addition to Chemkin, which is optimized for surface and gas phase reactions, the use of Gaussian DFT tool will also enable handling of liquid phase processes.

559047 Molecular Dynamics Simulation of Graphene Oxide-Enzyme Assembly for Gaseous Catalysis

Gaseous enzymatic catalysis holds great promise for chemical synthesis but is not yet applied due mainly to the water content dilemma. On one hand, an enzyme in dry state is stable but not active because of hindered conformational transition. On the other hand, a wetted enzyme may appear a high activity but is less stable because of water attack through hydrogen bond exchanging between surface amino residues with free water molecules in the moisture. We have proposed an idea of using graphene oxide (GO) to form enzyme aerogel, in which GO provides a hydrophilic microenvironment required for enzyme catalysis, and demonstrated its effectiveness in using *Candida antarctica* lipase B (CALB) as an example. In this study, we use molecular dynamics simulation to probe the assembly of GO and CALB and illustrate how GO stabilizes the active conformation of CALB in anhydrous status. Three kinds of GO sheets with different oxygen content were tested, in which a simulated annealing procedure was applied to accelerate the assembly process. It is shown that a higher oxygen content of GO intensifies its interaction with CALB via generating more intermolecular hydrogen bonds. The presence of GO reduces the intermolecular hydrogen bond between CALB and water, indicating the replacement of surface water molecules, terms as essential water, as observed by TGA analysis. Moreover,the presence of GO made the secondary structure of the CALB closer to its native form in the aqueous phase, this is favorable for the catalysis where the enzyme conformation transition is required. These simulation results offer a molecular insight into GO-CALB aerogel, which is essential to the design of GO-enzyme aerogel for gaseous enzymatic catalysis.

581360 The Development and Application of Novel Yolk@Shell Materials to CO_2 Recycling

The yolk@shell materials detailed here were designed to confront two key areas of catalytic deactivation: coking and sintering. Coking is known to occur in a number of ways: adsorption to the surface through a "monolayer-multilayer" mechanism[1]–[3], total encapsulation of the material that completely deactivates the catalyst [3] or fill micro- and mesopores to prevent or at least limit access of the reactants to the active sites [3]. Sintering, too, occurs in several ways: reduction in active surface area due to the growth of the catalytic phase, reduction in the amount of support area due to the collapse of the support phase in addition to the collapse of the porous structure [4].

To prevent these problems, the *mesoporous*-SiO_2 shell, within which our active core is held, provides a physical shield to mitigate sintering and acts as a barrier to prevent the blocking of the active sites through coke formation.

Our catalysts were tested in the dry reforming of methane (DRM) and Reverse Water Gas Shift (RWGS) reactions due to renewed interest in the "green" chemistry approach to produce added value products through chemical recycling CO_2. These reactions were also selected for their high temperature requirement that cause catalysts to be prone to sintering and coking mechanisms [5]–[8].

Herein, we report the successful synthesis of two Ni-based Yolk@shell catalysts: Ni/ZnO@*m*-SiO_2 for the DRM reaction and NiCo@SiO_2 towards the RWGS reaction. The NiZnO@*m*-SiO_2 catalyst was found to display considerable long-term activity over 143 hours, far surpassing standard supported metal nanoparticle materials tested under the same conditions. The post reaction characterization of the catalysts detailed no discernable sintering of the Ni nanoparticles inside the shell and SEM confirmed the successful protection of the core from carbon deposition.

Several NiCo@SiO_2 materials were produced with varied Ni content; 2, 5 and 10 wt%. These materials were applied to the methanation/RWGS reaction. Initial reactive testing has found that all these materials displayed considerable low-temperature activity towards the RWGS reaction. Interestingly, the results suggest that the Ni loading has a greater effect on the selectivity of the material as all the catalysts displayed similar levels of CO_2

determine the effect of property variation. Subsequent stability testing of the 2 wt% Ni catalyst, revealed that these materials are capable of significant long-term activity at 600°C, detailing no sign of deactivation over an 89-hour study.

In conclusion, the novel morphology of these materials affords long term protection from the reactive environments faced by all catalysts. The benefits of the morphology are clearly presented when considering the comparison of our materials to two reference materials (Ni/Al_2O_3 and Ni/SiO_2-Al_2O_3) with comparable Ni content in the DRM, clearly displaying long term isothermal performance and sintering resistance, while still providing substantial levels of CO_2 conversion. Further study regarding the NiCo@SiO_2 materials is expected to detail further enhancement and selectivity over existing catalysts at substantially lower temperatures.

[1] S. Niknaddaf, M. Soltani, A. Farjoo, and F. Khorasheh, "Modeling of Coke Formation and Catalyst Deactivation in Propane Dehydrogenation over a Commercial Pt-Sn/γ-Al2O3Catalyst," *Pet. Sci. Technol.*, vol. 31, no. 23, pp. 2451–2462, Dec. 2013.

[2] J. B. Butt and E. E. Petersen, *Activation, Deactivation, and Poisoning of Catalysts*. 1988.

[3] C. H. Bartholomew, "Mechanisms of catalyst deactivation," *Appl. Catal. A Gen.*, vol. 212, no. 1–2, pp. 17–60, 2001.

[4] C. H. Bartholomew, "Sintering Kinetics of Supported Metals: Perspectives from a Generalized Power Law Approach," *Stud. Surf. Sci. Catal.*, vol. 88, pp. 1–18, Jan. 1994.

[5] X. Zhao, H. Li, J. Zhang, L. Shi, and D. Zhang, "Design and synthesis of NiCe@m-SiO2 yolk-shell framework catalysts with improved coke- and sintering-resistance in dry reforming of methane," *Int. J. Hydrogen Energy*, vol. 41, no. 4, pp. 2447–2456, Jan. 2016.

[6] L. Pastor-Pérez, F. Baibars, E. Le Sache, H. Arellano-García, S. Gu, and T. R. Reina, "CO2 valorisation via Reverse Water-Gas Shift reaction using advanced Cs doped Fe-Cu/Al2O3catalysts," *J. CO2 Util.*, vol. 21, pp. 423–428, Oct. 2017.

[7] M. M. Nair, S. Kaliaguine, and F. Kleitz, "Nanocast $LaNiO_3$ Perovskites as Precursors for the Preparation of Coke-Resistant Dry Reforming Catalysts," *ACS Catal.*, vol. 4, no. 11, pp. 3837–3846, Nov. 2014.

[8] A. Cross *et al.*, "In Situ Preparation of Highly Stable Ni-Based Supported Catalysts by Solution Combustion Synthesis," *J. Phys. Chem. C*, vol. 118, no. 45, pp. 26191–26198, Nov. 2014.

569408 An Investigation into the Catalytic Cycle of Cytochrome P-450 Involving 1-n-Alkyl-3-Methylimidazolium Cations As Substrate

Ionic liquids have been projected as 'green' chemicals due to their environmentally benign characteristics like inherent low vapor pressure and flammability. As compared to conventional industrial solvents, their extremely low vapor pressures translate to the fact that they have negligible role in air emissions. These liquids can also be flexibly designed in order to tune their physical properties. Thousands of possible combinations of cations and anions are possible for such a design process. Out of the several ionic liquid classes, imidazolium-based ionic liquids have been one of the most widely utilized in diverse applications. Suitable variants of these types of ionic liquids have been applied to many processes in the chemical industry. Though being proved to be highly effective in driving complex industrial processes, experimental investigations have raised questions on their environmental degradability. Thus, it becomes imperative to include rational design into their synthesis. From an experimental point of view, considerable efforts have been put in this direction but molecular level details have not been explored in detail computationally. The present work aims to provide physical insight into the phenomena of ionic liquid biodegradability to aid in the future design of these solvents.

Cytochrome P-450 has been identified and widely studied for their role in oxidation of a wide variety of molecules in aerobic and anaerobic environments. Thus, it was deemed necessary to capture the effects of the cytochrome's active site molecule on imidazolium-based ($[C_n mim]^+$) cations to develop a computational framework for their biodegradability. For this, the enzymatic center of the P-450 molecule (heme) was modeled as an iron porphyrin molecule with an Fe-based center conjugated with the amino acid residue cysteine (FePCys) at its proximal side. The cations were included in the model as a potential substrate for the P-450 enzyme in complexation with the underlying porphyrin molecule. This interaction was modeled using DFT calculations by adapting a purely quantum mechanical framework at the M06 level of theory. To include the conformational effects, two different conformations of the ionic liquid cation, namely, tail up and tail down conformations, were considered in this work varying the 1-n-alkyl chain on the cation progressively along the homologous series (n =2,4,6,8,10). To provide insight into the dynamics of the substrate while binding to the active site molecule (FePCys) and further, different steps of the catalytic cycle were modeled using a synchronous approach. Result and discussion would describe the energetics and properties evaluated from the interaction between FePCys and ionic liquid cation from the binding, reduction and subsequent dioxygen insertion processes in the cycle. Also, key analysis of the interaction strength and thermodynamics of the system derived from each step would be included to provide further insight into the protein-ligand model.

569427 Design, Synthesis and Evaluation of Ir@Pt Bimetallic Catalysts for High Temperature Decomposition of so_3 to so_2 in the Hys Process for Thermochemical Water Splitting

One of the most promising new technologies for green energy production is development of an economic process for the solar driven Hybrid Sulfur (HyS) process. The key to success for this process is the high temperature, catalytic decomposition of sulfuric acid in the high temperature section of this process. After thermal volatilization and cracking of H_2SO_4 to form steam and SO_3, catalytic decomposition of SO_3 to SO_2 and O_2 must occur to regenerate SO_2 for the formation of H_2 from the electrolyzer portion of the process. Supported Pt catalysts are the most active of Group VIII metals for decomposition of SO_3 to SO_2 + ½ O_2. However, the reaction is conducted at temperatures = 800°C and conventional, supported Pt catalysts typically undergo sintering and loss of activity after only several hours at reaction conditions. Thus, the reaction and entire process becomes economically unattractive. Most of the external energy input (about 78%) comes in thermal form and is exchanged between the high temperature thermal source and the sulfuric acid section of the HyS process. Only 22% of the external input is in electric form to drive the low temperature electrolysis section. Thus, the efficiency of this thermochemical process is primarily affected by the performance of the high temperature catalytic decomposition section and having an efficient high temperature decomposition section is critical to reduce H_2 production costs. In addition, the decomposition section also affects the capital cost of the HyS chemical plant itself. Recent studies have demonstrated that the cost of the decomposition reactor represents about 53% of the overall HyS plant cost.

We have found that selective placement of Pt by electroless deposition, either as a partial or complete shell, on higher surface free energy (SFE) metal cores such as Ir (SFE = 3230 ergs/cm^2) results in active and stable Pt catalysts for SO_3 decomposition. The thermodynamic stabilization of the lower SFE Pt component (SFE = 2691 ergs/cm^2) on higher SFE Ir cores helps to constrain diffusion of Pt to form larger Pt aggregates. Evaluation data for long term decomposition of a gas stream containing up to 95% H_2SO_4 will be presented. Also discussed will be the enhanced stability and catalyst performance of non-oxidic supports such as BN, which maintains its surface area and pore structures at temperatures as high as 900°C.

571710 Solid-Solid Reactions: A Variation on a Classical Theme

All the classical models of solid-solid reacting systems, assume that the contact among reacting particles happen in continuous mode like solid-fluid reacting systems. Whereas, to precisely model these systems, only a few attempts made till date based on a rigorous account of the fact that the contact between reacting particles only occurs at a finite number of points on the surface of particles. In this work, we formulate a general contact-point based model for solid-solid reactions and simulate it using a finite element methodology on the COMSOL Multiphysics® platform. The model takes account the limited contact between two reactants at specific locations called as 'contact-points', also the unsteady-state behavior of the reaction, and spans the entire gamut of regimes from complete kinetic control to complete diffusion control. The diffusion-controlled case is particularly important, being a very likely scenario in solid-solid reactions, and also because a number of asymptotic cases for this case are available in the literature. The main assumptions for the study are isothermal conditions, only one (mobile) reactant species is diffusing into the matrix of other (stationary) reactant particles and stationary reactant particle size remains same during course of the reaction. The parameters involved in this models are Thiele Modulus (Th), density parameter (K) and contact points (N_{AB}). The rigorous formulated model extends over significant ranges of the above mentioned parameters.

Further the present model is compared with the most recent similar approximate Dalvi-Suresh (DS) model in this genre of contact point modelling. Dalvi-Suresh model is among the few which shows the correct asymptotic behavior in the limit of large number of contact points. However, the assumptions made in DS model, in order to keep the model mathematically tractable were not been validated previously in a rigorous way. So, in this work by comparing the results of our rigorously formulated model with the DS model, the applicability of the latter is studied, which proved that the simplified DS model can be applicable in the range of strong diffusion control (High Th) to moderate diffusion control only fails in case of reaction kinetic control (low Th) which is unlikely to happen in solid-solid reacting systems. However for the case of small density difference of the two reactants (low K), the assumption which treats the unsteady state behavior as a pseudo-steady state is limiting its accuracy for intermediate number of contact points.

Currently the work to plot the parametric space diagram for DS model is going on. Also the new model which accounts for the size change of reactant particle during the reaction due to formation of product is under study.

576556 Modeling of Transport-Kinetic Interactions in Various Design Configurations of Egg-Shell Catalyst Particles

Catalyst design is one of the most significant step in the development of a heterogeneous catalytic reactor process (Sie and Krishna, 1998). From a catalyst design perspective, specification of size, shape, and internal pore structure are interlinked parameters that have received notable attention by catalyst scientists and engineers since these properties affect the catalyst effectiveness factor, pressure drop, rate of catalyst attrition, catalyst mechanical integrity, catalyst life, amongst other key measures of performance. Determination of optimized catalyst design parameters is often a tedious trial-and-error process since tradeoffs exist in how the observed reaction rate and other measures of performance may be affected by altering one or both of these parameters.

Technical catalysts utilized to meet the economic norms of industrial processes are supplied as carrier-free, supported, or coated configurations. These catalysts are available in different sizes and traditional shapes, such as powders for slurry or fluidized bed reactors, and spheres, pellets, rings, extrudates, and honeycombs for fixed-bed reactors (LePage *et al.*, 2008). Methods commonly employed for catalyst preparation include deposition, precipitation and co-precipitation, gel formation and selective removal (Haber *et al.*, 1995). In many catalysts, the active metals may be distributed as a thin layer on a suitable support, which is often referred to as an egg-shell catalyst type.

Particle-scale transport-kinetic interactions, depending upon their magnitude, can affect the observed reaction rates for products and reactants. Significant transport effects are common in fixed-bed catalytic processes that employ catalyst particles with diameters from 5 to 15 mm (Sie and Krishna, 1998). Hence, it is important to utilize not only the exterior catalyst surface for deposition of active metals, but also the interior of the catalyst support to help reduce transport effects. In an effort to improve catalyst performance, non-traditional catalyst shapes with uniform metal distribution have been developed, such as multi-lobes, wagon wheels, miniliths and monoliths (Sie, 1993; Centi & Perathoner, 2003). In addition, egg-shell type catalysts with non-uniform metal distribution have been investigated (Lin and Chou, 1994; Bukur *et al.*, 2018).

The primary objective of this work is to compare catalyst performance for different design configurations of egg-shell catalyst particles using several oxidation and hydrogenation reactions of commercial importance to quantify the proof of concept. COMSOL Multiphysics® is used as the numerical engine for modeling the particle-scale transport-kinetic interactions. An oxidation test case reaction was utilized for modeling intraparticle diffusion and reaction for four different design configurations, namely 1. Egg-shell; 2. Egg-shell with a central opening; 3. Egg white; and 4. Egg yolk. Figures 1(a) to 1(d) show the concentration profiles oxidation test reaction reactant A for these configurations, respectively. Figures 1(e) to 1(h) show the particle concentration profiles for reactants (A, B) and products (C, D, and E) for these same catalyst design configurations. Higher conversions of reactant A were observed in egg-white and egg-yolk catalyst design configurations. Results for several test reaction cases provide guidance on future catalyst synthesis for optimization.

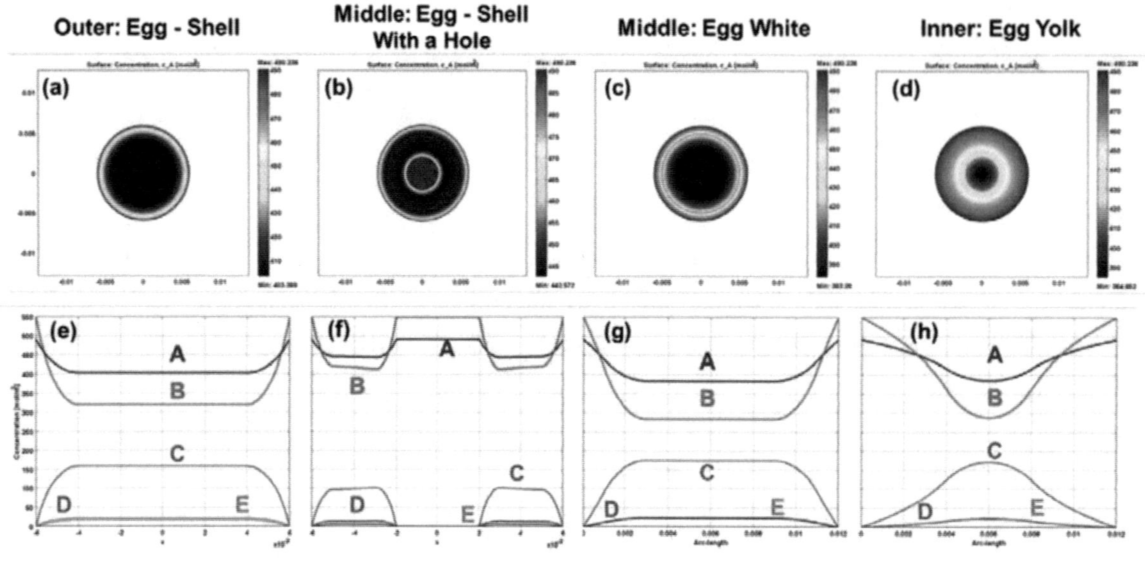

Figure 1. Concentration profiles for different design configuration of egg-shell catalyst particle.

References

Bukur, Dragomir B., Mandic, Miloš, Todic, Branislav, Nikacevic, Nikola. Pore diffusion effects on catalyst effectiveness and selectivity of cobalt based Fischer-Tropsch catalyst. Catalysis Today. 2018.

Cento, G., Perathoner, S. Integrated design for solid catalysts in multiphase reactions. CATTECH. 2003: 7(3): 78-89.

Haber, J., Block, J. H., Delmon, B. Manual of methods and procedures for catalyst characterization (Technical Report). Pure and Applied Chemistry. 1995: 67(8-9): 1257-1306.

LePage, J. F., Schlögl, R. , Wainwright, M. S., Schü, F. , Unger, K. , Ko, E. I., Jacobsen, H. , Kleinschmit, P. , Menon, R. G., Delmon, B. , Lee, K. , Misino, M. and Oyama, S. T. Preparation of solid catalysts: Section 2.1. In: Handbook of Heterogeneous Catalysis (eds. G. Ertl, H. Knözinger and J. Weitkamp), 2008: 57-66.

Lin, Tzong-Bin, Chou, Tse-Chuan. Selective hydrogenation of isoprene on eggshell and uniform palladium profile catalysts. Applied Catalysis A: General. 1994: 108(1): 7-19.

Sie, S.T. Intraparticle diffusion and reaction kinetics as factors in catalyst particle design, The Chemical Engineering Journal and the Biochemical Engineering Journal. 1993: 53(1): 1-11.

Sie, S.T., Krishna, R. Process Development and scale up: II. Catalyst design strategy. Reviews in Chemical Engineering. 1998: 14(3): 159-202.

571126 Mechanism of Effluence of Internal Substance from Pluronic Micelles By External Stimuli

Process intensification has attracted much attention in the field of chemical engineering since the 2000s. Stankiewicz and Moulijn defined process intensification as "Any chemical engineering development that leads to a substantially smaller, cleaner and more energy efficient technology," and it can be categorized into two classes: equipment and methods. In a general chemical industry, many reactors, mixture operations, and separation operations are needed to produce chemical product, and the design and maintenance become complex. On the other hand, high performance drug carrier such as polymer micelles has been investigated in the field of drug delivery system. In this study, the application of the drug carrier to a chemical reaction process was investigated from the view point of process intensification. Pluronic micelle and hydrophobic dye were used as a model reactant carrier and internal substances, respectively. The effects of ultrasonic irradiation on effluence of internal dye from Pluronic micelles were investigated. Moreover, the effects of radical scavenger addition on the effluence of dye was also investigated to consider a mechanism of effluence of internal substance from Pluronic micelles. For comparison, the effect of thermal stimulation on the release of dye was also investigated. A Pluronic micelle solution containing NKX-1595 was prepared by a dialysis method. The sample was irradiated with ultrasound for 10 min, and the effects of ultrasonic frequency, power, and radical scavenger addition on the effluence of the dye were investigated. The internal dye was rapidly released from the micelle at low ultrasonic frequencies, and radical scavenger addition was not influence the effluence of dye. Therefore, ultrasonic physical effect is considered one of the main factors. In addition, we investigated the effects of thermal stimulation on effluence of dye at several environmental temperatures. The internal dye was gradually released from the micelle when the environmental temperature was higher than 303 K.

576009 DESI-MS Based High Throughput Reaction Screening to Guide Bulk/Microfluidic Chemistry: A Correlation Analysis between Droplet and Bulk/Microfluidic Reaction Systems

High throughput screening (HTS) technologies are intensively used in various field of chemistry, from reaction pathway optimization to molecule discovery. Desorption electron ionization mass spectrometry (DESI-MS) is a recent technology that enables to carry out reactions in volumes as small as 50 nL with throughputs up to ~3000 reactions/h [1]. The essence of DESI-MS based HTS is that small volumes of reaction mixtures are spotted by a pipetting robot into high density (up to 6144 spots/plate) porous polytetrafluoroethylene (PTEF) based reaction plate, where each spot can represent and individual reaction. Then, the plate is analysed by the DESI-MS, which scans the whole ETFE plate surface by spraying it with a spray-solvent. The secondary droplets generated by the sprayer desorbs the reaction mixture from the plate, and these droplets are fed into the MS. Hence, the whole mass spectrum of each spot is collected, which enables to analyse the reaction mixture. The time the reagents spends in the droplet is in the order of milliseconds, in which the reactions happens. It is hypothesised that the reactions in the droplets are accelerated for the much reduced dimensions of droplets and their greatly increased surface areas [2] often by 10^4-10^6 times. This droplet acceleration enables to reach the throughputs of ~3000 reactions/h. However, the same droplet acceleration mechanisms raises significant questions when one tries to use the HTS data to guide batch/microfluidic reactions [3]: how the DESI-MS outcomes are expected to translate to larger volume scales, where there are no droplet effects anymore? This important question brings us to the topic of this work.

In order to analyse the correlation between the reactions observed in DESI and bulk, comparative studies were carried out for several representative reaction classes. For simplicity, in the abstract we present the correlation data for nucleophilic aromatic substitution reactions (S_NAr). S_NAr reactions are a versatile transformation in the modern organic chemistry arsenal and an important reaction class used for synthesizing pharmaceutically and biologically relevant molecules. The S_NAr reaction mechanism have been extensively investigated. It involves stepwise addition–elimination process wherein the first step involves nucleophilic attacks of the substrate to provide a Meisenheimer complex followed by the loss of the leaving group. The reaction typically involves an amine as the nucleophile. In this work, S_NAr reaction were carried out using amines and aryl halides in presence of bases for both DESI and bulk microtiter formats. The reactions used in the current analysis are published in the literature [3], the statistical interpretation of agreement between the DESI and bulk microliter reactions is in the focus of the current work.

Two S_NAr datasets are available: the first set (S_NAr1) involves 12 aryl halides, 8 amines, 2 reaction solvents and 4 basic conditions (3 bases and a 'no base' condition), which were carried out in DESI and in bulk microliter format at 150 °C with 15 h reaction time. The second dataset (S_NAr2) involves the same 11 aryl halides except one as S_NAr1, but 8 different bases, with one reaction solvent (from S_NAr1) and the same 4 basic conditions. In contrast with S_NAr1, the second dataset contains bulk reactions at 150 °C and 200 °C with 1,4 and 15 h reaction time. Hence, S_NAr1 covers broader space of reaction conditions, whereas S_NAr2 has richer bulk information content for the individual reactions. A simple yet effective comparison is provided by correlation plots, which is a representation of the product intensity of a given reaction recorded in the bulk as a function of product intensity in DESI. The intensity threshold, which is a critical intensity that has to be reached in order to count it as positive reaction – typically used in the DESI-MS based HTS technology - splits the correlation plot into four important quadrants. Q2 and Q3 are the regions of good agrement between DESI and bulk. The reactions in Q1 are positives in bulk and negative in DESI: in the context of DESI guided bulk reaction desing these reactions would have been missed. Q4 points are positives in DESI but negatives in bulk, which are misguiding, since these positive outcomes does not translates as potitives in bulk reaction.

In the S_NAR1 the correlation in the yes-no information space between the two reactor systems is greater than 70 %. Although, 22.5 % of points were false negatives (worked in bulk but not in DESI), and 3.6 % were false negatives. Depending on the nature of reactions and the applied bulk reaction conditions, the bulk outcome can often be both "yes" and "no". A correlation plot that compares the DESI outcome with the best bulk outcome based on the S_NAr2 data revealed that there are practically no false positives. Hence, in this case, the DESI does not misguide, and the true question to answer is: how to find the best reaction conditions for bulk and microfluidics (reaction time and temperature)? These questions can be addressed by the existing reaction engineering techniques, and therefore are out of objective of the current analysis. There are several surprising conclusions that can be drawn from the rich bulk information content of S_NAr2 reactions which will be discussed in details in the presentation. These are related to the thermal degradation as well as the importance of kinetic and thermodynamic control in the two reactor systems that can lead to either good agreement of discrepancy between the two platforms.

References:

[1] M. Wleklinski, B.P. Loren, C. Ferreira, Z. Jaman, L. Avramova, T.J.P. Sobreira, D.H. Thompson, R.G. Cooks, High Throughput Reaction Screening Using Desorption Electrospray Ionization Mass Spectrometry, Chem. Sci. 0 (2018) 1–7. doi:10.1039/C7SC04606E.

[2] X. Yan, R.M. Bain, R.G. Cooks, Organic Reactions in Microdroplets: Reaction Acceleration Revealed by Mass Spectrometry, Angew. Chemie - Int. Ed. 55 (2016) 12960–12972. doi:10.1002/anie.201602270.

[3] M. Wleklinski, C.E. Falcone, B.P. Loren, Z. Jaman, K. Iyer, H.S. Ewan, S.H. Hyun, D.H. Thompson, R.G. Cooks, Can Accelerated Reactions in Droplets Guide Chemistry at Scale?, European J. Org. Chem. 2016 (2016) 5480–5484. doi:10.1002/ejoc.201601270.

[4] Z. Jaman, D. L. Logsdon, B. Szilágyi, T. J. P. Sobreira, D. Aremu, L. Avramova, R. G. Cooks, D. H. Thompson, High Throughput Experimentation and Continuous Flow Validation of Nucleophilic Aromatic Substitution Reactions, manuscript is in preparation

577479 Understanding the Promotional Effects of Acids and Halides on Direct Synthesis of H_2O_2 and Their Effect on the Nature of Catalytically Active Pd

Direct synthesis ($H_2 + O_2 \rightarrow H_2O_2$) of H_2O_2 over supported Pd nanoparticles offers a promising alternative for H_2O_2 production to replace chlorinated oxidants in industrial processes instead of the current cost- and energy- intensive technology, the anthraquinone autoxiation (AO) process. However, direct synthesis suffers from low H_2O_2 selectivities (< 60%) on standard Pd catalysts [1]. Acids (e.g. HCl, H_2SO_4, H_3PO_4) and halides (KCl, KBr, NaBr) are often added to reaction mixture to improve the H_2O_2 selectivities, yet presence of acids can lead to significant rates of metal dissolution [2]. The dissolved Pd is oxidized to homogeneous Pd complexes which may catalyze H_2O_2 formation. Additionally, the fundamental role of these complex promoters in influencing catalysis is still not clear despite the vast body of literature reporting their effects on catalyst performance. Here, we combine kinetic and spectroscopic measurements to probe the relationships between rates, selectivities, and populations of homogeneous and heterogeneous Pd species to elucidate the nature of active species in direct synthesis. Furthur, we aim to understand the intrinsic manner in which these promoters improve the rates and selectivities in direct synthesis of H_2O_2 by conducting steady state rate measurements as a function of promoter concentrations (NaBr and H_3PO_4), H_2 and O_2 pressures and temperature.

To understand the nature of active sites, kinetic measurements were conducted in a semi batch reactor equipped with *operando* UV-Vis capabilities. Pd atoms primarily exist as suspended Pd^0 nanoparticles at steady state (>75% of total Pd) in absence of HCl regardless of the form of Pd introduced ($PdCl_2$, $PdSO_4$, $Pd(NO_3)_2$, Pd-SiO_2). Injection of HCl after reaction has achieved steady state increases the Pd^{2+} concentration due to oxidation of the Pd^0 nanoparticles to $PdCl_x(H_2O)_{(4-x)}^{(x-2)}$. When water is used as solvent, the H_2O_2 selectivities increase after the first HCl addition on Pd-SiO_2 and $Pd(NO_3)_2$ and decrease following a 2nd HCl injection. Yet for $PdCl_2$, the selectivity decreases after both HCl additions. H_2 and O_2 consumption rates decrease after each HCl addition which alludes to a loss of active sites. These observations are inconsistent with soluble Pd^{2+} species acting as the active catalyst for H_2O_2 formation. Rather, the suspended Pd nanoparticles are responsible for H_2O_2 formation. Chloride ions (in case of $PdCl_2$, the counter anion of the salt) can strongly bind to suspended Pd^0 nanoparticles and inhibit the cleavage of the critical O-O bond, hence reducing primary H_2O formation and H_2O_2 hydrogenation rates. Similar trends in H_2O_2 rates and selectivities were observed when methanol was used as solvent instead of water indicating that the form of the active species is independent of the choice of solvent.

Steady-state H_2O_2 and H_2O formation rates were measured within a fixed bed, plug flow reactor as functions of H_2 and O_2 pressure (10-400 kPa), temperature (275-295 K), and NaBr concentration ($0 - 10^{-3}$ M) in presence and absence of H_3PO_4 (10^{-3} M) on silica-supported ~7 nm Pd nanoparticles to probe the role of promoters in enhancing selectivities. H_2O_2 selectivities increase with increasing [Br^-] (15% in absence of Br^- to 40% at 10^{-4} M Br^-). However, increasing [Br^-] further to 10^{-3} M decreased the selectivity to 35% indicating that there is an optimum [Br^-] to achieve high selectivity. Furthermore, rates of H_2O_2 formation increase with increasing [Br^-] up to 10^{-4} M Br^-, but decline with further addition of Br^-. These observations suggest that bromide ions adsorb on the Pd clusters and reduce the number of active sites available for O_2 adsorption leading to a decrease in H_2O_2 formation rates and selectivities beyond 10^{-4} M [Br^-]. Addition of 10^{-3} M H_3PO_4 further improved the H_2O_2 selectivity to 77% at 10^{-4} M NaBr and to

~95% at 10^{-3} M NaBr. Addition of H_3PO_4 in presence of 10^{-4} M NaBr raises barriers for H_2O formation ($\Delta H(H_2O)$) (9±1 and 50±6 kJ mol^{-1} in absence and presence of H_3PO_4 respectively) much more than barriers for H_2O_2 formation ($\Delta H(H_2O_2)$) (1±1 and 9±1 kJ mol^{-1} in absence and presence of H_3PO_4 respectively) which partly explains the ~2 fold increase in selectivity in the presence of H_3PO_4 at 10^{-4} M NaBr. This result indicates that O-O binds more weakly to the active sites in the presence of both H_3PO_4 and NaBr and hence must overcome larger barrier for dissociation.

It is believed that the halides block sites responsible for O-O dissociation [3]. The results shown here suggest that bromide reduces electron backdonation to the $2\pi^*$ antibonding orbitals of O_2, resulting in weaker adsorption of O_2 to surfaces but also greater barriers for O-O bond rupture leading to enhanced H_2O_2 selectivities. We will attempt to elucidate the true role of halides by making comparisons of activation enthalpies across a range of concentrations. Finally, the role of acids will be tested by performing activation enthalpy measurement in varying concentrations of H_3PO_4 in absence of NaBr to deconvolute the effects of acids and halides. Hence, this detailed study on the promotional effect of acids and halides will help us understand the role of promoters in affecting catalysis separately as well as together which in turn will help us design more productive, selective and stable process for H_2O_2 formation by direct synthesis.

References

1. Flaherty, D.W., *ACS Catal.,* **2018**, *8*, 1520.
2. Chinta, S., Lunsford, J.H., *J. Catal*., **2004**, *225*, 249.
3. Bassler, P., Weidenbach, M., Goebbel, H., *Chem. Eng. Trans*., **2010**, *21*, 571.

577862 Studying Heterogeneous Cobalt Catalysts for Sulfite Oxidation to Compare with Theoretical Models

The sulfite oxidation reaction, where air is bubbled into a stirred liquid tank with dissolved sodium sulfite, can be used to determine liquid side volumetric mass transfer coefficients. Many of the assumptions that go into the calculations rely on appropriate reaction kinetics. Therefore, a catalyst can be used to appropriately use theoretical models. In this study, cobalt was immobilized on heterogeneous supports and compared with homogeneous analogs of cobalt as well as the theoretical models. It was determined that some of the key assumptions broke down when operated at various conditions, which should be avoided. This was especially true if comparing with uncatalyzed reactions. In this study we approached the reaction as a tool to teach undergraduate students the importance of assumptions as well as various techniques for analyzing gas/liquid reactions including using the effluent gas oxygen content to measure reaction progress. By using a heterogeneous catalyst, this experiment lends itself as an excellent opportunity to reinforce concepts learned in the classroom with real world applications which include concepts such as mass transfer, catalysis, and reaction engineering. After the reaction, the catalyst can be recovered and reused thus reducing waste disposal and catalyst loss compared to homogeneous catalyst use.

561744 Distinguishing Homogeneous and Heterogeneous Catalytic Pathway in Cuprous Oxide Nanoparticle-Catalyzed C-C Coupling Reactions

Carbon-carbon (C-C) coupling reactions are a diverse set of reactions that are used to produce a wide range of compounds including pharmaceuticals, aromatic-polymers, high performance materials and agrochemicals. The main drawback that limits the reactions though is the wide use of homogenous palladium catalyst. Homogenous palladium (Pd) catalyzed processes require numerous downstream operations to separate catalyst from the final product. Recently, metal and metal-oxide nanoparticle-based heterogeneous catalysts have emerged as attractive catalysts for these reactions. However, identification of the catalytic pathway remains a challenge in the field.

In this contribution, we show that by choosing appropriate reaction conditions, the reaction pathway can be controlled to occur via a truly homogeneous or heterogeneous pathway in Cu_2O-nanoparticle-catalyzed C-C coupling reactions. We utilized a combination of reactor study, flame atomic absorption spectroscopy, UV-Vis extinction spectroscopy, electrospray ionization mass spectrometry and transmission electron microscopy to characterize the role of homogenous Cu catalytic species under different reaction conditions. We also show that in-situ formed homogenous Cu complex can successfully catalyze (i) oxidative C-C homo-coupling of terminal alkynes, and (ii) C-C cross-coupling reactions between aryl halides and terminal alkyne. The molecular reaction mechanisms are proposed for both hetero and homo coupling activity and molecular structures for the intermediates are proposed. The mechanistic understanding from this work has the potential for the development of (a) an inexpensive copper based catalyst as a suitable replacement for expensive Pd based catalysts, and (b) a heterogeneous catalyst capable of exhibiting long term stability to develop energy efficient and environmentally friendly flow processes.

562662 Synthesis of Tetrapropylammonium Bromide (TPABr)

Tetrapropylammonium Bromide(TPABr) is a hygroscopic white-coloured salt used as a template for reaction. The catalyst like ZSM-5 and zeolite TS-1 can be prepared using the proposed product. As per the conventional techniques, it is extremely difficult to obtain with a higher yield. In the present work, a method for producing TPABr by reacting tripropylamine with Propyl bromide in a polar solvent at a temperature between 60°C and 160°C is carried out.

TPABr is prepared by mixing propyl bromide with tripropylamine in a polar solvent, in an inert atmosphere, under reflux conditions for a sufficient period of time to form a solution containing TPABr. After cooling the resulting solution, the solution was mixed with water, thus forming a mixture and recovering TPABr in high yield from the mixture. The reaction can be carried out at different time intervals and in presence of different polar solvents. The reaction can be optimized depending on the solvent used. Usually more the solvent taken leads to increase in the yield but recovery of solvent should be economical. Usually 20% to 5 time of reaction mixture is suggested as solvent quantity to be taken.

In the present work, the TPABr is synthesized at various conditions and evaluated with the successfulness through NMR. The developed and required TPABr is further characterized by NMR, FTIR, FE-SEM, TEM, XRD and Surface area.

565920 Bifurcation Analysis of Coupled Homogeneous-Heterogeneous Reactions in Monolith Reactors

Bifurcation analysis of coupled homogeneous-heterogeneous reactions in monoliths

Bhaskar Sarkar and Vemuri Balakotaiah

Chemical Engineering Department, University of Houston, Houston, TX 77024, USA

Keywords:

Bifurcation, homogeneous ignition, partial oxidation, OCM

Abstract:

Catalytic partial oxidation is an attractive technology for meeting future energy demands and production of intermediate chemicals. The models describing this process typically involve both catalytic and homogeneous reactions and strong species as well as thermal coupling. There have been numerous experimental and computational studies on these systems in the past twenty-five years. Most of these studies have used micro-kinetic models for homogeneous and catalytic reactions or focused on hydrodynamic aspects. Such analyses lead to models that are not amenable to a bifurcation analysis in the parameter space as they are computationally complex due to existence of exponentially thin boundary layers in space and/or time. Consequently, a clear picture of the essential/qualitative features of the process as the parameters are changed is not entirely clear. The objective of this study is to determine how the various operating and design parameters such as the space time, inlet fluid temperature, inlet mole ratio, pressure and channel hydraulic radius impact the bifurcation (ignition and extinction) behavior of coupled homogeneous-heterogeneous reactions in monoliths. We use the oxidative coupling of methane (OCM) as test example and use a hierarchy of reactor and chemistry models to determine their impact on the predicted bifurcation behavior.

First, we investigate a twelve step global kinetic model with seven reactions in heterogeneous phase (wall reactions) and five reactions in homogeneous phase using a two-phase 'Short Monolith' or 'Gauze Reactor Model'. We calculate various bifurcation diagrams and bifurcation sets for different state variables (gas temperature, solid temperature, conversion, selectivities etc.) vs. feed inlet temperature (T_{in}) and determine a complete phase diagram in the plane of oxygen to methane ratio and space time. Our calculations show that it is possible to obtain ~80% C_2 selectivity with ~20% CH_4 conversion for inlet CH_4/O_2 mole ratio of 6-8 and space time of 0.1-1s with La_2O_3/CaO catalyst. Higher CH_4/O_2 mole ratios lead to higher selectivities at the cost of lower CH_4 conversion and productivity. It is found that operations on mass transfer controlled regime result in higher selectivities but lead to lower conversion and require higher space times, thereby again lowering the productivity. Further, on the ignited branches the fluid exit temperatures are found to be low enough (~1000K) to conduct stable autothermal operation of OCM.

Next, we move towards a more realistic multi-scale coarse-grained model for the monolith reactor, which includes diffusion of species and heat through the washcoat (catalyst). Both the steady state and dynamic hysteresis curves are calculated with this low dimensional model using flat and parabolic velocity profiles. The qualitative bifurcation features of both the models are found to be similar and the coarse-grained model approaches the short monolith model in the limit of small washcoat thickness.

567485 Validated CFD Simulations of a Bench-Scale Fixed Bed Fischer-Tropsch Reactor

The Fischer-Tropsch synthesis (FTS), which converts syngas into liquid fuel, has a very important application and perspective. The fixed bed reactor (FBR) is one of the most competitive and mature reactors applied in small scale FTS, such as biomass to liquid fuel process (BTL). Compared to lab scale FBR, heat transfer is more important and the temperature profile along the catalyst bed cannot be ignored in a bench scale FBR. In this study, a bench scale reactor with 50 mm diameter and 1 000 mm length was specially designed for temperature measurement and employed in FTS. Four thermocouple sheaths were placed at different radial positions of the reactor. Each thermocouple can slide smoothly inside the sheath and measure the temperature in axially direction. Serial experiments were conducted at low temperature FTS conditions (2 MPa and 458K) with Co-based catalyst. The temperature distributions and reaction performance were investigated when changing the inert gas (nitrogen) content and gas hourly space velocity (GHSV) respectively.

Computational fluid dynamics (CFD) simulation is an efficient access for certain chemical operating unit designing, result prediction and operating condition optimization. A 3D pseudo-homogeneous FBR model including the oil bath on the shell side was set up and solved by Fluent in this study. Thermal equilibrium was assumed between the gas phase and the solid phase (catalyst bed). A widely used semi-empirical kinetic model was implemented to describe the FTS. When separately comparing the temperature profile (Figure 1) and reaction performance (CO conversion, CH_4 selectivity and C_{5+} hydrocarbon selectivity) from simulation results to experimental data, it showed a good agreement. Based on this validated model, numerical experiments can be conducted to give a precise guideline for further studies on optimization of FTS operating conditions or hot spot prediction.

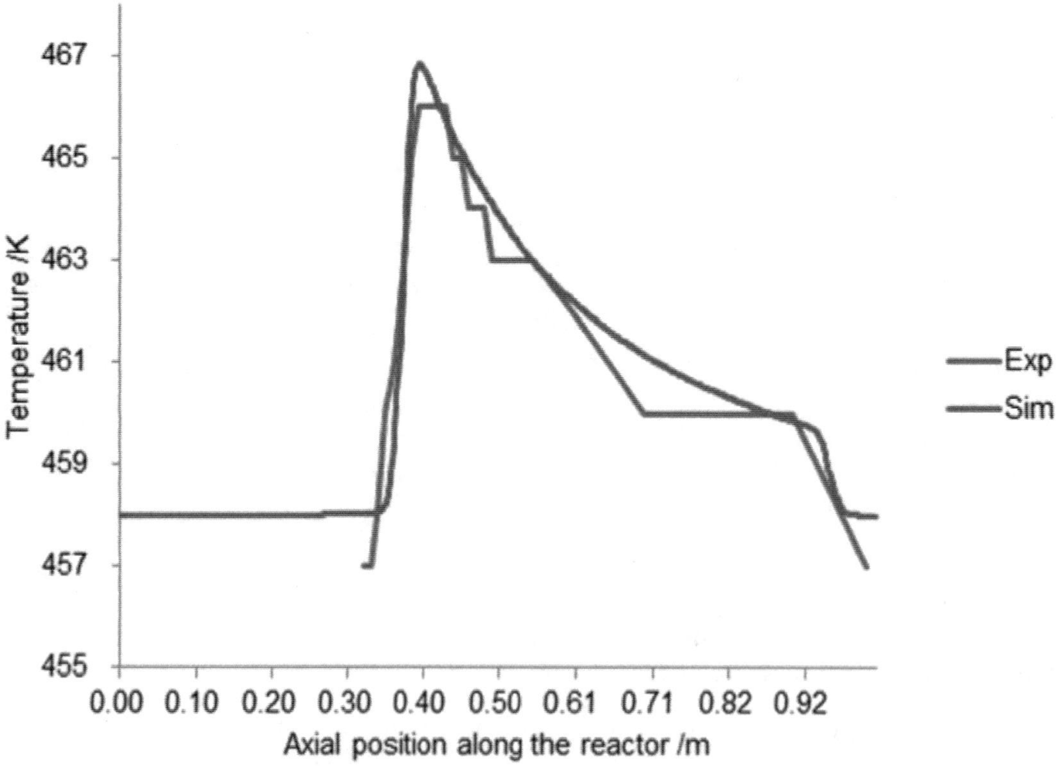

Figure 1 Temperature comparison of simulation results and experimental data

572672 Two-Dimensional Heterogeneous Reactor Model for an Exothermic Reaction Exhibiting Deactivation By Fouling: Modeling and Computational Strategy

Two-dimensional heterogeneous reactor model for an exothermic reaction exhibiting deactivation by fouling: modeling and computational strategy

Javier Ibáñez,[a] Jeroen Poissonnier,[a] and Joris Thybaut[a]

[a] *Laboratory for Chemical Technology, Ghent University, Ghent, B-9052, Belgium*

Abstract:

This contribution reports on a two-dimensional heterogeneous reactor model for a fixed bed catalytic reactor. The model considers temperature and concentration gradients in both phases. The gas-phase chemical reaction, is highly exothermic and presents particularities that cannot be tackled with the effectiveness factor approach. Notably, deactivation by fouling is considered, simultaneously addressing active site disappearance and pore blockage. The computational effort is alleviated employing efficient numerical integration algorithms and applying a quasi-steady state approach.

1. Introduction

The generally considered idealization of a packed bed reactor consists of a homogeneous model in which temperature and concentration gradients within all phases are considered negligible. For reactions presenting rapid kinetics and pronounced exothermicity, the gradients between phases become non-negligible. Under these circumstances, a two-dimensional heterogeneous reactor model becomes the appropriate choice. This more fundamental approach becomes especially relevant when catalyst deactivation takes place, since well-described local conditions are key to accurately simulate the evolution of catalytic activity.

In this work we aim to model the performance evolution for a packed bed reactor in which an exothermic reaction, represented as $A + B \leftrightarrow C$, with ΔH^{o} ~ 110.0 $kJ\cdot mol^{-1}$ is carried out. A deactivation of the pellet shell has been experimentally observed, in which the porosity of the catalyst is lost in the outer layers, whereas the core maintains its original state. Nonetheless, the plugged pores hinder the diffusion of the reactants and render the pellet inactive. In the following we describe a two-dimensional heterogeneous model, as described by de Wasch et al. [1], accounting for catalyst deactivation by fouling and pore blockage.

2. The modeling approach

The set of main features and assumptions considered in the heterogeneous packed bed reactor model are as follows:

1. Concentration and temperature gradients exist in the axial and radial directions on the fluid phase.
2. The chemical reaction occurs exclusively in the solid phase.
3. The solid phase consists of finite cylindrical pellets, presenting concentration and temperature gradients in the axial and radial directions.
4. The catalyst pellets are considered of negligible size with respect to the tubular reactor and are exposed to homogeneous boundary conditions.
5. The mass and heat transfer resistance at the fluid-solid interphase are considered.
6. The deactivation mechanism proceeds via coke deposition and is modeled by means of a decrease of the effective molecular diffusivity.

In order to derive the balances over the reactor and pellets, their spatial coordinates are discretized in concentric hollow cylinders, as depicted in Figure 1(a).

Particularly, the fluid phase continuity equations are as follows:

$$\frac{\partial(c_{i,f}u_s)}{\partial z} = D_{i,f}\frac{1}{r}\frac{\partial}{\partial r}\left(r\frac{\partial c_{i,f}}{\partial r}\right) - \frac{1-\varepsilon}{V_p}\int_0^{V_p} R_{i,s}dV \tag{1}$$

$$u_s\rho_f c_p\frac{\partial T_f}{\partial z} = \lambda_{er,f}\frac{\partial}{\partial r}\left(r\frac{\partial T_f}{\partial r}\right) - ha_v(T_f - T_s) \tag{2}$$

The boundary conditions applied correspond to those of the inlet at z=0, while the symmetry condition applies at r=0 and wall conditions at $r=R_t$:

$$c_{i,f} = c_{i,f}^0 \qquad T_f = T_f^0 \qquad at\ z = 0, \quad \forall r \tag{3}$$

$$\partial c_{i,f}/\partial r = 0 \qquad at\ r = 0\ and\ r = R_t \quad \forall z \tag{4}$$

$$\frac{\partial T_f}{\partial r} = -\frac{\alpha_w}{\lambda_{er,f}}(T - T_w) \qquad at\ r = R_t \quad \forall z \tag{5}$$

The solid phase mass and heat transfer occur solely by means of diffusion and are respectively described by the following expressions:

$$D_{i,eff}\left(\frac{\partial^2 c_{i,s}}{\partial z^2} + \frac{1}{r}\frac{\partial}{\partial r}\left(r\frac{\partial c_{i,s}}{\partial r}\right)\right) + R_{i,s} = 0 \tag{6}$$

$$\lambda_{er,s}\left(\frac{\partial^2 T_s}{\partial z^2} + \frac{1}{r}\frac{\partial}{\partial r}\left(r\frac{\partial T_s}{\partial r}\right)\right) - r\Delta H_{rx} = 0 \tag{7}$$

Neumann boundary conditions apply at the outer surface of the pellet, whereas symmetry conditions apply at r and z = 0.

$$\frac{\partial c_{i,s}}{\partial r} = -\frac{k_m}{D_{i,f}}(c_{i,s}^{surf} - c_{i,f}) \qquad at\ r = R_p, \quad \forall z \tag{8}$$

$$\frac{\partial T_s}{\partial r} = -\frac{h}{k_{f,s}}(c_{i,s}^{surf} - c_{i,f}) \qquad at\ r = R_p, \quad \forall z \tag{9}$$

$$\frac{\partial c_{i,s}}{\partial z} = -\frac{k_m}{D_{i,f}}(c_{i,s}^{surf} - c_{i,f}) \qquad at\ z = L/2, \quad \forall r \tag{10}$$

$$\frac{\partial T_s}{\partial z} = -\frac{h}{k_{f,s}}(c_{i,s}^{surf} - c_{i,f}) \qquad at\ z = L/2, \quad \forall r \tag{11}$$

The deactivation is considered to proceed by fouling and pore clogging, since a decrease of the net porosity is observed in the pellet shell. Such mechanism occurs in a set of consecutive steps: first, the coke precursor adsorbs on the active sites; second, the precursor is transformed into actual coke which further grows in the catalyst pores, hindering the diffusion of reactants and finally blocking their access. Theoretical approaches based on percolation theory have been proposed to model the evolution of the accessibility of the active sites as a function of coke content [2]. The success of these models is based upon a good description of the pore structure, which might only be known for purely crystalline catalysts. A more practical approach was proposed by Nevicato et al. [3]. A semi-empirical expression was developed based on a mechanistic description of the sequence of steps involved in the deactivation process. For the case of diffusion-limited reactions, the following expression could be derived:

$$\frac{D_{i,eff}}{D^0_{i,eff}} = \left[\frac{(k_e + k_f)\exp(-k_e C_{coke,s})}{k_f + k_e \exp(-(k_e + k_f)C_{coke,s})}\right]^2 \tag{12}$$

Finally, the following expressions are used to describe the main reaction and coke deposition rates:

$$r_{rx} = \frac{k_{rx}(C_A C_B - C_c/K_{eq})}{1 + K_B C_B} \tag{13}$$

$$r_{coke} = \frac{k_{coke}(C_A C_B)}{1 + K_B C_B} \tag{14}$$

By combining the above set of PDEs, the reactor model can be solved accounting for the effect of coke deposition in the catalyst deactivation. Depictions of the model outputs can be visualized in Figure 1(b-c), which illustrates a mapping of the concentration of A within a catalyst pellet and the evolution of the center-line axial temperature profiles with time on stream. The computational effort is effectively alleviated by automatically switching between the steady state solution and the deactivation function. In this approach, the reactor equations are solved at steady state and assumed to remain invariant for a finite amount of time. The deactivation functions are then computed at these invariant conditions, allowing to update the map of catalytic activity in the reactor. The automatic switching is based on the evolution of this activity map, which assures an accurate solution and saves computational effort.

3. Conclusions

A two-dimensional heterogeneous reactor model is here proposed. This model describes the time evolution of a packed bed reactor, considering interphase and intraparticle concentration and temperature gradients. The experimentally observed catalyst shell deactivation is described by means of a semi-empirical expression. This approach parts from a mechanistic basis, lumping the consecutive processes of active site coverage, growth of coke chains and pore blockage into a single expression. For the case of mass transfer limited reactions, the coke concentration can be linked with the effective diffusivity. This relation is here used to describe the evolution of the catalytic activity. Finally, the computational effort required for the complex transient modeling is minimized by automatically switching between the steady state solution and the deactivation function.

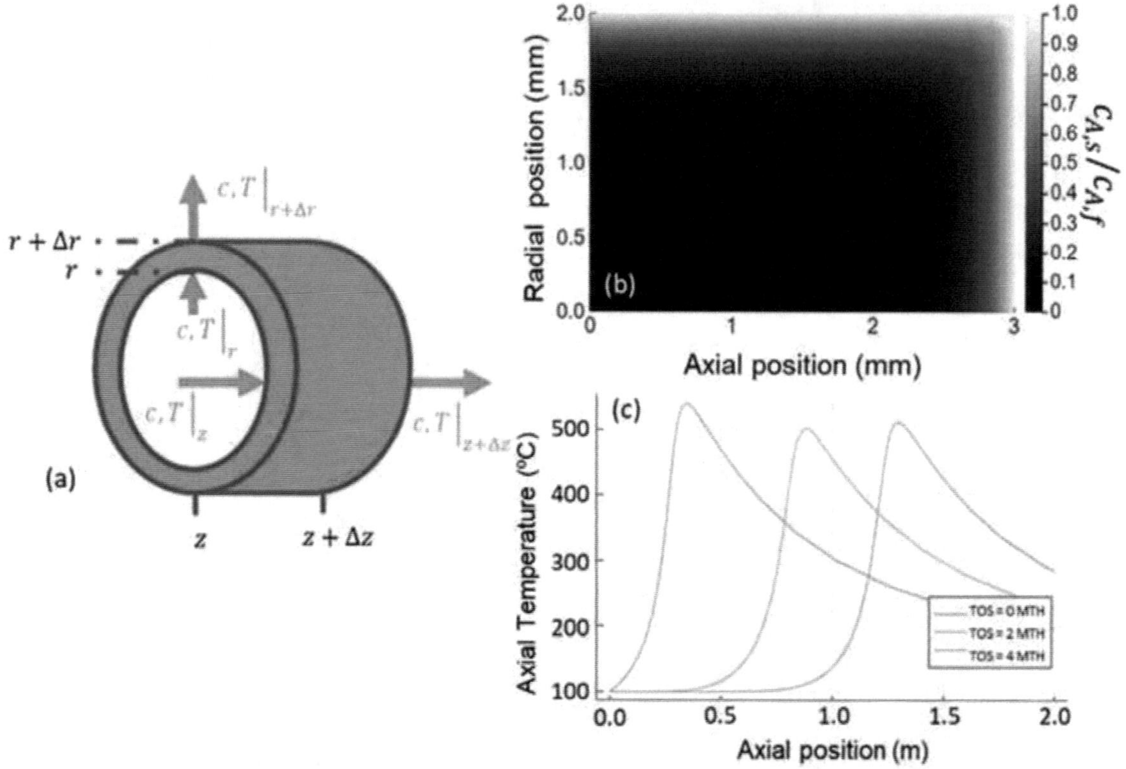

Figure 1. (a) Schema of the differential volume used for the derivation of the cylindrical pellet and reactor balances. **(b)** Heatmap showing the relative concentration of A inside a fresh catalyst pellet as a function of radial and axial position.**(c)** Time evolution of the center-line axial temperature profile over 4 months of time on stream.

Glossary

			Greek Letters		
c	concentration	$mol \cdot m^3$	α_w	wall heat transfer coefficient	$kJ \cdot m^{-2} \cdot s^{-1} \cdot K^{-1}$
c_p	specific heat	$kJ \cdot kg^{-1} \cdot K^{-1}$	ε	bed void fraction	-
D	diffusivity	$m^2 \cdot s^{-1}$	λ_{er}	effective radial thermal	$kJ \cdot m^{-1} \cdot s^{-1} \cdot K^{-1}$
h	film heat transfer coefficient	$kJ \cdot m^{-2} \cdot s^{-1} \cdot K^{-1}$		conductivity	
		-1			3

ΔH_{rx}	reaction enthalpy	kJ·mol		ρ	Density	kg·m
k_m	film mass transfer coefficient	m/s				
L	pellet length	m		Subscripts and superscripts		
r	radial position	m		i	component i	
r_i	reaction rate for reaction i	$mol \cdot m^{-3} \cdot s^{-1}$		f	fluid phase	
R	generation rate	$mol \cdot m^{-3} \cdot s^{-1}$		rx	chemical reaction	
R_p	pellet radius	m		s	solid phase	
T	temperature	K		w	heat exchanging wall	
u_s	superficial velocity	$m \cdot s^{-1}$				
V_p	pellet volume	m^3				
z	axial position	m				

4. References

1. de Wasch, A. P. & Froment, G. F. A two dimensional heterogeneous model for fixed bed catalytic reactors. *Chem. Eng. Sci.* **26,** 629–634 (1971).

2. Froment, G. . Modeling of catalyst deactivation. *Appl. Catal. A Gen.* **212,** 117–128 (2001).

3. Nevicato, D., Pitault, I., Forissier, M. & Bernard, J. R. The activity decay of cracking catalysts: chemical and structural deactivation by coke. *Stud. Surf. Sci. Catal.* **88,** 249–256 (1994).

557510 Modeling of Polymerization of Methyl Methacrylate in Homogeneous Systems As a Framework for Processes Improvements

The kinetics of acrylic polymerization is very complex in nature. Diffusion-limited problems such as gel and glass effects impact directly the polymerization kinetics and molar masses. Attempts to model these phenomena based on physical properties of the system have been presented, but models based on adjustable parameters (semi-empirical models) have shown a more pragmatical approach for prediction of molecular weight and conversion of such polymerizations.

In the present study, the polymerization of methyl methacrylate is selected and tested with recent kinetics models of the literature, either partially based on free-volume parameters, or based solely on empirical parameters. These models were extended to include additional reactions, such as depropagation, termination by combination and transfer to monomer, in a first attempt to generalize them to batch and semi-batch polymerizations in temperature range from 50°C to 140°C, under isothermal and non-isothermal conditions. The particular case of chain transfer agents was also tested. The molecular weight distribution was calculated with the method of moments.

A simulation program (Reactormodel) was developed for this study. New parameters were adjusted to fit experimental data taken from the literature using a simplex search method, applied to the normalized sum of squared errors between experimental and calculated values, generating two new models. These models were then compared with an extensive range of experiments, and showed improved prediction of polymerization parameters, when compared with the original models. The models were then submited to a sensitivity analysis, which qualitatively demonstrate their behavior and limitations in extrapolated conditions.

557260 Tributyl Citrate Production from Calcium Citrate: Solid-Liquid Reaction Kinetics

Plasticizers are chemical compounds used to enhance the flexibility, softness and durability of plastic materials. Phthalates are the most widely used plasticizer with large consumption in the manufacturing of PVC products. Recently, increasing concerns have appeared regarding the use of phthalates as additives in plastic for bottles, food packaging, toys and medical devices. As the plasticizers are not chemically bounded to the polymeric matrix, they can migrate from the polymeric materials. Thus, several studies have demonstrated that phthalates represent a risk for human health and the environment as a result of their toxicity and low-biodegradability. As a result of these problems, phthalates have been banned in many countries for sensitive applications, boosting the demand for non-toxic, biodegradable, and biobased plasticizers.

Among the variety of biobased plasticizers, citric acid esters have been considered promising phthalates substitutes because they offer a similar plasticizing performance. Among citrate esters, tributyl citrate (TBC) and acetyl tributil citrate (ATBC) have received a special attention, since they have been successfully used in many industrial applications. Nevertheless, currently production costs make TBC limits its potential to become a major player in the global plasticizers market. Regularly, TBC is synthetized through semi-batch esterification of citric acid with a large excess of butanol to overcome equilibrium limitations, using homogeneous acid catalyst and long residence times. As TBC production costs largely depend on raw materials costs, the use of lower costs feedstock would highly improve process economics.

During citric acid purification, there is a neutralization step where an intermediate calcium salt is precipitated from the culture broth. This stream is subjected to filtration, drying, water suspension, and acidification to recover the citric acid in aqueous solution. Later the acid is purified by ion exchange, adsorption, crystallization steps, filtration, and drying. Thus, it is expected that TBC production costs could be reduced by using the intermediate calcium solution as feedstock.

Then, the aim of this work is to explore the use of calcium citrate as raw material in the production of TBC. In the proposed process, dried calcium salt is dispersed in butanol and the insoluble mixture is acidified with sulfuric acid, allowing the simultaneous extraction and esterification of citric acid. This involves a solid-solid-liquid reaction where solid calcium citrate reacts with sulfuric acid to produce solid calcium sulfate, and citric acid in alcoholic solution. Then, dissolved citric acid reacts with butanol to produce butyl citrates owing to the presence of sulfuric acid (i.e. esterification catalyst). The water produced from esterification can be up taken by the solid calcium sulfate, thus shifting equilibrium to the esters side.

In order to evaluate the feasibility of the proposed process, a solid-solid-liquid reaction model was developed to describe the simultaneous acidification-esterification stage. Initially, the esterification kinetics of pure citric acid with butanol using sulfuric acid as catalyst was experimentally measured. A Box-Behnken experimental design was carried out, exploring the kinetic effects of reaction temperature, acid\alcohol feed ratio, and catalyst loading. In parallel, experiments were carried out to determinate the chemical equilibrium constant under the operating conditions.

Subsequently, the solid-liquid reaction experiments were performed in batch reactors using calcium citrate as reactant. The salt was dispersed in a solution of butanol and sulfuric acid. An experimental design was accomplished evaluating the kinetic effects of temperature and feed reactant ratio. Then, based upon the experimental observation of both reaction steps, a solid-solid-liquid reaction model was constructed to describe reaction mechanisms. Microscopy images help elucidating and developing the corresponding solid reaction model. Finally, the experimental data were used to regress a kinetic model and a good agreement with observations was obtained. Then, the regressed model can be further used for process design and up-scaling.

560395 Mechanism and Kinetics of Metal Dissolution in Nitric Acid

Mechanism and Kinetics of Metal Dissolution in Nitric Acid

Abstract

Dissolution of metals in acid is an important step in the production of metal salts, extraction of metals from its ore and extraction of metals from e-waste through the hydrometallurgical route. The r eaction mechanism and the stoichiometry of metal dissolution vary with the concentration of acid used and stirring speed. The dissolution of metals is reported to exhibit autocatalytic kinetics under certain operating conditions.

To understand the different physio-chemical mechanisms prevailing in the system, we focus on the dissolution of copper in nitric acid under controlled conditions for different nitric acid concentrations and stirring speeds. Specifically we focus on the autocatalytic behavior observed for low acid concentrations and high stirring speeds. When copper reacts with nitric acid then NO_x are released depending on the concentration of nitric acid.

During reaction, autocatalytic behavior is observed experimentally under low concentration and high stirring speeds as shown in Fig 1. When initial concentration of nitric acid is 4M then sigmoidal behavior is observed, confirming an autocatalytic effect. This autocatalytic behavior arises from the nitrous acid generated. This behavior is not observed at high concentration of nitric acid.

A reaction mechanism is developed which explains the generation of nitrous acid and autocatalytic effect behavior. The model also predicts that at high concentrations of nitric acid there is no autocatalytic effect observed.

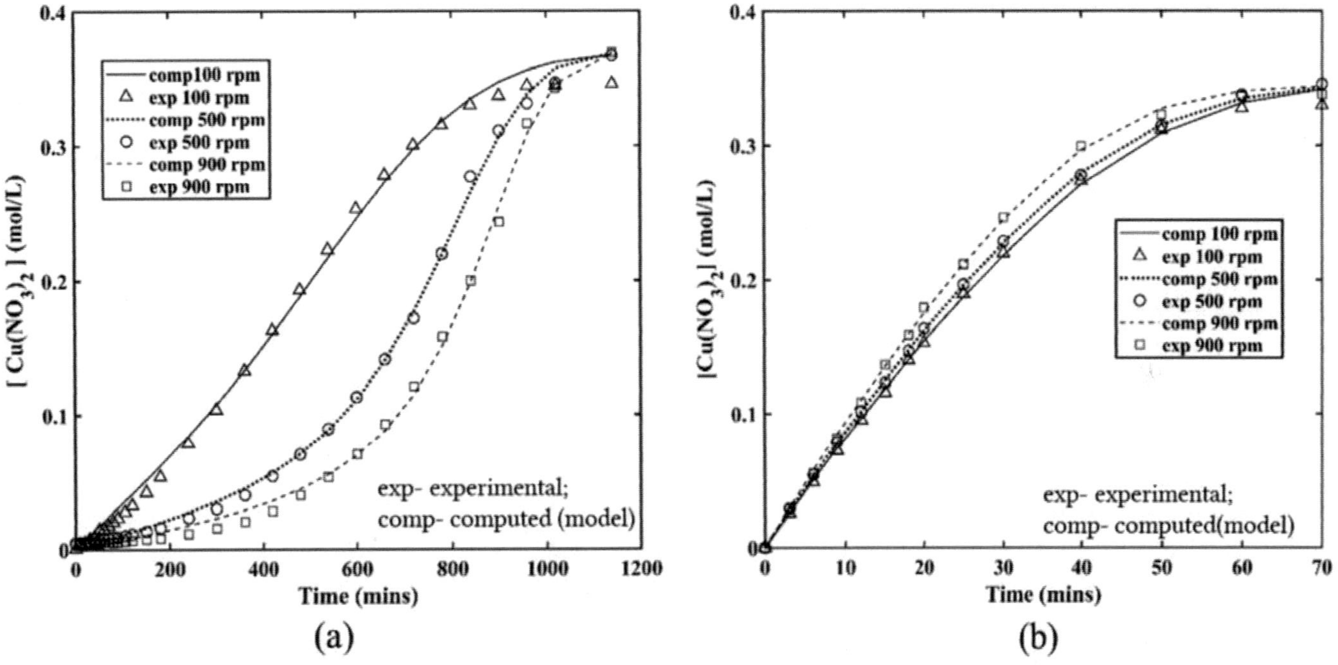

Fig 1. Comparison of experimental (marker) and model prediction (lines) for Cu dissolution

a) in 4M HNO_3 b) in 8M HNO_3

The mathematical model proposed incorporates the change in surface area of metallic copper during the dissolution process. This model considers multiple reactions and also accounts for change in volume in the reactor due to sampling. A heterogeneous model consists of Differential Algebraic Equations (DAEs) which considers both mass transfer and reaction is developed to predict the perfo rmance.

The kinetic parameters in the proposed model were estimated using a constrained optimization formulation based on particle swarm optimization technique. It is shown that model predictions fits well with the experimental data (see Fig 1). It is shown that two reactions are sufficient to describe the different features exhibited by the dissolution kinetics of copper.

Acknowledgment

This study was supported by Exploratory Research Grant Work, IIT Madras.

References

[1] U. Jadhav and H. Hocheng, "Hydrometallurgical Recovery of Metals from Large Printed Circuit Board Pieces," *Sci. Rep.*, vol. 5, no. 101, pp. 1–10, 2015.

[2] Levenspiel, O. *Chemical Reaction Engineering*; John Wiley & Sons: New York, 1999.

[3] Demir, H.; Özmetin, C.; Kocakerim, M. M.; Yapc, S.; Çopur, M. Determination of a Semi Empirical Kinetic Model for Dissolution of Metallic Copper Particles in HNO3 Solutions. *Chem. Eng. Process. Process Intensif.* 2004, *43* (8), 1095–1100.

[4] El-Chelkh, F. M.; Khalil, S. A.; Ei-Manguch, M. A.; Omar, H. A. New Stoichiometry for Copper Dissolution in Nitric Acid (the Authors Reply). J. Chem. Educ. 1987, 64 (12), 1070.

[5] Özmetin, C.; Þopur, M.; Yartasi, A.; Kocakerim, M. M.; Ozmetin, C.; Copur, M. Kinetic Investigation of Reaction between Metallic Silver and Nitric Acid Solutions. *Chem. Eng. Technol.* 2000, *23* (8), 707–711

569912 Dynamic Catalysis and Surface Resonance for Turnover Frequency Enhancement

Heterogeneous catalysts and their associated chemical reactors are key components of many chemical processes for petroleum production, fine chemical synthesis, and consumer products manufacturing. For each process, intensive reaction engineering has been employed to scale up, maximize product yields, and minimize production costs. Typical process and product development include varying temperature, pressure, catalyst selection, reactor type, and time/spacetime. For a given set of these variables, a wide range of reaction and catalyst classes have been found to obey a “volcano plot” relationship between reaction rate and several catalyst binding energy descriptors. This means that reaction engineering has a (currently) insurmountable catalytic speed limit, due to the trade-off between surface reaction rates and product desorption for a catalyst with static binding properties. In this work, we present several alternative handles for enhancing catalyst performance and chemical production rates. In this approach, catalyst binding properties are varied with time [1].

Using a CSTR model in Matlab, a model system with A → B and three elementary steps: (i) adsorption of A, (ii) surface reaction of A* → B*, and (iii) desorption of B was studied. The catalyst binding properties were varied using a square, sinusoidal, triangle, or saw-tooth waveform with a specified oscillation frequency (f_{osc} , [=] Hz) and amplitude (ΔU, [=] eV). The dynamic steady state rate was found to be a strong function of frequency and amplitude. For moderate amplitudes between 0.5-1.5 eV, there is a resonance frequency range between 10^3 -10^7 Hz where steady state rates level off at 100-10,000x the Sabatier maximum. With practicality in mind, the oscillation waveform was tested at the same frequency and amplitude to assess their effectiveness in rate enhancement. Square waveform were the highest performing but sinusoidal waveform still led to high rate enhancement. These new process handles of frequency, amplitude, and waveform allow for reaction engineering to greatly improve process performance.

[1] M. A. Ardagh, O. A. Abdelrahman, P. J. Dauenhauer, “Principles of Dynamic Heterogeneous Catalysis: Surface Resonance and Turnover Frequency Response” ChemRxiv Preprint, 2019. doi.org/10.26434/chemrxiv.7790009.v1

559861 Hierarchical "Nanoroll" like $MoS_2/Ti_3C_2T_x$ Hybrid with High Electrocatalytic Hydrogen Evolution Activity

Hydrogen evolution reaction (HER) is an efficient fossil fuel-free path to produce hydrogen, which is an attractive energy carrier to address the global environmental and energy crisis. Platinum (Pt)-based catalysts show brilliant catalytic performance for HER, but the high cost and scarcity severely limit their pervasive application. Molybdenum disulfide (MoS_2) is one of the most promising candidates, because its hydrogen adsorption free energy (ΔGH) is near the optimal value of 0 eV. The HER activity of MoS_2 is mainly derived from the edge sites, despite the possible contribution from the vacancies in the basal plane. Therefore, exposing more edge sites is an effective strategy to enhance the catalytic activity of MoS_2. The electrocatalytic performance of the MoS_2 is also determined by some kinetic factors, especially the electron transport between the MoS_2 and electrolyte. Therefore, conductive additives or supports, such as conductive carbon and graphene, are usually introduced to reduce the resistive loss and enhance the catalytic activity. Two-dimensional (2D) MXenes, such as $Ti_3C_2T_x$ (T_x de-notes surface functionalization, such as −O, −OH, and −F), is a new member of 2D materials. They show distinguished characteristics from graphene, such as the combination of hydrophilic surfaces and good electrical conductivity. Therefore, it may be used as a promising support to disperse catalysts. Meanwhile, MXenes has attracted much attention for HER reaction. Here, we report a simple strategy to synthesize the hierarchical "nanoroll" like $MoS_2/Ti_3C_2T_x$ hybrid by combining liquid nitrogen-freezing and subsequent annealing. The quick freezing of the $Ti_3C_2T_x$ nanosheets and ammonium tetrathiomolybdate mixture causes a sudden change in the strain of $Ti_3C_2T_x$, which leads to an interesting "nanoroll" like hierarchical structure. After annealing at H_2/Ar atmosphere, vertically aligned molybdenum sulfide (MoS_2) crystallites are in situ formed in and on the nanoroll like $Ti_3C_2T_x$. Notably, this hierarchical $MoS_2/Ti_3C_2T_x$ hybrid exhibits excellent HER catalytic activity with a small onset overpotential of 30 mV, and a more than 25-fold increase in the exchange current density compared with MoS_2 was observed. The methods described here may be applicable to the preparation of other MXene-based hybrids with novel hierarchical structures for a variety of applications.

570098 Copper-Carbon Nanomaterial Composite Coatings for Dramatic Improvements in Phase Change Heat Transfer

Phase change heat transfer is crucial to various engineering applications. Some prominent examples include distillation reboiler, nuclear reactor, high-powered electronic systems, and refrigeration. Boiling is a heat transfer process accompanied by phase change from liquid to vapor, subsequently, pool boiling involves the boiling of a stagnant liquid over a heated surface. This talk aims to examine the physical mechanisms of boiling heat transfer that are of prime importance to quantify the efficacy of the process. For various manifestations, the pool boiling performance of a surface is dictated by higher critical heat fluxes (CHF) and heat transfer coefficients.

This talk will examine various strategies to create multiscale surface-active engineered surfaces with tunable properties essentially roughness, porosity, hydrophilicity, wickability and wicking rates and their influence on the heat transfer properties. These surfaces were developed as benchmark studies, to produce a new class of hierarchical porous coatings to improve critical heat fluxes and heat transfer coefficients while facilitating morphology-induced mechanisms such as: wettability, contact angle hysteresis, roughness, and capillary wicking. In addition, the aging characteristics such as surface chemistry and morphology variablity resulting from repetitive chaotic boiling tests will also be discussed. High thermal conductivity along with improved hydrophilicity and wickability of the copper/carbon nanomaterial composites are attributed for the enhanced CHF. High-speed images revealed reduced bubble departure diameters and micro-size pores on the electrodeposited surface serving as nucleation sites. The increase in the bubble frequency and delayed formation of vapor blanket resulted in enhanced heat transfer properties.

574910 Investigation of Coal Plastic Composite Materials As a Sustainable Building Material Alternative

The wood plastic composite (WPC) market is significant, with a global value of $4.05 billion in 2015. The largest WPC market segments include building and construction (65%), automotive (18%), and electrical (8%) with the remainder associated with miscellaneous applications. Tremendous growth in global WPC materials production is expected with an 8% compound annual growth rate (CAGR), reaching $8.76 billion in 2023. The use of WPCs in construction applications, especially decking, is increasing tremendously. However, performance issues associated with high wood content engineered composites still exist.

Ohio University (OHIO) is developing coal plastic composite (CPC) materials for construction applications. CPCs may offer significant advantages including lower manufacturing energy intensity and costs, while providing a product with equivalent or superior properties to commercially available WPC products. Utilization of coal in plastic composite materials for use in construction applications is being studied, with coal filler content ranging from 40-70 wt. %. CPCs made with subbituminous and bituminous coals have been evaluated and compared to commercially available WPC decking materials. Mechanical testing results indicate CPCs possess higher flexural strength and comparable flexural modulus in comparison to commercial WPC decking products. CPCs also show flammability rating advantages as CPCs possess higher FIT and SIT in comparison to WPCs. Depending on the type of coal, the rate of burning of CPC samples ranged from 15-20 mm/min, whereas commercial WPC rate of burning ranged from 30-40 mm/min. With no antioxidant added to the CPCs, CPC OIT determined by differential scanning calorimetry (DSC) was comparable and in some cases higher than commercial WPCs.

Further, OHIO has developed Aspen Plus simulations and completed cradle-to-product life cycle analysis (LCA) studies for WPC and CPC products. Process simulation and LCA studies also indicate CPC products possess advantages in both manufacturing costs and greenhouse gas emissions associated with composite manufacturing. This presentation will discuss both bench-scale CPC R&D and recent results in scaling CPC manufacturing using commercial extrusion equipment.

577451 Colloidal, Nanoelectronic State Machines Based on 2D Materials As Smart Aerosolized Probes and Recorders

A here-to-fore unexplored property of two-dimensional (2D) electronic materials is their ability to graft modular electronic functionality onto colloidal particles so as to access local hydrodynamics in fluids to impart mobility, form composite materials, and enter spaces inaccessible to larger electronic systems. Herein, we demonstrate the design and fabrication of fully autonomous state machines (only 100x100x1 μm^3 in size) built onto a SU-8 particles powered by a 2D material-based photodiode. The on-board circuit connects a chemiresistor element and a memristor element, enabling the detection and storage of information after aerosolization, hydrodynamic propulsion to targets over 0.6 m away, and large area surface sensing of triethylamine, ammonia and aerosolized soot in inaccessible locations. An incorporated retro-reflector design allows for a facile position location using laser-scanning optical detection. Such microrobots may find widespread application as probes in confined environments, such as the human digestive tract, oil and gas conduits, chemical and biosynthetic reactors, and autonomous environmental sensors.

Reference: V.B. Koman, et al. Nature Nanotechnology volume 13, pages 819–827 (2018).

563902 Hexavalent Chromium Removal from Water Via Composite Nanofibers

Uniform electrospun polyacrylonitrile / iron (III) nitrate nanofibers with small diameters were successfully prepared by systematically optimizing the electrospinning parameters including polymer concentration, salt loading, applied voltage, tip to collector distance, and flow rate. These parameters are shown to affect the fiber morphology and the addition of iron (III) nitrate salt was found to be able to reduce the fiber size and required electrospinning voltage. The improved thermal properties of the composite fibers were investigated and shown by thermal gravimetric analysis (TGA) and differential scanning calorimetry (DSC). The composite nanofibers were further carbonized to carbon nanofibers to show their potential for hexavalent chromium removal from water. The removal percentage has been found to show a pH dependence that is related to the iron loading. This material has shown promising results towards an effective water purification method.

572376 Building Momentum for Sustainable Behaviors in Developing Regions Using Locally Managed Decentralized Circular Economy Principles

Despite the current threat from climate change, plastic collecting in the world's oceans, and the steady loss of biodiversity, the world continually fails to take action with regard to our rapidly changing ecosystem. Unfortunately, waiting on governments to act is no longer a viable option. Rapid change is needed and the pace of diplomacy is simply too slow. Democratic governments are reactionary and taking action to solve future problems is not a priority, even as the future draws ever closer. Change is in the hands of individuals and it our decisions and behaviors that will influence the future of our planet and our ability to inhabit it. Therefore, building momentum for sustainable behavior must begin with individuals. The neoliberal approach to environmental protection posits that individuals are motivated by rational self-interest and that economic incentives are necessary to achieve environmental goals. However, recent research suggests that monetary gain alone actually negatively impacts behavior, and often neglects the rural poor. As a result, models for projects designed to benefit the environment need more than just a monetary incentive, they must incorporate all three pillars of sustainability: environment, economy and society. One approach for building momentum for sustainable behavior with regard to municipal solid waste management, particularly in the developing world, is by implementing Locally Managed Decentralized Circular Economy (LMDCE) principles. This contribution will describe the role behavioral economics plays in the choices made by producers and consumers. The results of a case study on applying LMDCE principles in Kampala, Uganda to manage waste plastic accumulation by conversion to fuel oil will be presented.

575678 Supply Chain Optimization of a Locally-Managed Decentralized Solution for Conversion of Waste Plastic to Fuel in Developing Regions

Plastic waste accumulation is a global threat. Due to the lack of sufficient capital, infrastructure and knowledge pertaining to hazards associated with waste accumulation, plastic waste management is a serious issue in rural regions of developing countries. This research proposes the implementation of a locally-managed decentralized circular economy (LMDCE) for reusing and recycling waste plastic at its source of generation. A feasible LMDCE solution is conversion of waste plastic to fuel oil via thermal decomposition, utilizing appropriate technology.

This research presents an initial supply chain optimization case study of a small-scale, low-cost, LMDCE based waste plastic to fuel oil process – designed by the University of Kentucky – in Kampala, Uganda. Utilizing MATLAB computational tool, the optimum number, size, and locations of the waste plastic to fuel technology will be determined. The role of small entrepreneurs and waste pickers, amount of waste plastic completely removed from the ecosystem, and the sustainability of the proposed solution will further be discussed.

569345 Thermochemical Energy Storage Integration with Thermoelectric Power Generation

This paper will investigate the combination of thermochemical energy storage (TCES) with thermoelectric power generation for electricity production and storage in distributed settings. Thermoelectric generation (TEG) of electricity has been heavily studied as a potential method of utilizing solar or waste heat. For low peak power demands in distributed settings, TEGs have the advantage over conventional power generation cycles as they do not require moving parts, large capital investments, or long startup times. Already, thermoelectric generators (TEGs) are being marketed as accessories to camping or wood stoves to charge electronics without any moving parts. However, in many distributed settings, fuels are either unavailable, or in the case of biomass, may require enormous investments in time to gather. Thermal energy storage therefore becomes one possible method of providing the heat input to a TEG device. Although previously, sensible or latent energy storage methods have been coupled with TEG, the possibility of using thermochemical energy storage (TCES) has not been investigated. Although the overall energy efficiency of a TCES + TEG system is lower than a PV + battery storage system, it is expected that thermochemical energy storage offers much cheaper and scalable storage which, depending on the scale, could more than compensate for the cost of the TEG. Presented here is economic motivation for pursuing this unconventional and unlikely route to clean electricity and a preliminary experimental demonstration of a TCES+TEG concept.

571101 Techno-Economic Analysis and Life Cycle Assessment of Decentralized Preprocessing System for Fast Pyrolysis Biorefineries with Blended Feedstocks in the Southeastern USA

Biofuel production has a rapid growth in the past decades. Since 2002, the bioenergy portion contributed by biofuel has reached over five times in the United States and bioenergy was over 5% of the total U.S. energy consumption in 2013.[1] Rapid expansion of biofuel production results in the challenges in the traditional biomass supply chains that have large impacts on the economic effectiveness and production robustness of biorefineries. Traditional biorefinery systems with a single type of feedstock and centralized design may be vulnerable to variations in feedstock cost, quality, and availability. One possible solution is to introduce blended feedstocks and decentralized preprocessing facilities. Many Life Cycle Analysis (LCA) and Techno-Economic Analysis (TEA) have been developed to either investigate the benefits of decentralized biofuel supply chains[2] or blended feedstocks[3]. However, few of them have studied the potential economic and environmental trade-offs of using blended biomass with decentralized preprocessing sites, nor did they quantify the impacts of biomass quality and regional availability, and selection of preprocessing technologies on the overall life-cycle impacts.

This study addressed this knowledge gap by developing TEA and LCA models for biorefineries treating blended feedstocks (i.e., pine residues and switchgrass) with fast pyrolysis. "Depot", the decentralized preprocessing site, was introduced to preprocess biomass for size reduction, mixing, drying, and pelletization. Two different preprocessing technologies were explored: conventional pelleting processing (CPP) and high moisture pelleting process (HMPP).[4] Different scenarios were developed to investigate the impacts of blending ratio of pine and switchgrass, biorefinery and depot capacities, the number of depots, and preprocessing technologies. A Geospatial Information System (GIS) coupled with real-world geographic and biomass data was used to identify the locations of depots and biorefinery along with transportation distances. The TEA model calculated the Minimum Fuel Selling Price (MFSP) based on feedstock costs, the inputs from GIS, and the mass and energy balance in biorefinery from Aspen Plus process simulations. Those data were also used in combinations with the Life Cycle Inventory (LCI) data collected from literature to estimate the cradle-to-gate environmental footprints if different scenarios. The results of TEA and LCA were compared to further understand the trade-offs.

Preliminary results showed that MFSP highly depended on the biorefinery capacity, the blending ratio of feedstocks, and the process pathways of the depot. For biorefinery with a capacity from 1,000 to 2,500 oven dry ton (ODT)/day, MFSP decreased as the capacity increased. The impacts of blending ratio on MFSP were different for biorefineries with and without depots. For biorefinery without depot at 1,000 ODT/day, using a higher ratio of pine residues was more economically beneficial. As the capacity expands, the advantage of using pine residues gradually vanished. It was found that using more switchgrass at 2,000 and 2,500 ODT/day biorefineries resulted in lower MFSP. For biorefineries with depot, MFSP was lower for blending feedstocks than single feedstock type (pine residues). Additionally, introducing the decentralized depots showed the potential to lower the MFSP for biorefinery larger than 2,500 ODT/day. The results also showed that the HMPP preprocessing pathway had lower costs than CPP pathway.

Preliminary LCA results showed that the depot capacities or biorefinery capacities had relatively minor impacts on the life cycle primary energy consumption and Greenhouse Gas (GHG) emissions. Biorefineries, with or without depots, accounted for the major portion of primary energy consumptions. By using real biomass availability data, and investigating the impacts of variabilities in biomass components, preprocessing pathways, depot capacities, and biorefinery capacities, this study provides useful information for a variety of stakeholders for decision making related to investment and design of low-cost and large-scale biorefineries.

References

1. US EIA. Biofuels production drives growth in overall biomass energy use over past decade

https://www.eia.gov/todayinenergy/detail.php?id=15451. Accessed March 15, 2019.
2. Pérez, A. T. E., Camargo, M., Rincón, P. C. N., & Marchant, M. A. Key challenges and requirements for sustainable and industrialized biorefinery supply chain design and management: a bibliographic analysis. *Renewable and Sustainable Energy Reviews*. 2017, *69*, 350-359.
3. Akgul, O., Shah, N., & Papageorgiou, L. G. An optimisation framework for a hybrid first/second generation bioethanol supply chain. *Computers & Chemical Engineering*.2012, *42*, 101-114.
4. Lamers, P., Roni, M. S., Tumuluru, J. S., Jacobson, J. J., Cafferty, K. G., Hansen, J. K., Kenney, K., Teymouri, F. and Bals, B. Techno-economic analysis of decentralized biomass processing depots. *Bioresource technology*. 2015, *194*, 205-213.

576328 Ammonia Decomposition to Produce Hydrogen Using Chemical Looping Approaches: Considerations and Results for High Pressure Operation

A potential solution to reduce transportation costs for H_2 is to use ammonia (energy density: 4.25 kW/L). Ammonia (NH_3) is proposed to be an alternative H_2 carrier, due to its lower transportation costs and it is widely produced on an industrial scale using Haber-Bosch chemistry. From an end-user perspective, if ammonia is used as a H_2 carrier and if H_2 is still used as the primary energy source, there exists a technological need to convert NH_3 back to H_2 in an efficient way. OSU has proposed a thermochemical solids looping based NH_3 to H_2 (ATH) process that can operate at high thermal and H_2 production efficiencies. The ATH system can be operated as a two reactor system or a three reactor system, wherein the first the reducer reactor utilizes an intrinsic O_2 gradient driven by the reduction potential of NH_3 and a metal oxide Fe_2O_3/Fe_3O_4) to efficiently crack NH_3 to a mixture of N_2, H_2 and H_2O. The reduced metal-oxide from the reducer reactor is re-oxidized in a second reactor oxidizer using H_2O as the oxygen source. Depending on the system configuration, the three reactor system adds in a combustor reactor to re-oxidize the metal-oxides. Previous studies have focused on comparing the three reactor system to a two-reactor system and obtaining highest cold-gas efficiencies for each. The earlier chemical looping investigations quantified the thermodynamics of the chemical looping system for atmospheric pressure. So, while the earlier results are promising, it should be noted that various H_2 utilization applications require H_2 at pressures of ~30-50 bar.

This study seeks to quantify the thermodynamic feasibility associated with high-pressure chemical looping applications for Hydrogen production. This study initially quantifies the H_2 production potential of the chemical looping system for higher pressures under isothermal operating conditions. Identified optimal isothermal operating conditions are used for simulating a commercially relevant adiabatic reactor operation. Experimental results are used to provide kinetic parameters to limit design space of the chemical looping system obtained using thermodynamic simulations. The trade-off analysis shows individual and synergistic effects of variables like compression ratios, reactor size, hourly space velocities while using pinch and transshipment type technology for heat exchanger network synthesis. Preliminary results show that operating chemical looping system at higher pressures is feasible and the three reactor system maintains better performance than the two reactor chemical looping system at higher pressures.

562786 Synthesis and Performance Evaluation of New Bleach Activators

Tetraacetylethylenediamine (TAED) is used as a bleach activator in the detergent market. TAED is only used in powder formulations due to its instability in aqueous formulations and its reaction with bleaching agents (peroxide source) under storage conditions. To address the hydrolysis issue, two imide derivatives of TAED are synthesized. Due to the increased steric hindrance close to the imide group, these two molecules exhibited improved hydrolytic stability but still had similar perhydrolytic activity compared with TAED, suggesting their potential as bleach activators.

567378 Sustainable Product Design: A Life-Cycle Approach

A systematic framework is proposed for sustainable chemical product design. The product technical requirements are first identified as design constraints. Then, a base-case product is generated as a reference on top of which a more sustainable product is designed. To do so, the life cycle of base-case product (in particular, whether and how it should be recycled) is decided by using life cycle sustainability assessment (LCSA) or rule-based methods depending on the availability of life cycle inventory data. In addition, the hotspots (i.e., life cycle stages with major impact) are identified. Afterwards, to reduce the impact on hotspots, product design targets and design alternatives are generated using knowledge-base and heuristics. Lastly, LCSA or rule-based methods is applied to decide the most sustainable product from the generated product design alternatives. Two case studies – composite bumper beam and lithium ion battery – are provided to illustrate the framework.

571392 Combined Life Cycle Analysis and Environmental Risk Assessment As a Tool to Evaluate the Circular Economy of Plastics

There is currently over 7.3 billion tons of plastic in the Earth System, and it will grow to be 40 billion tons by 2050. Approximately 10 mega tons of plastics reach the oceans every year; 12 billion tons are projected to be in landfill by 2050. As these are released, the flux of plastics to surface waters, coastlines and oceans is amplified, with significant environmental, social and economic cost. In the global north, plastic ends up in the environment through leakages from waste collection, recycling and disposal, driven by the growth in single-use plastic, with a significant amount of egregious litter and informal disposal. Current understanding of the pathways and fate of plastic in the environment is poorly constrained, with only fragmentary evidence restricted to parts of the cycle. The problem is now part of public consciousness and is on the political agenda nationally and internationally. The work presented here is part of the £1m Redefining Single Use project at the University of Sheffield which aims to stimulate creative thinking across disciplines and explore novel ideas and solutions with potential to deliver more circular economic approaches to plastics manufacture and utilisation.

Overcoming the negative environmental impacts of our current use of plastics requires both a step change in thinking and innovative and transformative actions. To ensure we are not inadvertently causing more problems by solving one, sustainability analysis is essential. Here we combine the two analytical approaches which dominate environmental policy and decision-making to create a tool suitable for analysing the circular economy of plastics: environmental risk assessment (ERA) and life cycle assessment (LCA). Environmental risk assessment (ERA) couples information on the fate and behaviour of chemicals in the environment with information on the effects of chemical exposure on ecological systems to predict the probability that adverse environmental effects will occur (i.e. risk). Life cycle assessment (LCA) provides an integrative assessment of a product and process by considering the environmental and health effects from the extraction of raw materials to their recycling or disposal ('cradle to cradle/grave'). The framework of the new analysis tool will be discussed, along with an example of its use to analyse the use of plastics, and alternatives, in the supply chain of milk from farm to café.

577108 Sustainability Potential of Melamine Non-Woven Prepared As a High-Performance Material for Thermal and Acoustic Applications

Melamine Etherified Resin (MER) fibers, commercially known as smartMELAMINE, are modified melamine resin fibers obtained by melt-blown technology. A combination of organic, lightweight, fire-resistant, acoustic and thermal insulating properties of MER fibers in combination with excellent textile processability offers a wide range of potential industrial use. Results of flammability test of smartMELAMINE compared to commonly used synthetic fibers, showed that smartMELAMINE does not burn, shrink, melt or drip. These properties offer a great advantage for many technological uses, like a fire blocker, construction material, mobility applications (in automobiles, trains, buses, trucks and aircraft), for thermal protective clothing (workwear), heavy-duty military and firefighters gear, dust/hot gas filtration and various other industrial applications.

Various types of smartMELAMINE non-woven material can be produced with different molecular weight, grammage (30 – 600 gsm), material density (60 – 10 kg/m^3) and filament diameter (1 - 25 μm). In addition, obtained melamine non-woven can be chemically modified to obtain material with improved properties. Sustainability potential and recycling opportunities for the above-mentioned material in various stages of its life-cycle will be discussed.

Acknowledgments

Authors would like to acknowledge financial support from the Ministry of Education, Science and Sport of Republic of Slovenia and European Regional Development Fund for the project No. 5442-1/2018/106. This project has also received funding from the European Union's Horizon 2020 research and innovation programme under grant agreement No. 756081.

568330 Shaping Technology Innovation: A Multistage, Multiscale Sustainability Assessment and Decision Support Framework

Industries are experiencing a new wave of technology innovation largely due to the national endeavor for advanced manufacturing in the thrust areas of national importance. The accelerated innovations entail rapid transfer of emerging technologies into design and manufacturing of high performance products. Although the technologies become an engine of change and progress, the net profit brought to the environment and society could be questionable, if sustainability principles are not fully incorporated into technology development and application phases. Unfortunately, there have been nearly no systematic methodologies and effective tools to ensure the sustainability of technology innovations. This renders an urgent research need to develop methodologies and tools for inventors and organizations to evaluate potential sustainability risks of emerging technologies.

In this paper, we will introduce a framework for conducting comprehensive sustainability assessment and impact prediction of emerging technologies. The framework will be composed of a coherent set of concepts, propositions, principles, methodologies, and a tool that could assist organizations, researchers, and policy makers in shaping technology innovation for industrial sustainability. The framework has been tested through investigating new technologies for nanopaint design and nanopaint-based automotive coating manufacturing in highly sustainable process systems. Comprehensive sustainability performance comparison against current painting material and application will be presented.

---

* All correspondence should be addressed to Prof. Yinlun Huang (Phone: (313) 577-3771; E-mail: yhuang@wayne.edu).

571937 Estimating True Cost of Ionic Liquids Using Monetization

Due to the attractive properties of ionic liquids, they have found their way into many applications where they showed high potential in replacing existing chemicals and systems. However, rising concerns over their ecological impacts, e.g. toxicity and biodegradability, and high cost have limited their use. To address this issue, researchers have started to adopt the life cycle thinking for the design of greener and less expensive ionic liquids. However, having multiple metrics like cost and environmental impacts to compare the sustainability of ionic liquids can be challenging for decision making. This is especially true when different weighting factors can lead to different results in a multi-objective optimization problem. In this work, we apply monetization to estimate the true cost of ionic liquids by converting environmental units into a single monetary unit. Our results show that the monetized cost can be higher than or as high as the production cost. In this work, we also investigate the impact of atom economy, size of molecule and cation group on the monetized cost and its environmental contributions. We show how two ionic liquids of different atom economy can have the same cost of production but different monetized costs. We expect this to be a starting point for future studies targeting the design of more sustainable ionic liquids. For example, the impact of the different synthesis procedures of ionic liquids on both the cost and environmental impacts can be investigated. Also, how different anions and cation families affect the sustainability of other ionic liquids qualitatively and quantitavely is a question that need to be answered.

573592 Optimised Actions to Engage Cooperative Partnerships for an Effective Climate Change Mitigation

Climate change is one of the main challenges facing our generation which calls for urgent mitigation actions since otherwise global warning will continue to worsen. World leaders through international agendas have boosted the climate action since the earliest 80s and 90s, such as for example, with the establishment of United Nations Framework on Climate Change Conference and the Intergovernmental Panel on Climate Change, or the adoption of the Kyoto Protocol and more recently the ratification of the Paris Agreement. Unfortunately, despite the efforts made so far, carbon dioxide emissions continue to rise and perhaps more important, the overall pledged mitigation efforts presented by the countries fall short of meeting the long-term Paris goals.

Coordinating global efforts to fight against climate change remains a priority challenge, hindered by the need of satisfying so many conflicting interests and competing priorities in a fair manner. Traditional negotiations schemes have failed so far, leading to an impasse reflecting differences in the willingness to act with other parties. The backsliding of the United States Administration on climate change, with the rescission of the US Clean Power Plan and its intention to withdraw from the Paris Agreement, might trigger a dangerous domino effect among other countries following its intention to inaction. This situation calls for new initiatives and mechanisms to engage countries in a collective action shaping the transition towards a more sustainable future.

In this contribution, we present an energy system model, referred to as ERCOM as acronym of Emission Reduction Cooperation Model, which allows quantifying the benefits of cooperation by automated screening millions of alternatives for differing levels of cooperation, ultimately identifying the most cost-effective solutions for meeting a given CO_2 emission reduction target in the power sector. In essence, ERCOM is formulated as a mixed integer linear programming model which is capable of identifying the least cost pathway (i.e. optimal electricity mixes and electricity transmissions) so as to satisfy a given demand of electricity, where the binary variables are introduced to model whether a region cooperate or not. To demonstrate the capabilities of our tool, we employ the U.S. commitment to the Paris Agreement as a test bed, demonstrating that significant benefits can be attained with states collaborating to curb the CO_2 emissions thereby exploiting cost effective and low carbon regional advantages. In particular, our results show that, with all states cooperating, the cost of the electricity generation in the US could be reduced by 16% (i.e. US$46 billion per year) while simultaneously cutting carbon emissions by 70% below 2012 levels (i.e. 66 Mt CO_2 per year). It might be difficult to reach this full cooperative solution in practise, but our results show that, with only half of the states cooperating, these costs and emissions would be reduced by 10% and 68% respectively. Our findings show that pursuing climate change mitigation in cooperation can bring significant benefits not only for the climate but also for the U.S. economy. Unfortunately, the engagement of some regions might be compromised, since joint action entails an uneven distribution of efforts where states may be either benefited or penalised from cooperating. To circumvent this issue, we propose a compensation mechanism based on the fair allocation of collective gains following a equality principle which may be used as one option to spur cooperation since all states would emerge as benefited.

Overall, we argue that the approach we envisioned in this work may provide key ingredients for devising more effective schemes to overcome the current the impasse in climate change negotiations. Even if full cooperation remains elusive, our results demonstrate that the mere cooperation of a few parties may already lead to major environmental and economic benefits. The way forward to establish new partnerships for effective climate change mitigation will not be easy, but the mutual benefits attained will compensate any efforts made to foster deeper cooperation across regions.

576716 Achieving Campus Carbon Neutrality - Application of Sustainable Engineering Methods to Evaluate and Identify Technical and Ecological Solutions

In 2006, the American College & University Presidents' Climate Commitment (ACUPCC) was created as a public contract to meet the goal of carbon neutrality. Carbon neutrality is a term which means having no net greenhouse gas (GHG) emissions related to University activity through emission reductions and offsets. Currently, over 300 active signatories have committed a goal and target date for carbon neutrality before 2100.

Second Nature, the main sponsor of the ACUPCC, supports the Sustainability Indicator Management and Analysis Platform (SIMAP™) as a tool to calculate GHG emissions. SIMAP prepares GHG emissions inventories based on campus-provided inputs. However, the role of land use and carbon sequestration of campus property is disregarded in this tool. Projects which increase sequestration rates are treated as offsets, and according to Second Nature's *Carbon Markets & Offsets Guidance* should be the last priority for achieving carbon neutrality.

Many other models consider flows of carbon, or the carbon cycle, through sources and sinks. If we consider active sequestration, not carbon storage, as a sink, then we include a constant rate of carbon conversion into biomass across geographic landscapes. If our focus is only minimizing source emissions instead of balancing the cycle, we neglect solutions that enhance sinks.

SIMAP categorizes emissions into three scopes, where Scope 1 is the direct emissions from sources owned or controlled on campus. Similar to the on-site natural gas combustion or campus-owned transportation emissions, the sequestration occurs through campus-owned property and should be included as a campus-owned. Therefore, this research proposes that land use should be included into GHG emissions inventories as negative Scope 1 emissions. Currently, methane recovery in waste disposal is considered a negative Scope 3 emission. What does reframing land as carbon dioxide recovery look like in our sustainability analysis? What impacts could this have on campus sustainability and the pursuit of carbon neutrality?

Data has been collected through the process of writing the Climate Action Plan for The Ohio State University. It shows that carbon sequestration capacity can increase substantially through different scenarios of improved land management and land-use change. The potential of enhancing carbon sinks as a viable solution for University climate neutrality is apparent, resulting in land-use change and improved management scenarios which yield 3-6 times more carbon sequestration than the current landscape. Natural sequestration, which increases over time as biomass grows, is a solution which should be explored immediately, to reap the benefits as soon as possible.

However, how do we compare these results with the trade-offs of other technological and behavioral solutions? Many other necessary technological and behavioral options need to be considered such as: combined heat and power plants, increased building efficiency, on and off-site renewable energy, adjusted climate control settings, financed air travel policies, and greener commuting options. Reaching the goal of carbon neutrality requires drastic action, especially for large universities. Yet, navigating the number of options with constrained budgets and multiple stakeholders can cause paralysis in moving forward. To assist in informing University decision-makers of these trade-offs, optimization methods can be applied to compare the estimated carbon reductions and costs for any number of options, presenting multi-solution strategies that meet the needs of the University. Understanding financial, technological, spatial, and social constraints and dynamics of the carbon reduction strategies can inform the optimization problem to achieve the objective of carbon neutrality, assisting the development of the university's action plan.

Prioritizing land use as a last effort offset is a missed opportunity for colleges and universities and including optimization methods to develop plans and drive decisions is essential in realizing carbon neutrality in higher education. This research discusses some of the options and challenges that exist in reaching carbon neutrality for academic institutions, while also demonstrating a potential plan for a large

University.

577318 Attributional to Consequential Life Cycle Assessment – Steps Toward a Unified Framework

Life cycle assessment (LCA) using well established attributional methods is narrowly focused on determining the environmental impact and properties of a technology or product by portraying it as an isolated entity and linearly scaling it up or down in size to match its demand. It doesn't consider the introduction of this technology into society, its acceptance and the ripple effects that might occur due to such a technology in the real world. For example, introduction of a novel "green" technology might be obstructed due to lack of available resources, acceptance of its price by industries and consumers, and how competing technologies and products adapt themselves in response to its introduction into the market. Such a problem can be thought of as being analogous to a chemical engineering design problem in a laboratory (attributional LCA) versus a full-sized industry and its effect on the associated economy (real world impacts). In this sense, there are some gaps between engineering decisions based on technological excellence and market decisions based on the economy and human preferences.

One of the methods currently used to capture such cascading effects into the market and economy is Consequential LCA (CLCA). It accommodates the interaction between competing products in market, how prices affect the demand and supply of goods and also the incorporation of factor constraints. Even though around for two decades, the method has not matured fully. Different groups of researchers practice different flavors of applying CLCA, each with their own unique advantages and disadvantages. Broadly, the three methods are System expansion, Rectangular Choice-of-Technology, and Equilibrium modeling. System expansion and marginal technology based approach, first introduced by Weidema uses a hierarchical set of rules and decisions to determine competing products and technologies of the system in question and tries to determine in a logical fashion, how the consequences of incorporating this novel technology would propagate through the economy and society. Being a step-wise question and answer based approach, it is evident that the methods suffer from the problem of boundary cutoff and has a narrow scope of the extent of consequences considered. The Rectangular Choice-of-Technology (RCOT) framework has been developed to account for market effects as a form of economy and environmental resource (economy factor) constraints and multiple technology uses in analyzing economic systems. The RCOT framework could show gradual and partial adoption of alternative technologies to displace conventional ones as a consequence of market effects. However, the framework does not account for price changes of commodities due to changing market conditions. Equilibrium modelling is a different approach to CLCA, using the economic general equilibrium models to capture price effects, through supply demand elasticity values and microeconomic relations. They include microeconomic concepts such as price elasticities of supply and demand, market clearing constraints, production and consumption functions, maximization of profits of different agents in the model, such as government, investment agents, producers, household and foreign markets. However, the major drawback of using these models is that they are highly aggregated, making it difficult and uncertain to study the impact of a specific product or technology.

In this research work, we perform a broad comparison of these methods using the same case study and highlight the differences in scope, results and insights obtained from each method. We explore their advantages and disadvantages in detail. Such an exploration and comparison provides insight into the characteristics and benefits of each method. With the knowledge obtained from such an exploration, a CLCA based optimization framework that combines all the methods is suggested to give a novel multiscale solution. The equilibrium models in this framework will help in system expansion which reduces errors in system boundary selection while the marginal-based and RCOT approaches can be focused on processes where greater resolution and details are required. Along with that, the presence of third finer engineering scale enables designing of processes while performing CLCA, which is a novel approach in the Sustainable Process Design field that has typically used ALCA for determining environmental impacts.

Ref:

Ekvall, Tomas, and Bo P. Weidema. "System boundaries and input data in consequential life cycle inventory analysis." *The International Journal of Life Cycle Assessment* 9.3 (2004): 161-171.

580330 Application of Green-Accounting Methods for the Assessment of the Chicago Metropolitan Area's Sustainability Trends

Assessment of regional sustainability entails an integrated evaluation of trends in economic wellbeing and environmental indicators. We approach the measurement of sustainability, defined as non-declining utility or non-declining wealth, by resorting to green-accounting methods. This study is part of a multidisciplinary research project that deals with the quantitative assessment of urban sustainability at the US Environmental Protection Agency. In this regard, we explore the sustainability trends of the Chicago Metropolitan Area via two metrics rooted in economics, Green Net Metropolitan Product (GNMP) and Genuine Savings (GS). GNMP adjusts the traditional metric of economic wellbeing, the metropolitan Gross Domestic Product (GDP), by subtracting from it the values corresponding to the consumption of manmade capital and natural capital as well as the monetary values of damages caused by environmental pollution. GS measures changes in the total capital stock (or wealth), including manmade capital, human capital, and natural and environmental capital. Regional sustainability requires that GNMP be non-decreasing, and that GS be non-negative over time. We have estimated GNMP and GS for the Chicago Metropolitan Area (CMA) for the period, 1990-2015. Variation of similar methods have been applied in the literature for sustainability measurements of larger geographical entities, e.g. Heberling et al. (2012); Mota and Domingos (2013); and Wu and Heberling (2016). Environmental variables considered in our assessment are damage cost estimates due to air pollution, solid waste, scarcity rent due to water depletion, and value of ecosystem services of urban forests. In addition, we incorporate the value of time to account for the role of technological progress. With proper adaptation, these methods could be applied for the determination of sustainability of industrial processes and/or products. For example, Demeke et al. (2018) have shown that green-accounting methodologies can be coupled with Life-Cycle Assessment (LCA) to measure the sustainability of an industrial product.

Disclaimer: The views expressed in this abstract are those of the authors and do not necessarily represent the views or the policies of the U.S. Environmental Protection Agency.

579789 Sustainability Metrics for Managing and Processing Food Wastes

Food waste is a significant issue in the United States and globally. About 1/3 of the quantity of biomass that is produced to provide food is not consumed as food. This work addresses sustainability metrics related to the management and processing of food wastes. Food wastes may be used as raw materials to produce ethanol, biodiesel, methane, and fertilizer. Some are used as animal feed. Waste management options include incineration wastewater treatment, landfill disposal, and composting. Sustainability metrics include cost/income energy, carbon dioxide emissions, methane emissions, risk, and nutritional value. Waste minimization options are important as well. Several management options will be considered

563641 Sustainable Soluble Coffee Production Using Dynamic Membrane Processes

A study, in partnership with Nestlé Beverage USA, and U.S. Environmental Protection Agency, is underway to improve the sustainability profile for the manufacture of instant (soluble) coffee products. In this study, the application of membrane processes in the pre-concentration of coffee extract prior to spray/freeze drying has been found to be a promising alternative to energy-intensive separation methods such as evaporation. Such approach opens opportunities in the intensification of soluble coffee production in terms of energy-efficiency, water recovery, and waste reduction, thereby reducing the carbon footprint of the overall operations. In addition, the use of a novel dynamic vibratory separation process to address the limitations of traditional crossflow filtration, by imparting shear at the membrane surface, reducing gel formation and fouling, and improving flux was investigated. In this study, we highlight the benefits and limitations of the technology, and how laboratory investigations help guide industrial use and scale-up considerations; and the overall assessment of environmental impacts. Using reconstituted coffee extract solutions, we tested and screened various types of membranes to determine their performance in concentrating the samples. Parametric studies were also conducted in order to determine the effects of varying feed concentrations, transmembrane pressure, and surface shear (vibrational displacement) on permeate flux, and selectivity of filtration in terms of organic/inorganic rejection efficiencies. Scale-up trials from laboratory scale system were correlated to estimate design parameters for a commercial membrane filtration unit. Different cases involving the integration of membrane processes are compared with the conventional evaporative pre-concentration of coffee extract. As criteria for sustainability, the potential energy savings, operating costs reduction, economic feasibility, and resulting life cycle emissions profiles of the scaled-up operation are also presented. As a promising area for further investigation, our research shows opportunities in making soluble coffee and similar powdered beverage manufacturing operations greener.

570940 Sustainability Assessment: Uncertainty Handling and Impact on Decision Making

Sustainability assessment is a very complex appraisal method. It entails not only multidisciplinary aspects (environmental, economic and social), but also cultural and value-based elements. There have been various types of sustainability indicators for different types of systems and applications, and methodologies for indicator formulation strategy, data scaling, normalization, weighting, and aggregation. The practicing approaches for developing SA frameworks are: (a) a top-down approach – to enable experts/researchers to define and overall structure for achieving sustainability and subsequently indicators, (b) a bottom-up approach – to require participation of various stakeholders to understand the framework and propose key indicators, and (c) a mix of both. Each sustainability dimension (i.e., environmental, economic, or social) requires selection of a number of indicators to measure the performance of different aspects. This has posed a challenge: data/information uncertainty, which is pervasive in almost every phase of sustainability study. Examples include the available data about material or energy utilization, toxic/hazardous waste generation, and market fluctuation; the multifaceted makeup of the inter-entity dynamics, dependencies, and relationships; the prospect of forthcoming environmental policies, and the interrelationship among the triple-bottom-line aspects of sustainability, weighting methods, weights' values and aggregation methods.

In this presentation, we will discuss the characteristics of two classes of uncertainties, i.e., the aleatory and epistemic uncertainties that appear in various types of industrial sustainability problems. We will also describe four uncertainty processing methods: the probability bounds analysis method, the fuzzy arithmetic method, the information gap theory based method, and the interval parameter based method. The applicability of these methods in sustainability assessment at the product and process level will be discussed, and an application of these methods to geographically distributed biofuel manufacturing will be presented.

---

* All correspondence should be addressed to Prof. Yinlun Huang (Phone: (313) 577-3771; E-mail: yhuang@wayne.edu).

567869 Statistical Analysis of Streamlined Process Simulation-Based Gate-to-Gate Energy Use for Chemical Reactions and Separations

Chemical process energy is a key component of chemical life cycle inventory (LCI) data. Energy used in manufacturing of chemicals account for 50%-80% of the environmental impacts associated with chemical production. Current chemical LCIs are typically either dependent on aggregated plant-based data that are not specific to the target chemical, or on estimations based on simplified design calculations and/or use of proxy data. Such estimation techniques miss the process-level details and advanced thermodynamic calculations of chemical process simulation tools commonly used in chemical engineering design. In this work, a streamlined process simulation-based methodology for process energy data generation is used to generate gate-to-gate energy use data for manufacturing of organic chemicals. Chemical process is divided into three major blocks, with the reaction at the core, energy for pretreatment before the reaction, followed by energy required for separation and purification of the product. Chemical process simulator Aspen Plus is used to model and simulate 150 chemical processes based on this methodology. Potential process heat integration is analyzed using Aspen Energy Analyzer, which uses Pinch Analysis to provide an estimate of target heating and cooling utilities. The extracted energy data for each chemical process are used to provide a time-efficient method for predicting process heating and cooling requirements through multi-variate statistical methods such as regression analysis. Process design-based predictor variable such as heat of reaction, type of reaction (addition, substitution, etc.), and unit process (hydrogenation, nitration, etc.), and phase were used in the analysis. Heat of reaction has a strong and positive correlation with reactor heat duty, with a R^2 of 0.8. Separation operations on average require 46% (SEM = 4%) of the heating energy and 57% (SEM = 4%) of the cooling energy. The models from this study can be used for predicting process energy requirements for chemical production in early design stages, and for chemical LCI data generation in the absence of detailed industry data.

AUTHOR INDEX

AUTHOR INDEX

AUTHOR INDEX

AUTHOR INDEX

AUTHOR INDEX

AUTHOR INDEX

AUTHOR INDEX

AUTHOR INDEX

AUTHOR INDEX

AUTHOR INDEX